RECENT ADVANCES IN
REINFORCEMENT LEARNING

edited by

Leslie Pack Kaelbling
Brown University

Reprinted from
MACHINE LEARNING
An International Journal
Vol. 22, Nos. 1, 2 & 3
January/February/March 1996

KLUWER ACADEMIC PUBLISHERS
Boston / Dordrecht / London

Machine Learning

Volume 22, Nos. 1/2/3, January/February/March 1996

Special Issue on Reinforcement Learning
Guest Editor: Leslie Pack Kaelbling

Distributors for North America:
Kluwer Academic Publishers
101 Philip Drive
Assinippi Park
Norwell, Massachusetts 02061 USA

Distributors for all other countries:
Kluwer Academic Publishers Group
Distribution Centre
Post Office Box 322
3300 AH Dordrecht, THE NETHERLANDS

ISBN 978-1-4419-5160-1   e-ISBN 978-0-585-33656-5

Library of Congress Cataloging-in-Publication Data

A C.I.P. Catalogue record for this book is available
from the Library of Congress.

Machine Learning, 22, 5–6 (1996)

Editorial

THOMAS G. DIETTERICH   tgd@cs.orst.edu
Executive Editor

With this volume, *Machine Learning* begins its eleventh year of publication. The first ten years have been a very successful period for machine learning research and for the journal. One measure of our success is that for 1994 in the category of "Computer Science/Artificial Intelligence," *Machine Learning* was ranked seventh in citation impact (out of a total of 32 journals) by the Institute for Scientific Information. This reflects the many excellent papers that have been submitted over the years and the importance of these papers in subsequent research.

To inaugurate our second decade, we are pleased to announce that *Machine Learning* is now available in electronic form. The World Wide Web universal resource locator http://mlis.www.wkap.nl is the home page for *Machine Learning Online*. This provides a mixture of free and subscription-based services including

- Complete table of contents, index, and bibliography files (in bibtex format).

- Complete abstracts for all articles (in HTML and Postscript format)

- Full text of all articles published starting with Volume 15 (subscription required for access). Both HTML and postscript versions are available. In the HTML version, bibliography citations that refer to other papers published in *Machine Learning* are hyperlinked to those papers.

- A search engine for boolean searches of title, author, affiliation, keyword, bibliography entries, and publication year

- Electronic appendices. Authors can use this facility to make additional data and software available to readers of the journal

- Links to other machine learning information worldwide.

Individual and institutional (site license) electronic subscriptions are available. Subscriptions may be purchased for the usual printed journal, for electronic access, or for both methods. Mike Casey and the staff at Kluwer have put a tremendous amount of effort into *Machine Learning Online*, and I think you will find it an extremely valuable service. Jeff Schlimmer is supervising *Machine Learning Online* in the new position of Online Editor.

Finally, I am pleased to announce changes in the membership of the Editorial Board since my last editorial (Dietterich, 1994). At that time, the Editorial Board adopted a system of rotating 3-year terms for board members and for action editors. I want to thank the following editorial board members,

Bruce Buchanan, Gerry DeJong, John Holland, Yves Kodratoff, John McDermott, Ryszard Michalski, Paul Rosenbloom, Derek Sleeman, and Leslie Valiant

for their outstanding service to the journal. I particularly want to acknowledge the contribution of Ryszard Michalski who was one of the three original Editors of the journal (as well as one of the founders of our field). I also want to thank Paul Rosenbloom for his three years of service as an Action Editor for the journal.

To fill these vacancies, and also to enlarge the scope of the board, the Editorial Board elected the following new members who will serve until December 31, 1997:

David Aha, Avrim Blum, Lashon Booker, Sally Goldman, David Heckerman, Lisa Hellerstein, John Holland, Yann LeCun, Andrew Moore, Dean Pomerleau, Larry Rendell, Stuart Russell, Claude Sammut, Cullen Schaffer, and Ron Williams.

Each of these people has already worked hard for the journal as a reviewer or editor of a special issue. On behalf of the readership of *Machine Learning*, I thank them for their willingness to serve.

The editorial board also elected new Action Editors. The Action Editors are responsible for soliciting reviews and making all decisions regarding acceptance or rejection of manuscripts. This is the most difficult job in the journal. The terms of five action editors came to a close in the last year:

David Haussler, Kenneth DeJong, Leonard Pitt, Paul Utgoff and Ross Quinlan.

I want to extend my deepest thanks to the long hours and hard work that these action editors have contributed to the field and to the journal.

To fill these vacancies, the editorial board elected

David Haussler, Wolfgang Maass, Robert Holte, Michael Pazzani, and John Grefenstette

to serve as action editors until December 31, 1997. I am extremely pleased to have such a talented group of scientists serving in these important positions.

One final action was taken at the 1995 Editorial Board meeting. In recognition of our continuing interest in research on scientific discovery, Pat Langley was elected to a 3-year term as an action editor with responsibilities in this area. His term starts January 1, 1996. Pat was, of course, the founding Executive Editor of *Machine Learning*, and we are very pleased to have him once again supervising the review process.

References

Dietterich, T. G. (1994) Editorial: New editorial board members. *Machine Learning. 16,* 5.

Machine Learning, 22, 7–9 (1996)

Introduction

LESLIE PACK KAELBLING   lpk@cs.brown.edu
Computer Science Department, Brown University, Providence, RI, USA 02912-1910

This is the second special issue of *Machine Learning* on the subject of reinforcement learning. The first, edited by Richard Sutton in 1992, marked the development of reinforcement learning into a major component of the machine learning field. Since then, the area has expanded further, accounting for a significant proportion of the papers at the annual *International Conference on Machine Learning* and attracting many new researchers.

As the field grows, its boundaries become, perhaps necessarily, more diffuse. People are increasingly confused about what reinforcement learning *is*. I will take advantage of this guest editorial to outline one general conception of the field, and then to very briefly survey the current state of the art, including the papers in this issue.

1. What is reinforcement learning?

It is useful to think of reinforcement learning as a class of problems, rather than as a set of techniques. As Sutton said in the introduction to his special issue, "Reinforcement learning is the learning of a mapping from situations to actions so as to maximize a scalar reward or reinforcement signal." It is distinguished from supervised learning by the lack of a teacher specifying examples of the desired mapping and by the problem of maximizing reward over an extended period of time.

The most common techniques for solving reinforcement-learning problems are based on dynamic programming, developed in the operations research community (see texts by Puterman (1994) or Bertsekas (1995) for excellent overviews). They are based on the idea that an estimate of the utility of a state can be improved by looking ahead and using estimates of the utility of successor states. This is the basis of the temporal difference (TD) techniques (Sutton, 1988, Watkins, 1989). There are other methods, however, based on direct optimization of a policy (the paper by Moriarty and Miikkulainen in this issue illustrates the use of genetic algorithms for this purpose) or of the value function (the paper by Bradtke and Barto in this issue appplies least-squares methods), and it is important to consider these and others not yet invented, when referring to "reinforcement-learning techniques."

Robot control problems, such as navigation, pole-balancing, or juggling, are the canonical reinforcement-learning problems; but reinforcement-learning problems occur in many other situations. A particularly interesting set of applications for reinforcement learning occur in symbol-level learning (Dietterich, 1986). Tesauro's TD-Gammon system (Tesauro, 1995) is an example of one kind of symbol-level reinforcement learning. The system knows a complete model of backgammon initially and so could, in principle, simply compute the optimal strategy. However, this computation is intractable, so the model is used to generate experience, and a strategy is learned from that experience. What results is an extremely good approximate solution, which is focused, by the training experience, on the most typical situations of the game. Another example of symbol-level reinforcement learning is Zhang and Dietterich's scheduling system (Zhang & Dietterich, 1995). In this case, the problem of learning search-control rules in a problem solver is modeled as a reinforcement-learning problem; this model is much more appropriate than the typical explanation-based learning model, in which successful traces are thought of as providing instances for a supervised learning method (Dietterich & Flann, 1995).

2. State of the art

The problem of learning from temporally-delayed reward is becoming very well understood. The convergence of temporal-difference (TD) algorithms for Markovian problems with table look-up or sparse representations has been strongly established (Dayan & Sejnowksi, 1994, Tsitsiklis, 1994). The paper by Tsitsiklis and Van Roy provides convergence results for feature-based representations; results like these are crucial for scaling reinforcement-learning to large problems. Most convergence results for TD methods rely on the assumption that the underlying environment is Markovian; the paper by Schapire and Warmuth shows that, even for environments that are arbitrarily non-Markovian, a slight variant of the standard TD(λ) algorithm performs nearly as well as the best linear estimate of the value function.

One oft-heard complaint about the TD and Q-learning algorithms is that they are slow to propagate rewards through the state space. Two of the papers in this issue address this problem with traces. The paper by Singh and Sutton considers a new trace mechanism for TD and shows that it has some theoretical and empirical advantages over the standard mechanism. The technical note by Peng and Williams develops the use, suggested by Watkins, of traces in Q-learning.

Nearly all of the formal results for TD algorithms use the expected infinite-horizon discounted model of optimality; in this issue, two additional cases are considered. Mahadevan's paper explores the problem of finding policies that are optimal in the average case, considering both model-free and model-based methods. Heger's paper addresses dynamic programming and reinforcement learning in the minimax case, in which the agent should choose actions to optimize the worst possible result.

The problem of exploration in unknown environments is a crucial one for reinforcement learning. Although exploration is well-understood for the extremely simple case of k-armed bandit problems, this understanding does not extend to the exploration of more general environments. The paper by Koenig and Simmons considers the special case of exploration in multi-state environments with goals; it shows that the problem of finding the goal even once is potentially intractable, but that simple changes in representation can have a large impact on the complexity of the problem.

A crucial problem in reinforcement learning, as in other kinds of learning, is that of finding and using bias. Bias is especially crucial in reinforcement learning, because it plays a dual role: in addition to allowing appropriate generalizations to be made, it can guide the initial exploration in such a way that useful experience is gathered. The paper by Maclin and Shavlik allows humans to provide bias in the form of "advice" to their reinforcement-learning system; this advice is added to a neural-network representation of the value function and can be adjusted based on the agent's experience.

The papers in this issue represent a great deal of progress on problems of reinforcement learning. There is still, of course, a great deal of work remaining to be done. In particular, there are still important questions of scaling up, of exploration in general environments, of other kinds of bias, and of learning control policies with internal state. These problems, as well as others, are the subject of much current research. The future of reinforcement learning is exciting and challenging, and I hope you find this issue informative and inspiring.

References

Bertsekas, Dimitri P., (1995). *Dynamic Programming and Optimal Control.* Athena Scientific, Belmont, Massachusetts. Volumes 1 and 2.

Dayan, Peter & Sejnowski, Terrence J., (1994). TD(λ) converges with probability 1. *Machine Learning*, 14(3).

Dietterich, Thomas G., (1986). Learning at the knowledge level. *Machine Learning*, 1(3):287–315.

Dietterich, Thomas G. & Flann, Nicholas S., (1995). Explanation-based learning and reinforcement learning: A unified view. In *Proceedings of the Twelfth International Conference on Machine Learning*, pages 176–184, Tahoe City, California. Morgan Kaufmann.

Puterman, Martin L., (1994). *Markov Decision Processes.* John Wiley & Sons, New York.

Sutton, Richard S., (1988). Learning to predict by the method of temporal differences. *Machine Learning*, 3(1):9–44.

Tesauro, Gerald, (1995). Temporal difference learning and TD-Gammon. *Communications of the ACM*, pages 58–67.

Tsitsiklis, John N., (1994). Asynchronous stochastic approximation and Q-learning. *Machine Learning*, 16(3).

Watkins, C. J. C. H., (1989). *Learning from Delayed Rewards.* PhD thesis, King's College, Cambridge.

Zhang, Wei & Dietterich, Thomas G., (1995). A reinforcement learning approach to job-shop scheduling. In *Proceedings of the Fourteenth International Joint Conference on Artificial Intelligence*, pages 1114–1120, Montreal, Canada. Morgan Kaufmann.

Machine Learning, 22, 11–32 (1996)

Efficient Reinforcement Learning through Symbiotic Evolution

DAVID E. MORIARTY AND RISTO MIIKKULAINEN   moriarty,risto@cs.utexas.edu

Department of Computer Sciences, The University of Texas at Austin. Austin, TX 78712

Editor: Leslie Pack Kaelbling

Abstract. This article presents a new reinforcement learning method called SANE (Symbiotic, Adaptive Neuro-Evolution), which evolves a population of neurons through genetic algorithms to form a neural network capable of performing a task. Symbiotic evolution promotes both cooperation and specialization, which results in a fast, efficient genetic search and discourages convergence to suboptimal solutions. In the inverted pendulum problem, SANE formed effective networks 9 to 16 times faster than the Adaptive Heuristic Critic and 2 times faster than Q-learning and the GENITOR neuro-evolution approach without loss of generalization. Such efficient learning, combined with few domain assumptions, make SANE a promising approach to a broad range of reinforcement learning problems, including many real-world applications.

Keywords: Neuro-Evolution, Reinforcement Learning, Genetic Algorithms, Neural Networks.

1. Introduction

Learning effective decision policies is a difficult problem that appears in many real-world tasks including control, scheduling, and routing. Standard supervised learning techniques are often not applicable in such tasks, because the domain information necessary to generate the target outputs is either unavailable or costly to obtain. In reinforcement learning, agents learn from signals that provide some measure of performance and which may be delivered after a sequence of decisions have been made. Reinforcement learning thus provides a means for developing profitable decision policies with minimal domain information. While reinforcement learning methods require less *a priori* knowledge than supervised techniques, they generally require a large number of training episodes and extensive CPU time. As a result, reinforcement learning has been largely limited to laboratory-scale problems.

This article describes a new reinforcement learning system called SANE (Symbiotic, Adaptive Neuro-Evolution), with promising scale-up properties. SANE is a novel neuro-evolution system that can form effective neural networks quickly in domains with sparse reinforcement. SANE achieves efficient learning through *symbiotic evolution*, where each individual in the population represents only a partial solution to the problem: complete solutions are formed by combining several individuals. In SANE, individual neurons are evolved to form complete neural networks. Because no single neuron can perform well alone, the population remains diverse and the genetic algorithm can search many different areas of the solution space concurrently. SANE can thus find solutions faster, and to harder problems than standard neuro-evolution systems.

An empirical evaluation of SANE was performed in the inverted pendulum problem, where it could be compared to other reinforcement learning methods. The learning speed and generalization ability of SANE was contrasted with those of the single-layer Adaptive Heuristic Critic (AHC) of Barto et al. (1983), the two-layer Adaptive Heuristic Critic of Anderson (1987, 1989), the Q-learning method of Watkins and Dayan (1992), and the GENITOR neuro-evolution system of Whitley et al. (1993). SANE was found to be considerably faster (in CPU time) and more efficient (in training episodes) than the two-layer AHC, Q-learning, and GENITOR implementations. Compared to the single-layer AHC, SANE was an order of magnitude faster even though it required more training episodes. The generalization capability of SANE was comparable to the AHC and GEN-ITOR and was superior to Q-learning. An analysis of the final populations verifies that SANE finds solutions in diverse, unconverged populations and can maintain diversity in prolonged evolution. SANE's efficient search mechanism and resilience to convergence should allow it to extend well to harder problems.

The body of this article is organized as follows. After a brief review of neuro-evolution in section 2, section 3 presents the basic idea of symbiotic evolution. Section 4 describes the SANE method and its current implementation. The empirical evaluation of SANE in the inverted pendulum problem is presented in section 5, followed by an empirical analysis of the population dynamics in 6. Section 7 discusses related work, and section 8 briefly describes other tasks where SANE has been effectively applied and outlines future areas of future research.

2. Neuro-Evolution

Genetic algorithms (Holland 1975; Goldberg 1989) are global search techniques patterned after Darwin's theory of natural evolution. Numerous potential solutions are encoded in strings, called *chromosomes*, and evaluated in a specific task. Substrings, or *genes*, of the best solutions are then combined to form new solutions, which are inserted into the population. Each iteration of the genetic algorithm consists of solution evaluation and recombination and is called a *generation*. The idea is that structures that led to good solutions in previous generations can be combined to form even better solutions in subsequent generations.

By working on a legion of solution points simultaneously, genetic algorithms sample many different areas of the solution space concurrently. Such parallel search can be very advantageous in multimodal (multi-peaked) search spaces that contain several good but suboptimal solutions. Unlike gradient methods, which perform a point-to-point search and must search each peak sequentially, genetic algorithms may evolve several distinct groups of solutions, called *species*, that search multiple peaks in parallel. Speciation can create a quicker, more efficient search as well as protect against convergence at false peaks. However, for speciation to emerge the population must contain a diverse collection of genetic material, which prevents convergence to a single solution.

Since genetic algorithms do not require explicit credit assignment to individual actions, they belong to the general class of reinforcement learning algorithms. In genetic algorithms, the only feedback that is required is a general measure of proficiency for each potential solution. Credit assignment for each action is made implicitly, since poor solutions generally choose poor individual actions. Thus, which individual actions are most responsible for a good/poor solution is irrelevant to the genetic algorithm, because by selecting against poor solutions, evolution will automatically select against poor actions.

Recently there has been much interest in evolving artificial neural networks with genetic algorithms (Belew et al., 1990; Jefferson et al., 1991; Kitano, 1990; Koza and Rice, 1991; Nolfi and Parisi, 1991; Schaffer et al., 1992, Whitley et al., 1990). In most applications of neuro-evolution, the population consists of complete neural networks and each network is evaluated independently of other networks in the population. During evolution, the population converges towards a single dominant network. Such convergence is desirable if it occurs at the global optimum, however, often populations *prematurely converge* to a local optimum. Instead of multiple parallel searches through the encoding space, the search becomes a random walk using the mutation operator. As a result, evolution ceases to make timely progress and neuro-evolution is deemed pathologically slow.

Several methods have been developed to prevent premature convergence including fitness sharing [Goldberg & Richardson, 1987], adaptive mutation (Whitley et al., 1990), crowding [Dejong, 1975], and local mating [Collins & Jefferson, 1991]. Each of these techniques limits convergence through external operations that are often computationally expensive or produce a less efficient search. In the next section, a new evolutionary method will be presented that maintains diverse populations without expensive operations or high degrees of randomness.

3. Symbiotic Evolution

Normal evolutionary algorithms operate on a population of full solutions to the problem. In symbiotic evolution, each population member is only a *partial solution*. The goal of each individual is to form a partial solution that can be combined with other partial solutions currently in the population to build an effective full solution. For example in SANE, which applies the idea of symbiotic evolution to neural networks, the population consists of individual neurons, and full solutions are complete neural networks. Because single neurons rely on other neurons in the population to achieve high fitness levels, they must maintain a symbiotic relationship.

The fitness of an individual partial solution can be calculated by summing the fitness values of all possible combinations of that partial solution with other current partial solutions and dividing by the total number of combinations. Thus, an individual's fitness value reflects the average fitness of the full solutions in which the individual participated. In practice, however, evaluating all possible full solutions is intractable. The average fitness values are therefore approximated by evaluating n random subsets of partial solutions (n full solutions).

Partial solutions can be characterized as *specializations*. Instead of solving the entire problem, partial solutions specialize towards one aspect of the problem. For example, in an animal classification task one specialization may learn to recognize a mammal, while another specialization may learn to recognize a reptile. Whereas alone each specialization forms a poor classification system, conjunctively such specializations can form a complete animal classification system. Specialization will emerge because (1) individual fitness values are based on the performance of full solutions, and (2) individuals cannot delineate full solutions.

Specialization ensures diversity which prevents convergence of the population. A single partial solution cannot "take over" a population since to achieve high fitness values, there must be other specializations present. If a specialization becomes too prevalent, its members will not always be combined with other specializations in the population. Thus, redundant partial solutions do not always receive the benefit of other specializations and will incur lower fitness evaluations. Evolutionary pressures are therefore present to select against members of dominant specializations. This is quite different from standard evolutionary approaches, which always converge the population, hopefully at the global optimum, but often at a local one. In symbiotic evolution, solutions are found in diverse, *unconverged* populations.

Different specializations optimize different objective functions. In the animal classification example, recognizing mammals is different from recognizing reptiles. Evolution will, in effect, conduct separate, parallel searches in each specialization. This concurrent, divide-and-conquer approach creates a faster, more efficient search, which allows the population to discover better solutions faster, and to more difficult problems.

4. The SANE Implementation

SANE employs symbiotic evolution on a population of neurons that interconnect to form a complete neural network. More specifically, SANE evolves a population of hidden neurons for a given type of architecture such as a 2-layer-feedforward network (2 layers of weights). The basic steps in one generation of SANE are as follows (table 1): During the evaluation stage, random subpopulations of neurons of size ζ are selected and combined to form a neural network. The network is evaluated in the task and assigned a score, which is subsequently added to each selected neuron's fitness variable. The process continues until each neuron has participated in a sufficient number of networks. The average fitness of each neuron is then computed by dividing the sum of its fitness scores by the number of networks in which it participated. The neurons that have a high *average* fitness have cooperated well with other neurons in the population. Neurons that do not cooperate and are detrimental to the networks they form receive low fitness scores and are selected against.

Once each neuron has a fitness value, crossover operations are used to combine the chromosomes of the best-performing neurons. Mutation at low levels introduces genetic material that may have been missing from the initial population or lost during crossover operations. In other words, mutation is used only as an insurance policy against missing genetic material, not as a mechanism to create diversity.

Each neuron is defined in a bitwise chromosome that encodes a series of connection definitions, each consisting of an 8-bit label field and a 16-bit weight field. The value of the label determines where the connection is to be made. The neurons connect only to the input and the output layer, and every specified connection is connected to a valid unit. If the decimal value of the label, D, is greater than 127, then the connection is

Table 1. The basic steps in one generation of SANE.

1. Clear all fitness values from each neuron.
2. Select ζ neurons randomly from the population.
3. Create a neural network from the selected neurons.
4. Evaluate the network in the given task.
5. Add the network's score to each selected neuron's fitness variable.
6. Repeat steps 2-5 a sufficient number of times.
7. Get each neuron's average fitness score by dividing its total fitness values by the number of networks in which it was implemented.
8. Perform crossover operations on the population based on the average fitness value of each neuron.

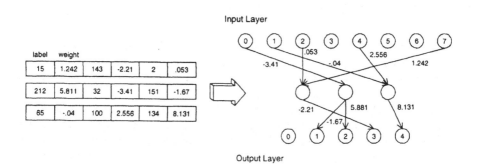

Figure 1. Forming a simple 8 input, 3 hidden, 5 output unit neural network from three hidden neuron definitions. The chromosomes of the hidden neurons are shown to the left and the corresponding network to the right. In this example, each hidden neuron has 3 connections.

made to output unit D mod O, where O is the total number of output units. Similarly, if D is less than or equal to 127, then the connection is made to input unit D mod I, where I is the total number of input units. The weight field encodes a floating point weight for the connection. Figure 1 shows how a neural network is formed from three sample hidden neuron definitions.

Once each neuron has participated in a sufficient number of networks, the population is ranked according to the average fitness values. A mate is selected for each neuron in the top quarter of the population by choosing a neuron with an equal or higher average fitness value. A one-point crossover operator is used to combine the two neurons' chromosomes, creating two offspring per mating. The offspring replace the worst-performing neurons in the population. Mutation at the rate of 0.1% is performed on the new offspring as the last step in each generation.

Selection by rank is employed instead of the standard fitness-proportionate selection to ensure a bias towards the best performing neurons. In fitness-proportionate selection, a string s is selected for mating with probability f_s/F, where f_s is the fitness of string s and F is the average fitness of the population. As the average fitness of the strings increase, the variance in fitness decreases [Whitley, 1994]. Without sufficient variance

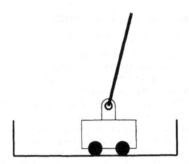

Figure 2. The cart-and-pole system in the inverted pendulum problem. The cart is pushed either left or right until it reaches a track boundary or the pole falls below 12 degrees.

between the best and worst performing strings, the genetic algorithm will be unable to assign significant bias towards the best strings. By selecting strings based on their overall rank in the population, the best strings will always receive significant bias over the worst strings even when performance differences are small.

The current implementation of SANE has performed well, however, SANE could be implemented with a variety of different neuron encodings and even network architectures that allow recurrency. More advanced encodings and evolutionary strategies may enhance both the search efficiency and generalization ability and will be a subject of future research.

5. Empirical Evaluation

To evaluate SANE, it was implemented in the standard reinforcement learning problem of balancing a pole on a cart, where its learning speed and generalization ability could be compared to previous reinforcement learning approaches to this problem.

5.1. The Inverted Pendulum Problem

The inverted pendulum or pole-balancing problem is a classic control problem that has received much attention in the reinforcement learning literature (Anderson, 1989; Barto et al., 1983; Michie and Chambers, 1968; Whitley et al., 1993). A single pole is centered on a cart (figure 2), which may move left or right on a horizontal track. Naturally, any movements to the cart tend to unbalance the pole. The objective is to push the cart either left or right with a fixed-magnitude force such that the pole remains balanced and the track boundaries are avoided. The controller receives reinforcement only after the pole has fallen, which makes this task a challenging credit assignment problem for a reinforcement learning system.

The controller is afforded the following state information: the position of the cart (ρ), the velocity of the cart ($\dot{\rho}$), the angle of the pole (θ), and the angular velocity of the pole ($\dot{\theta}$). At each time step, the controller must resolve which direction the cart is to be pushed. The cart and pole system can be described with the following second-order equations of motion:

$$\ddot{\theta}_t = \frac{mg\ sin\theta_t - cos\theta_t[F_t + m_p l \dot{\theta}_t^2\ sin\theta_t]}{(4/3)ml - m_p l\ cos^2\theta_t}, \quad \ddot{\rho}_t = \frac{F_t + m_p l[\dot{\theta}_t^2\ sin\theta_t - \ddot{\theta}_t\ cos\theta_t]}{m},$$

where

ρ: The position of the cart.
$\dot{\rho}$:   The velocity of the cart.
θ: The angle of the pole.
$\dot{\theta}$:   The angular velocity of the pole.
l: The length of the pole = 0.5 m.
m_p: The mass of the pole = 0.1 kg.
m: The mass of the pole and cart = 1.1 kg.
F: The magnitude of force = 10 N.
g: The acceleration due to gravity = 9.8.

Through Euler's method of numerical approximation, the cart and pole system can be simulated using discrete-time equations of the form $\theta(t+1) = \theta(t) + \tau\dot{\theta}(t)$, with the discrete time step τ normally set at 0.02 seconds. Once the pole falls past 12 degrees or the cart reaches the boundary of the 4.8 meter track, the trial ends and a reinforcement signal is generated. The performance of the controller is measured by the number of time steps in which the pole remains balanced. The above parameters are identical to those used by Barto et al. (1983), Anderson (1987), and Whitley et al. (1993) in this problem.

5.2.   Controller Implementations

Five different reinforcement-learning methods were implemented to form a control strategy for the pole-balancing problem: SANE, the single-layer Adaptive Heuristic Critic (AHC) of Barto et al. (1983), the two-layer AHC of Anderson (1987, 1989), the Q-learning method of Watkins and Dayan (1992), and the GENITOR system of Whitley et al. (1993). The original programs written by Sutton and Anderson were used for the AHC implementations, and the simulation code developed by Pendrith (1994) was used for the Q-learning implementation. For GENITOR, the system was reimplemented as described in [Whitley et al., 1993]. A control strategy was deemed successful if it could balance a pole for 120,000 time steps.

5.2.1.   SANE

SANE was implemented to evolve a 2-layer network with 5 input, 8 hidden, and 2 output units. Each hidden neuron specified 5 connections giving each network a total of 40 connections. The number of hidden neurons was chosen so that the total number of connections was compatible with the 2-layer AHC and GENITOR implementations. Each network evaluation consisted of a single balance attempt where a sequence of control decisions were made until the pole fell or the track boundaries were reached. Two hundred networks were formed and evaluated per generation, which allowed each neuron to participate in 8 networks per generation on average. The input to the network consisted of the 4 state variables $(\theta, \dot{\theta}, \rho, \dot{\rho})$, normalized between 0 and 1 over the following ranges:

$$
\begin{aligned}
\rho: \quad & (-2.4, 2.4) \\
\dot{\rho}: \quad & (-1.5, 1.5) \\
\theta: \quad & (-12°, 12°) \\
\dot{\theta}: \quad & (-60°, 60°)
\end{aligned}
$$

To make the network input compatible with the implementations of Whitley et al. (1993) and Anderson (1987), an additional bias input unit that is always set to 0.5 was included.

Each of the two output units corresponded directly with a control choice (left or right). The output unit with the greatest activation determined which control action was to be performed. The output layer, thus, represented a ranking of the possible choices. This approach is quite different from most neuro-control architectures, where the activation of an output unit represents a probability of that action being performed (Anderson, 1987; Barto et al., 1983; Whitley et al., 1993). For example, in many applications a decision of "move right" with activation 0.9 would move right only 90% of the time. Probabilistic output units allow the network to visit more of the state space during training, and thus incorporate a more global view of the problem into the control policy [Whitley et al., 1993]. In the SANE implementation, however, randomness is unnecessary in the decision process since there is a large amount of state space sampling through multiple combinations of neurons.

5.2.2.   The Adaptive Heuristic Critic

The Adaptive Heuristic Critic is one of the best-known reinforcement learning methods, and has been shown effective in the inverted pendulum problem. The AHC framework consists of two separate networks: an *action network* and an *evaluation network*. The action network receives the current problem state and chooses an appropriate control action. The evaluation network receives the same input, and evaluates or critiques the current state. The evaluation network is trained using the temporal difference method [Sutton, 1988] to predict the expected outcome of the current trial given the current state and the action network's current decision policy. The differences in predictions between consecutive states provide effective credit assignment to individual actions selected by the action network. Such credit assignment is used to train the action network using a standard supervised learning algorithm such as backpropagation.

Two different AHC implementations were tested: A single-layer version (Barto et al., 1983) and a two-layer version [Anderson, 1987]. Table 2 lists the parameters for each method. Both implementations were run directly from pole-balance simulators written

Table 2. Implementation parameters for each method.

	1-AHC	2-AHC	QL		GENITOR	SANE
Action Learning Rate (α):	1.0	1.0		Population Size:	100	200
Critic Learning Rate (β):	0.5	0.2	0.2	Mutation Rate:	Adaptive	0.1%
TD Discount Factor (γ):	0.95	0.9	0.95	Chromosome Length:	35 (floats)	120 (bits)
Decay Rate (λ):	0.9	0		Subpopulation (ζ):		8

by Sutton and Anderson, respectively. The learning parameters, network architectures, and control strategy were thus chosen by Sutton and Anderson and presumably reflect parameters that have been found effective.

Since the state evaluation function to be learned is non-monotonic (Anderson, 1989) and single-layer networks can only learn linearly-separable tasks, Barto et al. (1983) discretized the input space into 162 nonoverlapping regions or "boxes" for the single-layer AHC. This approach was first introduced by Michie and Chambers (1968), and it allows the state evaluation to be a linear function of the input. Both the evaluation and action network consist of one unit with a single weight connected to each input box. The output of the unit is the inner product of the input vector and the unit's weight vector, however, since only one input box will be active at one time, the output reduces to the weight corresponding to the active input box.

In the two-layer AHC, discretization of the input space is not necessary since additional hidden units allow the network to represent any non-linear discriminant function. Therefore, the same continuous input that was used for SANE was also used for the two-layer AHC. Each network (evaluation and action) in Anderson's implementation consists of 5 input units (4 input variables and one bias unit set at 0.5), 5 hidden units, and one output unit. Each input unit is connected to every hidden unit and to the single output unit. The two-layer networks are trained using a variant of backpropagation [Anderson, 1989]. The output of the action network is interpreted as the probability of choosing that action (push left or right) in both the single and two-layer AHC implementations.

5.2.3. Q-learning

Q-learning (Watkins, 1989; Watkins and Dayan 1992) is closely related to the AHC and is currently the most widely-studied reinforcement learning algorithm. In Q-learning, the Q-function is a predictive function that estimates the expected return from the current state and action pair. Given accurate Q-function values, called Q values, an optimal decision policy is one that selects the action with the highest associated Q value (expected payoff) for each state. The Q-function is learned through "incremental dynamic programming" [Watkins & Dayan, 1992], which maintains an estimate $\hat{Q}$ of the Q values and updates the estimates based on immediate payoffs and estimated payoffs from subsequent states.

Our Q-learning simulations were run using the simulation code developed by Pendrith (1994), which employs one-step updates as described by Watkins and Dayan (1992). In this implementation, the Q-function is a look-up table that receives the same discretized input that Barto et al. created for the single-layer AHC. Actions on even-numbered steps are determined using the stochastic action selector described by Lin (1992). The action on odd-numbered steps is chosen deterministically according to the highest associated Q-value. Pendrith (1994) found that such interleaved exploration and exploitation greatly improves Q-learning in the pole-balancing domain. Our experiments confirmed this result: when interleaving was disabled, Q-learning was incapable of learning the pole-balancing task.

5.2.4. GENITOR

The motivation for comparing SANE to GENITOR is twofold. GENITOR is an advanced genetic algorithm method that includes external functions for ensuring population diversity. Diversity is maintained through *adaptive mutation*, which raises the mutation rate as the population converges (section 7.1). Comparisons between GENITOR's and SANE's search efficiency thus test the hypothesis that symbiotic evolution produces an efficient search without reliance on additional randomness. Since GENITOR has been shown to be effective in evolving neural networks for the inverted pendulum problem [Whitley et al., 1993], it also provides a state-of-the-art neuro-evolution comparison.

GENITOR was implemented as detailed in [Whitley et al., 1993] to evolve the weights in a fully-connected 2-layer network, with additional connections from each input unit to the output layer. The network architecture is identical to the two-layer AHC with 5 input units, 5 hidden units and 1 output unit. The input to the network consists of the same normalized state variables as in SANE, and the activation of the output unit is interpreted as a probabilistic choice as in the AHC.

5.3. Learning-Speed Comparisons

The first experiments compared the time required by each algorithm to develop a successful network. Both the number of pole-balance attempts required and the CPU time expended were measured and averaged over 50 simulations. The number of balance attempts reflects the number of training episodes required. The CPU time was included because the number of balance attempts does not describe the amount of overhead each algorithm incurs. The CPU times should be treated as ballpark estimates because they are sensitive to the implementation details. However, the CPU time differences found in these experiments are large enough to indicate real differences in training time among the algorithms. Each implementation was written in C and compiled using the cc compiler on an IBM RS6000 25T workstation with the -O2 optimization flag. Otherwise, no special effort was made to optimize any of the implementations for speed.

The first comparison (table 3) was based on the static start state of Barto et al. (1983). The pole always started from a centered position with the cart in the middle of the track. Neither the pole nor the cart had any initial velocity. The second comparison (table 4), was based on the random start states of Anderson (1987, 1989) and Whitley et al. (1993). The cart and pole were both started from random positions with random initial

Table 3. The CPU time and number of pole balance attempts required to find a successful network starting from a centered pole and cart in each attempt. The number of pole balance attempts refers to the number of training episodes or "starts" necessary. The numbers are computed over 50 simulations for each method. A training failure was said to occur if no successful network was found after 50,000 attempts. The differences in means are statistically significant ($p < .01$) except the number of pole balance attempts between Q-learning and GENITOR.

Method	CPU Seconds				Pole Balance Attempts				Failures
	Mean	Best	Worst	SD	Mean	Best	Worst	SD	
1-layer AHC	130.6	17	3017	423.6	232	32	5003	709	0
2-layer AHC	99.1	17	863	158.6	8976	3963	41308	7573	4
Q-learning	19.8	5	99	17.9	1975	366	10164	1919	0
GENITOR	9.5	4	45	7.4	1846	272	7052	1396	0
SANE	5.9	4	8	0.6	535	70	1910	329	0

velocity. The positions and velocities were selected from the same ranges that were used to normalize the input variables, and could specify a state from which pole balancing was impossible. With random initial states, a network was considered successful if it could balance the pole from any single start state.

5.3.1. Results

The results show the AHCs to require significantly more CPU time than the other approaches to discover effective solutions. While the single-layer AHC needed the lowest number of balance attempts on average, its long CPU times overshadow its efficient learning. Typically, it took over two minutes for the single-layer AHC to find a successful network. This overhead is particularly large when compared to the genetic algorithm approaches, which took only five to ten seconds. The two-layer AHC performed the poorest, exhausting large amounts of CPU time and requiring at least 5, but often 10 to 20, times more balance attempts on average than the other approaches.

The experiments confirmed Whitley's observation that the AHC trains inconsistently when started from random initial states. Out of the 50 simulations, the single-layer AHC failed to train in 3 and the two-layer AHC failed in 14. Each unsuccessful simulation was allowed to train for 50,000 pole balance attempts before it was declared a failure. The results presented for the AHC in tables 3 and 4 are averaged over the successful simulations, excluding the failures.

Q-learning and GENITOR were comparable across both tests in terms of mean CPU time and average number of balance attempts required. The differences in CPU times between the two approaches are not large enough to discount implementation details, and when started from random start states the difference is not statistically significant. Both Q-learning and GENITOR were close to an order of magnitude faster than the AHCs and incurred no training failures.

SANE expended one half of the CPU time of Q-learning and GENITOR on average and required significantly fewer balance attempts. Like Q-learning and GENITOR, SANE

Table 4. The CPU time and number of pole balance attempts required to find a successful network starting from random pole and cart positions with random initial velocities. The differences in means between Q-learning and GENITOR are not significant ($p < .01$); the other mean differences are.

Method	CPU Seconds				Pole Balance Attempts				
	Mean	Best	Worst	SD	Mean	Best	Worst	SD	Failures
1-layer AHC	49.4	14	250	52.6	430	80	7373	1071	3
2-layer AHC	83.8	13	311	61.6	12513	3458	45922	9338	14
Q-learning	12.2	4	41	7.8	2402	426	10056	1903	0
GENITOR	9.8	4	54	7.9	2578	415	12964	2092	0
SANE	5.2	4	9	1.1	1691	46	4461	984	0

found solutions in every simulation. In addition, the time required to learn the task varied the least in the SANE simulations. When starting from random initial states, 90% of the CPU times (in seconds) fall in the following ranges:

$$
\begin{aligned}
\text{1-layer AHC:} & \quad [17, 136] \\
\text{2-layer AHC:} & \quad [17, 124] \\
Q\text{-learning:} & \quad [4, 20] \\
\text{GENITOR:} & \quad [4, 17] \\
\text{SANE:} & \quad [4, 6]
\end{aligned}
$$

Thus, while the AHC can vary as much as 2 minutes among simulations and Q-learning and GENITOR about 15 seconds, SANE consistently finds solutions in 4 to 6 seconds of CPU time, making it the fastest and most consistent of the learning methods tested in this task.

5.3.2. Discussion

The large CPU times of the AHC are caused by the many weight updates that they must perform after every action. Both the single and two-layer AHCs adjust every weight in the neural networks after each activation. Since there are thousands of activations per balance attempt, the time required for the weight updates can be substantial. The Q-learning implementation reduces this overhead considerably by only updating a single table entry after every step; however, these continuous updates still consume costly CPU cycles. Neither SANE nor GENITOR require weight updates after each activation, and do not incur these high overhead costs.

Note that the Q-function can be represented efficiently as a look-up table only when the state space is small. In a real-world application, the enormous state space would make explicit representation of each state impossible. Larger applications of Q-learning are likely to use neural networks (Lin 1992), which can learn from continuous input values in an infinite state space. Instead of representing each state explicitly, neural networks form internal representations of the state space through their connections and weights, which allows them to generalize well to unobserved states. Like the AHC, a neural network implementation of Q-learning would require continuous updates of all neural network weights, which would exhaust considerably more CPU time than the table look-up implementation.

Both the single-layer AHC and Q-learning had the benefit of a presegmented input space, while the two-layer AHC, GENITOR, and SANE methods received only undifferentiated real values of the state variables. Barto et al. (1983) selected the input boxes according to prior knowledge of the "useful regions" of the input variables and their compatibility with the single-layer AHC. This information allowed the single-layer AHC to learn the task in the least number of balance attempts. The input partitioning, however, did not extend well to the Q-learner, which required as many pole-balance attempts as the methods receiving real-valued inputs.

Interestingly, the results achieved with GENITOR were superior to those reported by Whitley et al. (1993). This disparity is probably caused by the way the input variables were normalized. Since it was unclear what ranges Whitley et al. (1993) used for normalization, the input vectors could be quite different. On average, our implementation of GENITOR required only half of the attempts, which suggests that Whitley et al. may have normalized over an overly broad range.

The comparison between SANE and GENITOR confirms our hypothesis that symbiotic evolution can perform an efficient genetic search without relying on high mutation rates. It appears that in GENITOR, the high mutation rates brought on through adaptive mutation may be causing many disruptions in highly-fit schemata (genetic building blocks), resulting in many more network evaluations required to learn the task.

5.4. Generalization Comparisons

The second battery of tests explored the generalization ability of each network. Networks that generalize well can transfer concepts learned in a subset of the state space to the rest of the space. Such behavior is of great benefit in real-world tasks where the enormous state spaces make explicit exploration of all states infeasible. In the pole balancing task, networks were trained until a network could balance the pole from a single start state. How well these networks could balance from other start states demonstrates their ability to generalize.

One hundred random start states were created as a test set for the final network of each method. The network was said to successfully balance a start state if the pole did not fall below 12° within 1000 time steps. Table 5 shows the generalization performance over 50 simulations. Since some initial states contained situations from which pole balancing was impossible, the best networks were successful only 80% of the time.

Generalization was comparable across the AHCs and the genetic algorithm approaches. The mean generalization of the Q-learning implementation, however, was significantly lower than those of the single-layer AHC, GENITOR, and SANE. This disparity is likely due to the look-up table employed by the Q-learner. In the single-layer AHC, which uses the same presegmented input space as the Q-learner, all weights are updated after visiting a single state, allowing it to learn a smoother approximation of the control function. In

Table 5. The generalization ability of the networks formed with each method. The numbers show the percentage of random start states balanced for 1000 time steps by a fully-trained network. Fifty networks were formed with each method. There is a statistically significant difference ($p < .01$) between the mean generalizations of Q-learning and those of the single-layer AHC, GENITOR, and SANE. The other differences are not significant.

Method	Mean	Best	Worst	SD
1-layer AHC	50	76	2	16
2-layer AHC	44	76	5	20
Q-learning	41	61	13	11
GENITOR	48	81	2	23
SANE	48	81	1	25

Q-learning, only the weight (i.e. the table value) of the currently visited state is updated, preventing interpolation to unvisited states.

Whitley et al. (1993) speculated that an inverse relationship exists between learning speed and generalization. In their experiments, solutions that were found in early generations tended to have poorer performance on novel inputs. Sammut and Cribb (1990) also found that programs that learn faster often result in very specific strategies that do not generalize well. This phenomenon, however, was not observed in the SANE simulations. Figure 3 plots the number of network evaluations incurred before a solution was found against its generalization ability for each of the 50 SANE simulations. As seen by the graph, no correlation appears to exist between learning speed and generalization. These results suggest that further optimizations to SANE will not restrict generalization.

6. Population Dynamics in Symbiotic Evolution

In section 3, we hypothesized that the power of the SANE approach stems from its ability to evolve several specializations concurrently. Whereas standard approaches converge the population to the desired solution, SANE forms solutions in diverse, unconverged populations. To test this hypothesis, an empirical study was conducted where the diversity levels of populations evolved by SANE were compared to those of an otherwise identical approach, but one that evolved a population of networks. Thus, the *only* difference between the two approaches was the underlying evolutionary method (symbiotic vs. standard). Whereas in SANE each chromosome consisted of 120 bits or one neuron definition, in the standard approach each chromosome contained 960 bits or 8 (the value of ζ) neuron definitions. All other parameters including population size, mutation rate, selection strategy, and number of networks evaluated per generation (200) were identical.

The comparisons were performed in the inverted pendulum problem starting from random initial states. However, with the standard parameter settings, (section 5.1), SANE found solutions so quickly that diversity was not even an issue. Therefore, the pendulum length was extended to 2.0 meters. With a longer pole, the angular acceleration of the pole $\ddot{\theta}$ is increased, because the pole has more mass and the pole's center of mass is farther away from the cart. As a result, some states that were previously recoverable

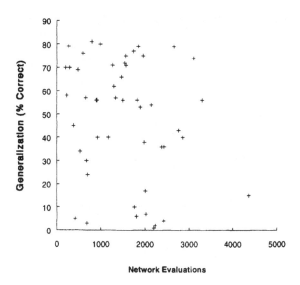

Figure 3. A plot of the learning speed versus generalization for the final networks formed in the 50 SANE simulations. The learning speed is measured by the number of network evaluations (balance attempts) during the evolution of the network, and the generalization is measured by the percentage of random start states balanced for 1000 time steps. The points are uniformly scattered indicating that learning speed does not affect generalization.

no longer are. The controllers receive more initial states from which pole balancing is impossible, and consequently require more balance attempts to form an effective control policy. A more difficult problem to learn prolongs the evolution and thereby makes the population more susceptible to diversity loss.

Ten simulations were run using each method. Once each simulation established a successful network, the diversity of the population, Φ, was measured by taking the average Hamming distance between every two chromosomes and dividing by the length of the chromosome:

$$\Phi = \frac{2 \sum_{i=1}^{n} \sum_{j=i+1}^{n} H_{i,j}}{n(n-1)l},$$

where n is the population size, l is the length of each chromosome, and $H_{i,j}$ is the Hamming distance between chromosomes i and j. The value Φ represents the probability that a given bit at a specific position on one chromosome is different from a bit at the same position on a different chromosome. Thus, a random population would have $\Phi = 0.5$ since there is a 50% probability that any two bits in the same position differ.

Figure 4 shows the average population diversity Φ as a function of each generation. A significant loss of diversity occurred in the standard approach in early generations as the populations quickly converged. After only 50 generations, 75% of any two chromosomes were identical ($\Phi = 0.25$). After 100 generations, 95% of two chromosomes were the

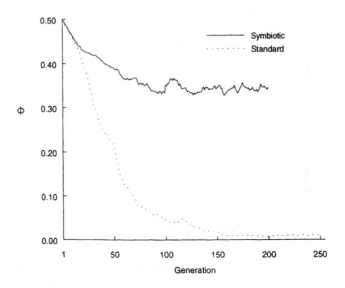

Figure 4. The average population diversity Φ after each generation. The diversity measures were averaged over 10 simulations using each method. The symbiotic method maintained high levels of diversity, while the standard genetic algorithm quickly converged.

same. In the symbiotic approach, the diversity level decreased initially but reached a plateau of 0.35 around generation 100. The symbiotic diversity level never fell below 0.32 in any simulation. SANE was able to form solutions in every simulation, while the standard approach found solutions in only 3. SANE found its solutions between 10 and 201 generations (67 on average), with an average final diversity Φ = 0.38. The three solutions found by the standard approach were at generations 69, 76, and 480, with an average diversity of 0.14. The failed simulations were stopped after 1000 generations (i.e. 200,000 pole-balance attempts).

These results confirm the hypothesis that symbiotic evolution establishes solutions in diverse populations and can maintain diversity in prolonged evolution. Whereas evolving full solutions caused the population to converge and fail to find solutions, the symbiotic approach always found a solution and in an unconverged population. This cooperative, efficient, genetic search is the hallmark of symbiotic evolution and should allow SANE to extend to more difficult problems.

7. Related Work

Work most closely related to SANE can be divided into two categories: genetic reinforcement learning and coevolutionary genetic algorithms.

7.1. Genetic Reinforcement Learning

Several systems have been built or proposed for reinforcement learning through genetic algorithms, including both symbolic and neural network approaches. The SAMUEL system [Grefenstette et al., 1990] uses genetic algorithms to evolve rule-based classifiers in sequential decision tasks. Unlike most classifier systems where genetic algorithms evolve individual rules, SAMUEL evolves a population of classifier systems or "tactical plans" composed of several action rules. SAMUEL is a model-based system that is designed to evolve decision plans offline in a simulation of the domain and then incrementally add the current best plan to the actual domain. SAMUEL has been shown effective in several small problems including the evasive maneuvers problem [Grefenstette et al., 1990] and the game of cat-and-mouse [Grefenstette, 1992], and in more recent work, SAMUEL has been extended to the task of mobile robot navigation [Grefenstette & Schultz, 1994]. The main difference between SAMUEL and SANE lies in the choice of representation. Whereas SAMUEL evolves a set of rules for sequential decision tasks, SANE evolves neural networks. The interpolative ability of neural networks should allow SANE to learn tasks quicker than SAMUEL, however, it is easier to incorporate pre-existing knowledge of the task into the initial population of SAMUEL.

GENITOR (Whitley and Kauth, 1988; Whitley, 1989) is an "aggressive search" genetic algorithm that has been shown to be quite effective as a reinforcement learning tool for neural networks. GENITOR is considered aggressive because it uses small populations, large mutation rates, and rank-based selection to create greater variance in solution space sampling. In the GENITOR neuro-evolution implementation, each network's weights are concatenated in a real-valued chromosome along with a gene that represents the crossover probability. The crossover allele determines whether the network is to be mutated or whether a crossover operation is to be performed with a second network. The crossover allele is modified and passed to the offspring based on the offspring's performance compared to the parent. If the offspring outperforms the parent, the crossover probability is decreased. Otherwise, it is increased. Whitley refers to this technique as adaptive mutation because it tends to increase the mutation rate as populations converge.

There are several key differences between GENITOR and SANE. The main difference is that GENITOR evolves full networks and requires extra randomness to maintain diverse populations, whereas SANE ensures diversity by building it into the evaluation itself. As demonstrated in the pole-balancing simulations, GENITOR's high mutation rates can lead to a less efficient search than SANE's symbiotic approach. Another difference between SANE and Whitley's approach lies in the network architectures. In the current implementation of GENITOR [Whitley et al., 1993], the network architecture is fixed and only the weights are evolved. The implementor must resolve *a priori* how the network should be connected. In SANE, the topology of the network evolves together with the weights, granting more freedom to the genetic algorithm to manifest useful neural structures.

7.2. Coevolutionary Genetic Algorithms

Symbiotic evolution is somewhat similar to implicit fitness sharing or co-adaptive genetic algorithms (Smith et al., 1993; Smith and Gray, 1993). In their immune system model, Smith et al. (1993) evolved artificial antibodies to recognize or match artificial antigens. Since each antibody can only match one antigen, a diverse population of antibodies is necessary to effectively guard against a variety of antigens. The co-adaptive genetic algorithm model, however, is based more on competition than cooperation. Each antibody must compete for survival with other antibodies in the subpopulation to recognize the given antigen. The fitness of each individual reflects how well it matches its opposing antigen, not how well it cooperates with other individuals. The antibodies are thus not dependent on other antibodies for recognition of an antigen and only interact implicitly through competition. Horn et al. (1994) characterize this difference as weak cooperation (co-adaptive GA) vs. strong cooperation (symbiotic evolution). Since both approaches appear to have similar effects in terms of population diversity and speciation, further research is necessary to discover the relative strengths and weaknesses of each method.

Smith (1994) has recently proposed a method where a learning classifier system (LCS) can be mapped to a neural network. Each hidden node represents a classifier rule that must compete with other hidden nodes in a winner-take-all competition. Like SANE, the evolution in the LCS/NN is performed on the neuron level instead of at the network level. Unlike SANE, however, the LCS/NN does not form a complete neural network, but rather relies on a gradient descent method such as backpropagation to "tune" the weights. Such reliance precludes the LCS/NN from most reinforcement learning tasks where sparse reinforcement makes gradient information unavailable.

Potter and De Jong have developed a symbiotic evolutionary strategy called Cooperative Coevolutionary Genetic Algorithms (CCGA) and have applied it to both neural network and rule-based systems (Potter and De Jong, 1995a; Potter et al., 1995b). The CCGA evolves partial solutions much like SANE, but distributes the individuals differently. Whereas SANE keeps all individuals in a single population, the CCGA evolves specializations in distinct subpopulations or *islands*. Members of different subpopulations do not interbreed across subpopulations, which eliminates haphazard, destructive recombination between dominant specializations, but also removes information-sharing between specializations.

Evolving in distinct subpopulations places a heavier burden on *a priori* knowledge of the number of specializations necessary to form an effective complete solution. In SANE, the number and distribution of the specializations is determined implicitly throughout evolution. For example, a network may be given eight hidden neurons but may only require four *types* of hidden neurons. SANE would evolve four different specializations and redundantly select two from each for the final network. While two subpopulations in the CCGA could represent the same specialization, they cannot share information and therefore are forced to find the redundant specialization independently. Potter and De Jong (1995a) have proposed a method that automatically determines the number of partial solutions necessary by incrementally adding random subpopulations. This approach appears promising, and motivates further research comparing the single population and incremental subpopulation approaches.

8. Extending SANE

SANE is a general reinforcement learning method that makes very few domain assumptions. It can thus be implemented in a broad range of tasks including real-world decision tasks. We have implemented SANE in two such tasks in the field of artificial intelligence: value ordering in constraint satisfaction problems and focusing a minimax search (Moriarty and Miikkulainen, 1994a, 1994b).

Value ordering in constraint satisfaction problems is a well-studied task where problem-general approaches have performed inconsistently. A SANE network was used to decide the order in which types or classes of cars were assigned on an assembly line, which is an NP-complete problem. The network was implemented in a chronological backtrack search and the number of backtracks incurred determined each network's score. The final SANE network required 1/30 of the backtracks of random value ordering and 1/3 of the backtracks of the commonly-used maximization-of-future-options heuristic.

In the second task, SANE was implemented to focus minimax search in the game of Othello. SANE formed a network to decide which moves from a given board situation are promising enough to be evaluated. Such decisions can establish better play by effectively hiding bad states from the minimax search. Using the powerful evaluation function from Bill [Lee & Mahajan, 1990], the SANE network was able to generate better play while examining 33% fewer board positions than a normal, full-width minimax search using the same evaluation function.

Future work on SANE includes applying it to larger real-world domains with multiple decision tasks. Possible tasks include local area network routing and scheduling, robot control, elevator control, air and automobile traffic control, and financial market trading. Since SANE makes few domain assumptions, it should be applicable in each of these domains as well as many others. An important question to be explored in such domains is: can SANE simultaneously evolve networks for separate decision tasks? For example, in a local area network, can neurons involved in priority queuing be simultaneously evolved with neurons for packet routing? Evolving neurons to form many different networks should not be any different than for a single network, since even in a single network SANE must develop neurons that specialize and serve very different roles. To evolve multiple networks, the input layers and output layers of each network could be concatenated to form a single, multi-task network. Which input units are activated and which output units are evaluated would depend on which decision task was to be performed. Since a hidden neuron could establish connections to any input or output unit, it could specialize its connections to a single decision task or form connections between sub-networks that perform different tasks. Such inter-network connections could produce interesting interactions between decision policies.

A potential key advantage of symbiotic evolution not yet fully explored is the ability to adapt quickly to changes in the environment. Often in control applications, fluctuations in the domain may require quick adaptation of the current decision policy. In gradient or point-to-point searches, adaptation can be as slow as retraining from a random point. Similarly in standard genetic algorithms which converge the population to a single solution, the lack of diverse genetic material makes further traversals of the solution space extremely slow. In SANE, however, the population does not converge. SANE's population should therefore remain highly adaptive to any changes in the fitness landscape.

While we believe that symbiotic evolution is a general principle, applicable not only to neural networks but to other representations as well, not all representations may be compatible with this approach. Symbiosis emerges naturally in the current representation of neural networks as collections of hidden neurons, but preliminary experiments with other types of encodings, such as populations of individual network connections, have been unsuccessful [Steetskamp, 1995]. An important facet of SANE's neurons is that they form complete input to output mappings, which makes every neuron a primitive solution in its own right. SANE can thus form subsumption-type architectures (Brooks, 1991), where certain neurons provide very crude solutions and other neurons perform higher-level functions that fix problems in the crude solutions. Preliminary studies in simple classification tasks have uncovered some subsumptive behavior among SANE's neurons. An important focus for future research will be to further analyze the functions of evolved hidden neurons and to study other symbiotic-conducive representations.

9. Conclusion

SANE is a new reinforcement learning system that achieves fast, efficient learning through a new genetic search strategy called symbiotic evolution. By evolving individual neurons, SANE builds effective neural networks quickly in diverse populations. In the inverted pendulum problem, SANE was faster, more efficient, and more consistent than the earlier AHC approaches, Q-learning, and the GENITOR neuro-evolution system. Moreover, SANE's quick solutions do not lack generalization as suggested by Whitley et al. (1993). Future experiments will extend SANE to more difficult real-world problems where the ability to perform multiple concurrent searches in an unconverged population should allow SANE to scale up to previously unachievable tasks.

Acknowledgments

Thanks to Mark Pendrith and Claude Sammut for providing their implementation of Q-learning for pole balancing and to Rich Sutton and Charles Anderson for making their AHC simulation code available. This research was supported in part by the National Science Foundation under grant #IRI-9504317.

References

Anderson, C. W. (1987). Strategy learning with multilayer connectionist representations. Technical Report TR87-509.3, GTE Labs, Waltham, MA.

Anderson, C. W. (1989). Learning to control an inverted pendulum using neural networks. *IEEE Control Systems Magazine*, 9:31–37.

Barto, A. G., Sutton, R. S., & Anderson, C. W. (1983). Neuronlike adaptive elements that can solve difficult learning control problems. *IEEE Transactions on Systems, Man, and Cybernetics*, SMC-13:834–846.

Belew, R. K., McInerney, J., & Schraudolph, N. N. (1991). Evolving networks: Using the genetic algorithm with connectionist learning. In Farmer, J. D., Langton, C., Rasmussen, S., and Taylor, C., editors, *Artificial Life II*. Reading, MA: Addison-Wesley.

Brooks, R. A. (1991). Intelligence without representation. *Artificial Intelligence*, 47:139–159.

Collins, R. J., & Jefferson, D. R. (1991). Selection in massively parallel genetic algorithms. In *Proceedings of the Fourth International Conference on Genetic Algorithms*, 249–256. San Mateo, CA: Morgan Kaufmann.

De Jong, K. A. (1975). *An Analysis of the Behavior of a Class of Genetic Adaptive Systems*. PhD thesis, The University of Michigan, Ann Arbor, MI.

Goldberg, D. E. (1989). *Genetic Algorithms in Search, Optimization and Machine Learning*. Reading, MA: Addison-Wesley.

Goldberg, D. E., & Richardson, J. (1987). Genetic algorithms with sharing for multimodal function optimization. In *Proceedings of the Second International Conference on Genetic Algorithms*, 148–154. San Mateo, CA: Morgan Kaufmann.

Grefenstette, J., & Schultz, A. (1994). An evolutionary approach to learning in robots. In *Proceedings of the Machine Learning Workshop on Robot Learning, Eleventh International Conference on Machine Learning*. New Brunswick, NJ.

Grefenstette, J. J. (1992). An approach to anytime learning. In *Proceedings of the Ninth International Conference on Machine Learning*, 189–195.

Grefenstette, J. J., Ramsey, C. L., & Schultz, A. C. (1990). Learning sequential decision rules using simulation models and competition. *Machine Learning*, 5:355–381.

Holland, J. H. (1975). *Adaptation in Natural and Artificial Systems: An Introductory Analysis with Applications to Biology, Control and Artificial Intelligence*. Ann Arbor, MI: University of Michigan Press.

Horn, J., Goldberg, D. E., & Deb, K. (1994). Implicit niching in a learning classifier system: Nature's way. *Evolutionary Computation*, 2(1):37–66.

Jefferson, D., Collins, R., Cooper, C., Dyer, M., Flowers, M., Korf, R., Taylor, C., & Wang, A. (1991). Evolution as a theme in artificial life: The genesys/tracker system. In Farmer, J. D., Langton, C., Rasmussen, S., and Taylor, C., editors, *Artificial Life II*. Reading, MA: Addison-Wesley.

Kitano, H. (1990). Designing neural networks using genetic algorithms with graph generation system. *Complex Systems*, 4:461–476.

Koza, J. R., & Rice, J. P. (1991). Genetic generalization of both the weights and architecture for a neural network. In *International Joint Conference on Neural Networks*, vol. 2, 397–404. New York, NY: IEEE.

Lee, K.-F., & Mahajan, S. (1990). The development of a world class Othello program. *Artificial Intelligence*, 43:21–36.

Lin, L.-J. (1992). Self-improving reactive agents based on reinforcement learning, planning, and teaching. *Machine Learning*, 8(3):293–321.

Michie, D., & Chambers, R. A. (1968). BOXES: An experiment in adaptive control. In Dale, E., and Michie, D., editors, *Machine Intelligence*. Edinburgh, UK: Oliver and Boyd.

Moriarty, D. E., & Miikkulainen, R. (1994a). Evolutionary neural networks for value ordering in constraint satisfaction problems. Technical Report AI94-218. Department of Computer Sciences, The University of Texas at Austin.

Moriarty, D. E., & Miikkulainen, R. (1994b). Evolving neural networks to focus minimax search. In *Proceedings of the Twelfth National Conference on Artificial Intelligence*, 1371-1377. Seattle, WA: MIT Press.

Nolfi, S., & Parisi, D. (1992). Growing neural networks. *Artificial Life III*. Reading, MA: Addison-Wesley.

Pendrith, M. (1994). On reinforcement learning of control actions in noisy and non-Markovian domains. Technical Report UNSW-CSE-TR-9410. School of Computer Science and Engineering, The University of New South Wales.

Potter, M., & De Jong, K. (1995a). Evolving neural networks with collaborative species. In *Proceedings of the 1995 Summer Computer Simulation Conference*. Ottawa, Canada.

Potter, M., De Jong, K., & Grefenstette, J. (1995b). A coevolutionary approach to learning sequential decision rules. In *Proceedings of the Sixth International Conference on Genetic Algorithms*. Pittsburgh, PA.

Sammut, C., & Cribb, J. (1990). Is learning rate a good performance criterion for learning? In *Proceedings of the Seventh International Conference on Machine Learning*, 170–178. Morgan Kaufmann.

Schaffer, J. D., Whitley, D., & Eshelman, L. J. (1992). Combinations of genetic algorithms and neural networks: A survey of the state of the art. In *Proceedings of the International Workshop on Combinations of Genetic Algorithms and Neural Networks (COGANN-92)*. Baltimore, MD.

Smith, R. E. (1994). Is a learning classifier system a type of neural network? *Evolutionary Computation*, 2(1).

Smith, R. E., Forrest, S., & Perelson, A. S. (1993). Searching for diverse, cooperative populations with genetic algorithms. *Evolutionary Computation*, 1(2):127–149.

Smith, R. E., & Gray, B. (1993). Co-adaptive genetic algorithms: An example in Othello strategy. Technical Report TCGA 94002, Department of Engineering Science and Mechanics, The University of Alabama.

Steetskamp, R. (1995) Explorations in symbiotic neuro-evolution search spaces. Masters Stage Report, Department of Computer Science, University of Twente. The Netherlands.

Sutton, R. S. (1988). Learning to predict by the methods of temporal differences. *Machine Learning*, 3:9–44.

Syswerda, G. (1991). A study of reproduction in generational and steady-state genetic algorithms. In Rawlings, G., editor, *Foundations of Genetic Algorithms*, 94–101. San Mateo, CA: Morgan-Kaufmann.

Watkins, C. J. C. H. (1989). *Learning from Delayed Rewards*. PhD thesis, University of Cambridge, England.

Watkins, C. J. C. H., & Dayan, P. (1992). Q-learning. *Machine Learning*, 8(3):279–292.

Whitley, D. (1989). The GENITOR algorithm and selective pressure. In *Proceedings of the Third International Conference on Genetic Algorithms*. San Mateo, CA: Morgan Kaufman.

Whitley, D. (1994). A genetic algorithm tutorial. *Statistics and Computing*, 4:65–85.

Whitley, D., Dominic, S., Das, R., & Anderson, C. W. (1993). Genetic reinforcement learning for neurocontrol problems. *Machine Learning*, 13:259–284.

Whitley, D., & Kauth, J. (1988). GENITOR: A different genetic algorithm. In *Proceedings of the Rocky Mountain Conference on Artificial Intelligence*, 118–130. Denver, CO.

Whitley, D., Starkweather, T., & Bogart, C. (1990). Genetic algorithms and neural networks: Optimizing connections and connectivity. *Parallel Computing*, 14:347–361.

Received October 20, 1994
Accepted February 24, 1995
Final Manuscirpt October 6, 1995

Machine Learning, 22, 33–57 (1996)

Linear Least–Squares Algorithms for Temporal Difference Learning

STEVEN J. BRADTKE   bradtke@gte.com
GTE Data Services, One E Telecom Pkwy, DC B2H, Temple Terrace, FL   33637

ANDREW G. BARTO   barto@cs.umass.edu
Dept. of Computer Science, University of Massachusetts, Amherst, MA   01003-4610

Editor: Leslie Pack Kaelbling

Abstract.   We introduce two new temporal difference (TD) algorithms based on the theory of linear least–squares function approximation. We define an algorithm we call Least–Squares TD (LS TD) for which we prove probability–one convergence when it is used with a function approximator linear in the adjustable parameters. We then define a recursive version of this algorithm, Recursive Least–Squares TD (RLS TD). Although these new TD algorithms require more computation per time–step than do Sutton's TD(λ) algorithms, they are more efficient in a statistical sense because they extract more information from training experiences. We describe a simulation experiment showing the substantial improvement in learning rate achieved by RLS TD in an example Markov prediction problem. To quantify this improvement, we introduce the *TD error variance* of a Markov chain, σ_{TD}, and experimentally conclude that the convergence rate of a TD algorithm depends linearly on σ_{TD}. In addition to converging more rapidly, LS TD and RLS TD do not have control parameters, such as a learning rate parameter, thus eliminating the possibility of achieving poor performance by an unlucky choice of parameters.

Keywords: Reinforcement learning, Markov Decision Problems, Temporal Difference Methods, Least–Squares

1. Introduction

The class of temporal difference (TD) algorithms (Sutton, 1988) was developed to provide reinforcement learning systems with an efficient means for learning when the consequences of actions unfold over extended time periods. They allow a system to learn to predict the total amount of reward expected over time, and they can be used for other prediction problems as well (Anderson, 1988, Barto, et al., 1983, Sutton, 1984, Tesauro, 1992). We introduce two new TD algorithms based on the theory of linear least–squares function approximation. The recursive least–squares function approximation algorithm is commonly used in adaptive control (Goodwin & Sin, 1984) because it can converge many times more rapidly than simpler algorithms. Unfortunately, extending this algorithm to the case of TD learning is not straightforward.

We define an algorithm we call Least–Squares TD (LS TD) for which we prove probability–one convergence when it is used with a function approximator linear in the adjustable parameters. To obtain this result, we use the instrumental variable approach (Ljung & Söderström, 1983, Söderström & Stoica, 1983, Young, 1984) which provides a way to handle least–squares estimation with training data that is noisy on both the input and output observations. We then define a recursive version of this algorithm, Recursive Least–Squares TD (RLS TD). Although these new TD algorithms require more computation per time step than do Sutton's TD(λ) algorithms, they are more efficient in a statistical sense because they extract more information from training experiences. We describe a simulation experiment showing the substantial improvement in learning rate achieved by RLS TD in an example Markov prediction problem. To quantify this improvement, we introduce the *TD error variance* of a Markov chain, σ_{TD}, and experimentally conclude that the convergence rate of a TD algorithm depends linearly on σ_{TD}. In addition to converging more rapidly, LS TD and RLS TD do not have control parameters, such as a learning rate parameter, thus eliminating the possibility of achieving poor performance by an unlucky choice of parameters.

We begin in Section 2 with a brief overview of the policy evaluation problem for Markov decision processes, the class of problems to which TD algorithms apply. After describing the TD(λ) class of algorithms and the existing convergence results in Sections 3 and 4, we present the least–squares approach in Section 5. Section 6 presents issues relevant to selecting an algorithm, and Sections 7 and 8 introduce the TD error variance and use it to quantify the results of a simulation experiment.

2. Markov Decision Processes

TD(λ) algorithms address the policy evaluation problem associated with discrete–time stochastic optimal control problems referred to as Markov decision processes (MDPs). An MDP consists of a discrete–time stochastic dynamic system (a controlled Markov chain), an immediate reward function, R, and a measure of long–term system performance. Restricting attention to finite–state, finite–action MDP's, we let X and A respectively denote finite sets of states and actions, and P denote the state transition probability function. At time step t, the controller observes the current state, x_t, and executes an action, a_t, resulting in a transition to state x_{t+1} with probability $P(x_t, x_{t+1}, a_t)$ and the receipt of an immediate reward $r_t = R(x_t, x_{t+1}, a_t)$. A (stationary) *policy* is a function $\mu : X \to A$ giving the controller's action choice for each state.

For each policy μ there is a *value function*, V^μ, that assigns to each state a measure of long–term performance given that the system starts in the given state and the controller always uses μ to select actions. Confining attention to the *infinite–horizon discounted* definition of long–term performance, the value function for μ is defined as follows:

$$V^\mu(x) = E_\mu[\sum_{k=0}^{\infty} \gamma^k r_k | x_0 = x],$$

where γ, $0 \leq \gamma < 1$, is the discount factor and E_μ is the expectation given that actions are selected via μ. (In problems in which one can guarantee that there will exist some finite time τ such that $r_k = 0$ for $k \geq \tau$, then one can set $\gamma = 1$.) The objective of the MDP is to find a policy, μ^*, that is optimal in the sense that $V^{\mu^*}(x) \geq V^\mu(x)$ for all $x \in X$ and for all policies μ.

Computing the evaluation function for a given policy is called *policy evaluation*. This computation is a component of the policy iteration method for finding an optimal policy, and it is sometimes of interest in its own right to solve prediction problems, the perspective taken by Sutton (Sutton, 1988). The evaluation function of a policy must satisfy the following consistency condition: for all $x \in X$:

$$V^\mu(x) = \sum_{y \in X} P(x, y, \mu(x))[R(x, y, \mu(x)) + \gamma V^\mu(y)].$$

This is a set of $|X|$ linear equations which can be solved for V^μ using any of a number of standard direct or iterative methods when the functions R and P are known. The TD(λ) family of algorithms apply to this problem when these functions are not known. Since our concern in this paper is solely in the problem of evaluating a fixed policy μ, we can omit reference to the policy throughout. We therefore denote the value function V^μ simply as V, and we omit the action argument in the functions R and P. Furthermore, throughout this paper, by a Markov chain we always mean a finite–state Markov chain.

3.   The TD(λ) Algorithm

Although any TD(λ) algorithm can be used with a lookup–table function representation, it is most often described in terms of a parameterized function approximator. In this case, V_t, the approximation of V at time step t, is defined by $V_t(x) = f(\theta_t, \phi_x)$, for all $x \in X$, where θ_t is a parameter vector at time step t, ϕ_x is a feature vector representing state x, and f is a given real–valued function differentiable with respect to θ_t for all ϕ_x. We use the notation $\nabla_{\theta_t} V_t(x)$ to denote the gradient vector at state x of V_t as a function of θ_t.

Table 1. Notation used in the discussion of the TD(λ) learning rule.

x, y, z	states of the Markov chain
x_t	the state at time step t
r_t	the immediate reward associated with the transition from state x_t to x_{t+1}; $r_t = R(x_t, x_{t+1})$.
$\bar{r}$	the vector of expected immediate rewards; $\bar{r}_x = \sum_{y \in X} P(x, y) R(x, y)$; $\bar{r}_t = \sum_{y \in X} P(x_t, y) R(x_t, y)$
S	the vector of starting probabilities
X'	the transpose of the vector or matrix X
V	the true value function
ϕ_x	the feature vector representing state x
ϕ_t	the feature vector representing state x_t. $\phi_t = \phi_{x_t}$.
Φ	the matrix whose x-th row is ϕ_x.
π_x	the proportion of time that the Markov chain is expected to spend in state x
Π	the diagonal matrix $\mathrm{diag}(\pi)$
θ^*	the true value function parameter vector
θ_t	the estimate of θ^* at time t
$V_t(x)$	the estimated value of state x using parameter vector θ_t
$\alpha_{\eta(x_t)}$	the step–size parameter used to update the value of θ_t
$\eta(x_t)$	the number of transitions from state x_t up to time step t.

Using additional notation, summarized in Table 1, the TD(λ) learning rule for a differentiable parameterized function approximator (Sutton, 1988) updates the parameter vector θ_t as follows:

$$\theta_{t+1} = \theta_t + \alpha_{\eta(x_t)} \left[R_t + \gamma V_t(x_{t+1}) - V_t(x_t) \right] \sum_{k=1}^{t} \lambda^{t-k} \nabla_{\theta_t} V_t(x_k)$$

$$= \theta_t + \alpha_{\eta(x_t)} \Delta \theta_t,$$

where

$$\Delta \theta_t = \left[R_t + \gamma V_t(x_{t+1}) - V_t(x_t) \right] \sum_{k=1}^{t} \lambda^{t-k} \nabla_{\theta_t} V_t(x_k)$$

$$= \left[R_t + \gamma V_t(x_{t+1}) - V_t(x_t) \right] \Sigma_t$$

and

$$\Sigma_t = \sum_{k=1}^{t} \lambda^{t-k} \nabla_{\theta_t} V_t(x_k). \tag{1}$$

Notice that $\Delta \theta_t$ depends only on estimates, $V_t(x_k)$, made using the latest parameter values, θ_t. This is an attempt to separate the effects of changing the parameters from the effects of moving through the state space. However, when V_t is not linear in the parameter vector θ_t and $\lambda \neq 0$, the sum Σ_t cannot be formed in an efficient, recursive manner. Instead, it is necessary to remember the x_k and to explicitly compute $\nabla_{\theta_t} V_t(x_k)$ for all $k \leq t$. This is necessary because if V_t is nonlinear in θ_t, $\nabla_{\theta_t} V_t(x_k)$ depends on θ_t. Thus, Σ_t cannot be defined recursively in terms of Σ_{t-1}. Because recomputing Σ_t in this manner at every time step can be expensive, an approximation is usually used. Assuming that the step–size parameters are small, the difference between θ_t and θ_{t-1} is also small. Then an approximation to Σ_k can be defined recursively as

$$\Sigma_t = \lambda \Sigma_{t-1} + \nabla_{\theta_t} V_t(x_t). \tag{2}$$

If V *is* linear in θ, then (2) can be used to compute Σ_t exactly. No assumptions about the step–size parameters are required, and no approximations are made.

We will be concerned in this paper with function approximators that are linear in the parameters, that is, functions that can be expressed as follows: $V_t(x) = \phi'_x \theta_t$. where ϕ'_x denotes the transpose of ϕ_x so that $\phi'_x \theta_t$ is the inner product of ϕ_x and θ_t. In this case, (2) becomes

$$\Sigma_t = \lambda \Sigma_{t-1} + \phi_t,$$

so that (1) simplifies to

$$\Sigma_t = \sum_{k=1}^{t} \lambda^{t-k} \phi_k.$$

```
1    Select θ₀.
2    Set t = 0.
3    for n = 0 to ∞ {
4         Choose a start state xₜ according to the start–state probabilities given by S.
5         Set Δₙ = 0.
6         while xₜ is not an absorbing state {
7              Let the state change from xₜ to xₜ₊₁ according to the Markov chain transition
              probabilities.
8              Set Δₙ = Δₙ + Δθₜ, where Δθₜ is given by (3).
9              t = t + 1.
10        }
11        Update the parameters at the end of trial number n: θₙ₊₁ = θₙ + αₙΔₙ.
12   }
```

Figure 1. Trial–based TD(λ) for absorbing Markov chains. A trial is a sequence of states generated by the Markov chain, starting with some initial state and ending in an absorbing state. The variable n counts the number of trials. The variable k counts the number of steps within a trial. The parameter vector θ is updated only at the end of a trial.

4.   Previous Convergence Results for TD(λ)

Convergence of a TD(λ) learning rule depends on the state representation, $\{\phi_x\}_{x \in X}$, and the form of the function approximator. Although TD(λ) rules have been used successfully with function approximators that are nonlinear in the parameter vector θ, most notably the use of a multi–layer artificial neural network in Tesauro's backgammon programs (Tesauro, 1992), convergence has only been proven for cases in which the value function is represented as a lookup table or as a linear function of θ when the feature vectors are linearly independent.[1] Sutton (Sutton, 1988) and Dayan (Dayan, 1992) proved parameter convergence in the mean under these conditions, and Dayan and Sejnowski (Dayan & Sejnowski, 1994) proved parameter convergence with probability 1 under these conditions for TD(λ) applied to absorbing Markov chains in a *trial–based* manner, i.e., with parameter updates only at the end of every trial. A trial is a sequence of states generated by the Markov chain, starting with some initial state and ending in an absorbing state. The start state for each trial is chosen according to a probability distribution S. Figure 1 describes this algorithm. Since parameter updates take place only at the end of each trial, $\Delta\theta_t$ must be defined somewhat differently from above:

$$\Delta\theta_t = \left[R_t + \gamma\theta'_n\phi_{t+1} - \theta'_n\phi_t \right] \sum_{k=1}^{t} \lambda^{t-k}\phi_k. \tag{3}$$

where n is the trial number and t is the time step. The parameter vector θ_n is held constant throughout trial n, and is updated only at the end of each trial.

Less restrictive theorems have been obtained for the TD(0) algorithm by considering it as a special case of Watkins' (Watkins, 1989) Q-learning algorithm. Watkins and Dayan (Watkins & Dayan, 1992), Jaakkola, Jordan, and Singh, (Jaakkola, et al., 1994), and Tsitsiklis (Tsitsiklis, 1993) note that since the TD(0) learning rule is a special case of Q-learning, their probability–one convergence proofs for Q-learning can be used to show that *on–line* use of the TD(0) learning rule (i.e., not trial–based) with a lookup–table function representation converges to V with probability 1. Bradtke (Bradtke, 1994) extended Tsitsiklis' proof to show that on–line use of TD(0) with a function approximator that is linear in the parameters and in which the feature vectors are linearly independent also converges to V with probability 1.

Bradtke also proved probability–one convergence under the same conditions for a normalized version of TD(0) that he called NTD(0) (Bradtke, 1994). Bradtke also defined the NTD(λ) family of learning algorithms, which are normalized versions of TD(λ). As with similar learning algorithms, the size of the input vectors ϕ_x can cause instabilities in TD(λ) learning until the step–size parameter, α, is reduced to a small enough value. But this can make the convergence rate unacceptably slow. The NTD(λ) family of algorithms addresses this problem. Since we use NTD(λ) in the comparative simulations presented below, we define it here.

The NTD(λ) learning rule for a function approximator that is linear in the parameters is

$$\theta_{t+1} = \theta_t + \alpha_t(x_t) \left[R_t + \gamma \theta_t' \phi_{t+1} - \theta_t' \phi_t \right] \sum_{k=1}^{t} \lambda^{t-k} \frac{\phi_k}{\epsilon + \phi_k' \phi_k}, \tag{4}$$

where ϵ is some small, positive number. If we know that all of the ϕ_t are non–zero, then we can set ϵ to zero. The normalization does not change the directions of the updates; it merely bounds their size, reducing the chance for unstable behavior.

5. A Least–Squares Approach to TD Learning

The algorithms described above require relatively little computation per time step, but they use information rather inefficiently compared to algorithms based on the least–squares approach. Although least–squares algorithms require more computation per time step, they typically require many fewer time steps to achieve a given accuracy than do the algorithms described above. This section describes a derivation of a TD learning rule based on least–squares techniques. Table 2 summarizes the notation we use in this section.

5.1. Linear Least–Squares Function approximation

This section reviews the basics of linear least–squares function approximation, including instrumental variable methods. This background material leads in the next section to a least–squares TD algorithm. The goal of linear least–squares function approximation is

Table 2. Notation used throughout this section in the discussion of Least–Squares algorithms.

Ψ	$\Psi : \Re^n \longrightarrow \Re$, the linear function to be approximated
ω_t	$\omega_t \in \Re^n$, the observed input at time step t
ψ_t	$\psi_t \in \Re$, the observed output at time step t
η_t	$\eta_t \in \Re$, the observed output noise at time step t
$\hat{\omega}_t$	$\hat{\omega}_t = \omega_t + \zeta_t$, the noisy input observed at time t
ζ_t	$\zeta_t \in \Re^n$, the input noise at time step t
$Cor(x, y)$	$Cor(x, y) = E\{xy'\}$, the correlation matrix for random variables x and y
ρ_t	$\rho_t \in \Re^n$, the instrumental variable observed at time step t

to linearly approximate some function $\Psi : \Re^n \longrightarrow \Re$ given samples of observed inputs $\omega_t \in \Re^n$ and the corresponding observed outputs $\psi_t \in \Re$. If the input observations are not corrupted by noise, then we have the following situation:

$$
\begin{aligned}
\psi_t &= \Psi(\omega_t) + \eta_t \\
&= \omega_t'\theta^* + \eta_t,
\end{aligned}
\tag{5}
$$

where θ^* is the vector of true (but unknown) parameters and η_t is the output observation noise.

Given (5), the least–squares approximation to θ^* at time t is the vector θ_t that minimizes the quadratic objective function

$$
J_t = \frac{1}{t} \sum_{k=1}^{t} [\psi_k - \omega_k'\theta_t]^2 .
$$

Taking the partial derivative of J_t with respect to θ_t, setting this equal to zero and solving for the minimizing θ_t gives us the tth estimate for θ^*,

$$
\theta_t = \left[\frac{1}{t} \sum_{k=1}^{t} \omega_k \omega_k' \right]^{-1} \left[\frac{1}{t} \sum_{k=1}^{t} \omega_k \psi_k \right] .
\tag{6}
$$

The following lemma, proved in ref. (Young, 1984), gives a set of conditions under which θ_t as defined by (6) converges in probability to θ^*:

LEMMA 1 *If the correlation matrix $Cor(\omega, \omega)$ is nonsingular and finite, and the output observation noise η_k is uncorrelated with the input observations ω_k, then θ_t as defined by (6) converges in probability to θ^*.*

Equation (5) models the situation in which observation errors occur only on the output. In the more general case, the input observations are also noisy. Instead of being able to directly observe ω_t, we can only observe $\hat{\omega}_t = \omega_t + \zeta_t$, where ζ_t is the input observation noise vector at time t. This is known as an *errors–in–variables* situation (Young, 1984). The following equation models the errors–in–variables situation:

$$
\begin{aligned}
\psi_t &= \Psi(\omega_t) + \eta_t \\
&= \Psi(\hat{\omega}_t - \zeta_t) + \eta_t \\
&= \hat{\omega}_t'\theta^* - \zeta_t'\theta^* + \eta_t.
\end{aligned}
\tag{7}
$$

The problem with the errors–in–variables situation is that we cannot use $\hat{\omega}_t$ instead of ω_t in (6) without violating the conditions of Lemma (1). Substituting $\hat{\omega}_t$ directly for ω_t in (6) has the effect of introducing noise that is dependent upon the current state. This introduces a bias, and θ_t no longer converges to θ^*. One way around this problem is to introduce *instrumental variables* (Ljung & Söderström, 1983, Söderström & Stoica, 1983, Young, 1984). An instrumental variable, ρ_t, is a vector that is correlated with the true input vectors, ω_t, but that is uncorrelated with the observation noise, ζ_t. The following equation is a modification of (6) that uses the instrumental variables and the noisy inputs:

$$\theta_t = \left[\frac{1}{t}\sum_{k=1}^{t}\rho_k\hat{\omega}_k'\right]^{-1}\left[\frac{1}{t}\sum_{k=1}^{t}\rho_k\psi_k\right]. \tag{8}$$

The following lemma, proved in ref. (Young, 1984), gives a set of conditions under which the introduction of instrumental variables solves the errors–in–variables problem.

LEMMA 2 *If the correlation matrix $Cor(\rho, \omega)$ is nonsingular and finite, the correlation matrix $Cor(\rho, \zeta) = 0$, and the output observation noise η_t is uncorrelated with the instrumental variables ρ_t, then θ_t as defined by (8) converges in probability to θ^*.*

5.2. *Algorithm LS TD*

Here we show how to use the instrumental variables method to derive an algorithm we call Least–Squares TD (LS TD), a least–squares version of the TD algorithm. The TD algorithm used with a linear–in–the–parameters function approximator addresses the problem of finding a parameter vector, θ^*, that allows us to compute the value of a state x as $V(x) = \phi_x'\theta^*$. Recall that the value function satisfies the following consistency condition:

$$\begin{aligned}
V(x) &= \sum_{y\in X} P(x,y)[R(x,y) + \gamma V(y)] \\
&= \sum_{y\in X} P(x,y)R(x,y) + \gamma \sum_{y\in X} P(x,y)V(y) \\
&= \bar{r}_x + \gamma \sum_{y\in X} P(x,y)V(y),
\end{aligned}$$

where $\bar{r}_x$ is the expected immediate reward for any state transition from state x. We can rewrite this in the form used in Section 5.1 as

$$\begin{aligned}
\bar{r}_x &= V(x) - \gamma \sum_{y\in X} P(x,y)V(y) \\
&= \phi_x'\theta^* - \gamma \sum_{y\in X} P(x,y)\phi_y'\theta^*
\end{aligned}$$

$$= (\phi_x - \gamma \sum_{y \in X} P(x, y)\phi_y)'\theta^*, \tag{9}$$

for every state $x \in X$. Now we have the same kind of problem that we considered in Section 5.1. The scalar output, $\bar{r}_x$, is the inner product of an input vector, $\phi_x - \gamma \sum_{y \in X} P(x, y)\phi_y$, and the true parameter vector, θ^*.

For each time step t, we therefore have the following equation that has the same form as (5):

$$r_t = (\phi_t - \gamma \sum_{y \in X} P(x_t, y)\phi_y)'\theta^* + (r_t - \bar{r}_t), \tag{10}$$

where r_t is the reward received on the transition from x_t to x_{t+1}. $(r_t - \bar{r}_t)$ corresponds to the noise term η_t in (5). The following lemma, proved in Appendix A, establishes that this noise term has zero mean and is uncorrelated with the input vector $w_t = \phi_t - \gamma \sum_{y \in X} P(x_t, y)\phi_y$:

LEMMA 3 *For any Markov chain, if x and y are states such that $P(x, y) > 0$, with $\eta_{xy} = R(x, y) - \bar{r}_x$ and $w_x = (\phi_x - \gamma \sum_{y \in X} P(x, y)\phi_y)$, then $E\{\eta\} = 0$, and $Cor(w, \eta) = 0$.*

Therefore, if we know the state transition probability function, P, the feature vector ϕ_x, for all $x \in X$, if we can observe the state of the Markov chain at each time step, and if $Cor(w, w)$ is nonsingular and finite, then by Lemma (1) the algorithm given by (6) converges in probability to θ^*.

In general, however, we do not know P, $\{\phi_x\}_{x \in X}$, or the state of the Markov chain at each time step. We assume that all that is available to define θ_t are ϕ_t, ϕ_{t+1} and r_t. Instrumental variable methods allow us to solve the problem under these conditions. Let $\hat{w}_t = \phi_t - \gamma\phi_{t+1}$, and $\zeta_t = \gamma \sum_{y \in X} P(x_t, y)\phi_y - \gamma\phi_{t+1}$. Then we can observe

$$\begin{aligned} \hat{w}_t &= \phi_t - \gamma\phi_{t+1} \\ &= (\phi_t - \gamma \sum_{y \in X} P(x_t, y)\phi_y) + (\gamma \sum_{y \in X} P(x_t, y)\phi_y - \gamma\phi_{t+1}) \\ &= w_t + \zeta_t, \end{aligned}$$

with $w_t = \phi_t - \gamma \sum_{y \in X} P(x_t, y)\phi_y$ and $\zeta_t = \gamma \sum_{y \in X} P(x_t, y)\phi_y - \gamma\phi_{t+1}$. We see, then, that the problem fits the errors–in–variables situation. Specifically, the following equation in the form of (7) is equivalent to the consistency condition (9):

$$r_t = (\phi_t - \gamma\phi_{t+1})'\theta^* - (\gamma \sum_{y \in X} P(x_t, y)\phi_y - \gamma\phi_{t+1})'\theta^* + (r_t - \bar{r}_t).$$

Following Section 5.1, we introduce an instrumental variable, ρ_t, to avoid the asymptotic bias introduced by errors–in–variables problems. The following lemma, which is proved in Appendix A, shows that $\rho_t = \phi_t$ is an instrumental variable because it is uncorrelated with the input observation noise, ζ_t, defined above:

```
1    Set t = 0.
2    repeat forever {
3        Set xt to be a start state selected according to the probabilities given by S.
4        while xt is not an absorbing state {
5            Let the state change from xt to xt+1 according to the Markov chain
             transition probabilities.
6            Use (11) to define θt.
7            t = t + 1.
8        }
9    }
```

Figure 2. Trial–based LS TD for absorbing Markov chains. A trial is a sequence of states that starts at some start state, follows the Markov chain as it makes transitions, and ends at an absorbing state.

LEMMA 4 *For any Markov chain, if (1) x and y are states such that $P(x,y) > 0$; (2) $\zeta_{xy} = \gamma \sum_{z \in X} P(x,z)\phi_z - \gamma\phi_y$; (3) $\eta_{xy} = R(x,y) - \bar{r}_x$; and (4) $\rho_x = \phi_x$, then (1) $Cor(\rho, \eta) = 0$; and (2) $Cor(\rho, \zeta) = 0$.*

Using ϕ_x as the instrumental variable, we rewrite (8) to obtain the LS TD algorithm:

$$\theta_t = \left[\tfrac{1}{t} \sum_{k=1}^{t} \phi_k(\phi_k - \gamma\phi_{k+1})' \right]^{-1} \left[\tfrac{1}{t} \sum_{k=1}^{t} \phi_k r_k \right]. \tag{11}$$

Figure 2 shows how (8) can be used as part of a trial–based algorithm to find the value function for an absorbing Markov chain. Figure 3 shows how (8) can be used as part of an algorithm to find the value function for an ergodic Markov chain. Learning takes place on–line in both algorithms, with parameter updates after every state transition. The parameter vector θ_t is not well defined when t is small since the matrix $\left[\tfrac{1}{t} \sum_{k=1}^{t} \phi_k(\phi_k - \gamma\phi_{k+1})' \right]$ is not invertible.

The LS TD algorithm has some similarity to an algorithm Werbos (Werbos, 1990) proposed as a linear version of his Heuristic Dynamic Programming (Lukes, et al., 1990, Werbos, 1987, Werbos, 1988, Werbos, 1992). However, Werbos' algorithm is not amenable to a recursive formulation, as is LS TD, and does not converge for arbitrary initial parameter vectors, as does LS TD. See ref. (Bradtke, 1994).

It remains to establish conditions under which LS TD converges to θ^*. According to Lemma 2, we must establish that $Cor(\rho, \omega)$ is finite and nonsingular. We take this up in the next section.

5.3. *Convergence of Algorithm LS TD*

In this section we consider the asymptotic performance of algorithm LS TD when used on–line to approximate the value functions of absorbing and ergodic Markov chains. The following lemma, proved in Appendix A, starts the analysis by expressing $\theta_{\text{LSTD}} \stackrel{\text{def}}{=} \lim_{t \to \infty} \theta_t$, the limiting estimate found by algorithm LS TD for θ^*, in a convenient form.

```
1   Set t = 0.
2   Select an arbitrary initial state, x_0.
3   repeat forever {
4       Let the state change from x_t to x_{t+1} according to the Markov chain transi-
        tion probabilities.
5       Use (11) to define θ_t.
6       t = t + 1.
7   }
```

Figure 3. LS TD for ergodic Markov chains.

LEMMA 5 *For any Markov chain, when (1) θ_t is found using algorithm LS TD; (2) each state $x \in X$ is visited infinitely often; (3) each state $x \in X$ is visited in the long run with probability 1 in proportion π_x; and (4) $[\Phi'\Pi(I - \gamma P)\Phi]$ is invertible, where Φ is the matrix of whose x–th row is ϕ_x, and Π is the diagonal matrix $\mathrm{diag}(\pi)$, then*

$$\theta_{\mathrm{LSTD}} = [\Phi'\Pi(I - \gamma P)\Phi]^{-1}[\Phi'\Pi\bar{r}]$$

with probability 1.

The key to using Lemma 5 lies in the definition of π_x: the proportion of time that the Markov chain is expected to spend over the long run in state x. Equivalently, π_x is the expected proportion of state transitions that take the Markov chain out of state x. For an ergodic Markov chain, π_x is the invariant, or steady–state, distribution associated with the stochastic matrix P (Kemeny & Snell, 1976). For an absorbing Markov chain, π_x is the expected number of visits out of state x during one transition sequence from a start state to a goal state (Kemeny & Snell, 1976). Since there are no transitions out of a goal state, $\pi_x = 0$ for all goal states. These definitions prepare the way for the following two theorems. Theorem 1 gives conditions under which LS TD as used in Figure 2 will cause θ_{LSTD} to converge with probability 1 to θ^* when applied to an absorbing Markov chain. Theorem 2 gives conditions under which LS TD as used in Figure 3 will cause θ_{LSTD} to converge with probability 1 to θ^* when applied to an ergodic Markov chain.

THEOREM 1 (CONVERGENCE OF LS TD FOR ABSORBING MARKOV CHAINS)
When using LS TD as described in Figure 2 to estimate the value function for an absorbing Markov chain, if (1) S is such that there are no inaccessible states; (2) $R(x, y) = 0$ whenever both $x, y \in \mathcal{T}$, the set of absorbing states; (3) the set of feature vectors representing the non-absorbing states, $\{\phi_x \mid x \in \mathcal{N}\}$, is linearly independent; (4) $\phi_x = 0$ for all $x \in \mathcal{T}$; (5) each ϕ_x is of dimension $m = |\mathcal{N}|$; and (6) $0 \leq \gamma \leq 1$; then θ^* is finite and the asymptotic parameter estimate found by algorithm LS TD, θ_{LSTD}, converges with probability 1 to θ^* as the number of trials (and state transitions) approaches infinity.

Different conditions are required in the absorbing and ergodic chain cases in order to meet the conditions of Lemma 5. The conditions required in Theorem 1 are generalizations of the conditions required for probability 1 convergence of TD(λ) for absorbing Markov chains. The conditions required in Theorem 2 are much less restrictive, though the discount factor γ must be less than 1 to ensure that the value function is finite.

THEOREM 2 (CONVERGENCE OF LS TD FOR ERGODIC MARKOV CHAINS)
When using LS TD as described in Figure 3 to estimate the value function for an ergodic Markov chain, if (1) the set of feature vectors representing the states, $\{\phi_x \mid x \in X\}$, is linearly independent; (2) each ϕ_x is of dimension $N = |X|$; (3) $0 \leq \gamma < 1$; then θ^* is finite and the asymptotic parameter estimate found by algorithm LS TD, θ_{LSTD}, converges with probability 1 to θ^* as the number of state transitions approaches infinity.

Theorems 1 and 2 provide convergence assurances for LS TD similar to those provided by Tsitsiklis (Tsitsiklis, 1993) and Watkins and Dayan (Watkins & Dayan, 1992) for the convergence of TD(0) using a lookup–table function approximator.

Proof of Theorem 1: Condition (1) implies that, with probability 1, as the total number of state transitions approaches infinity, the number of times each state $x \in X$ is visited approaches infinity. Since this is an absorbing chain, we have with probability 1 that the states are visited in proportion π as the number of trials approaches infinity. Therefore, by Lemma 5, we know that with probability 1

$$\theta_{\text{LSTD}} = [\Phi'\Pi(I - \gamma P)\Phi]^{-1} [\Phi'\Pi\bar{r}],$$

assuming that the inverse exists.

Conditions (3), (4), and (5) imply that Φ has rank m, with row x of Φ consisting of all zeros for all $x \in T$. Condition (1) implies that Π has rank m. Row x of Π consists of all zeros, for all $x \in T$. Φ has the property that if all rows corresponding to absorbing states are removed, the resulting submatrix is of dimensions $(m \times m)$ and has rank m. Call this submatrix A. Π has the property that if all rows and columns corresponding to absorbing states are removed, the resulting submatrix is of dimensions $(m \times m)$ and has rank m. Call this submatrix B. $(I - \gamma P)$ has the property that if all rows and columns corresponding to absorbing states are removed, the resulting submatrix is of dimensions $(m \times m)$ and has rank m (Kemeny & Snell, 1976). Call this submatrix C. It can be verified directly by performing the multiplications that $[\Phi'\Pi(I - \gamma P)\Phi] = [A'BCA]$. Therefore, $[\Phi'\Pi(I - \gamma P)\Phi]$ is of dimensions $(m \times m)$ and has rank m. Thus, it is invertible.

Now, (9) can be rewritten using matrix notation as

$$\bar{r} = (I - \gamma P)\Phi\theta^*. \tag{12}$$

This, together with conditions (2) and (6), implies that θ^* is finite. Finally, substituting (12) into the expression for θ_{LSTD} gives us

$$\begin{aligned} \theta_{\text{LSTD}} &= [\Phi'\Pi(I - \gamma P)\Phi]^{-1} [\Phi'\Pi(I - \gamma P)\Phi] \theta^* \\ &= \theta^*. \end{aligned}$$

Thus, θ_{LSTD} converges to θ^* with probability 1. ∎

Proof of Theorem 2: Since this is an ergodic chain, as t approaches infinity we have with probability 1 that the number of times each state $x \in X$ is visited approaches infinity. We also have with probability 1 that the states are visited in the long run in proportion π. Ergodicity implies that $\pi_x > 0$ for all $x \in X$. Therefore, Π is invertible. Condition (3) implies that $(I - \gamma P)$ is invertible. Conditions (1) and (2) imply that Φ is invertible. Therefore, by Lemma 5, we know that with probability 1

$$\theta_{\text{LSTD}} = [\Phi'\Pi(I - \gamma P)\Phi]^{-1} [\Phi'\Pi\bar{r}].$$

Condition (3) together with Equation (12) imply that θ^* is finite. And, as above, substituting (12) into the expression for θ_{LSTD} gives

$$\begin{aligned}
\theta_{\text{LSTD}} &= [\Phi'\Pi(I - \gamma P)\Phi]^{-1} [\Phi'\Pi(I - \gamma P)\Phi] \theta^* \\
&= \theta^*.
\end{aligned}$$

Thus, θ_{LSTD} converges to θ^* with probability 1. ∎

5.4. Algorithm RLS TD

Algorithm LS TD requires the computation of a matrix inverse at each time step. This means that LS TD has a computational complexity of $O(m^3)$, assuming that the state representations are of length m. We can use Recursive Least–Squares (RLS) techniques (Goodwin & Sin, 1984, Ljung & Söderström, 1983, Young, 1984) to derive a modified algorithm, Recursive Least–Squares TD (RLS TD), with computational complexity of $O(m^2)$. The following equation set specifies algorithm RLS TD:

$$e_t = R_t - (\phi_t - \gamma\phi_{t+1})'\theta_{t-1} \tag{13}$$

$$C_t = C_{t-1} - \frac{C_{t-1}\phi_t(\phi_t - \gamma\phi_{t+1})'C_{t-1}}{1 + (\phi_t - \gamma\phi_{t+1})'C_{t-1}\phi_t} \tag{14}$$

$$\theta_t = \theta_{t-1} + \frac{C_{t-1}}{1 + (\phi_t - \gamma\phi_{t+1})'C_{t-1}\phi_t} e_t\phi_t. \tag{15}$$

Notice that (15) is the TD(0) learning rule for function approximators that are linear in the parameters, except that the scalar step–size parameter has been replaced by a gain matrix. The user of an RLS algorithm must specify θ_0 and C_0. C_t is the tth sample estimate of $\frac{1}{t}\text{Cor}(\rho, \hat{\omega})^{-1}$, where ρ and $\hat{\omega}$ are defined as in Section 5.2. C_0^{-1} is typically chosen to be a diagonal matrix of the form βI, where β is some large positive constant. This ensures that C_0, the initial guess at the correlation matrix, is approximately 0, but is invertible and symmetric positive definite.

The convergence of RLS TD requires the same conditions as algorithm LS TD, plus one more. This is Condition A.1, or equivalently, Condition A.2:

Condition A.1: $\left[C_0^{-1} + \sum_{k=1}^{t} \rho_k\hat{\omega}_k'\right]$ must be non–singular for all times t.

Condition A.2: $[1 + \hat{\omega}_t'C_{t-1}\rho_t] \neq 0$ for all times t.

Under the assumption that the conditions A.1 and A.2 are maintained, we have that

$$C_t = \left[C_0^{-1} + \sum_{k=1}^{t} \rho_k \hat{\omega}_k' \right]^{-1}$$

and that

$$
\begin{aligned}
\theta_t &= \left[C_0^{-1} + \sum_{k=1}^{t} \rho_k \hat{\omega}_k' \right]^{-1} \left[C_0^{-1} \theta_0 + \sum_{k=1}^{t} \rho_k \psi_k \right] \\
&= \left[\frac{1}{t} C_0^{-1} + \frac{1}{t} \sum_{k=1}^{t} \rho_k \hat{\omega}_k' \right]^{-1} \left[\frac{1}{t} C_0^{-1} \theta_0 + \frac{1}{t} \sum_{k=1}^{t} \rho_k \psi_k \right].
\end{aligned}
$$

If the conditions A.1 and A.2 are not met at some time t_0, then all computations made thereafter will be polluted by the indeterminate or infinite values produced at time t_0. The non–recursive algorithm LS TD does not have this problem because the computations made at any time step do not depend directly on the results of computations made at earlier time steps.

5.5. Dependent or Extraneous Features

The value function for a Markov chain satisfies the equation

$$V = \left[I - \gamma P \right]^{-1} \bar{r}.$$

When using a function approximator linear in the parameters, this means that the parameter vector θ must satisfy the linear equation

$$\Phi \theta = \left[I - \gamma P \right]^{-1} \bar{r}. \tag{16}$$

In this section, the rows of Φ consist only of the feature vectors representing the non–absorbing states, and V only includes the values for the non–absorbing states. This is not essential, but it makes the discussion much simpler. Let $n = |\mathcal{N}|$ be the number of non–absorbing states in the Markov chain. Matrix Φ has dimension $n \times m$, where m is the length of the feature vectors representing the states.

Now, suppose that rank$(\Phi) = m < n$. Dayan (Dayan, 1992) shows that in this case trial–based TD(λ) (Figure 1) converges to $[\Phi'\Pi(I - \gamma P)\Phi]^{-1} [\Phi'\Pi\bar{R}]$ for $\lambda = 0$. This is the same result we achieved in Lemma 5, since $m = $ rank(Φ) if and only if $[\Phi'\Pi(I - \gamma P)\Phi]$ is invertible[2]. The proofs of Theorems 1 and 2 show convergence of θ_{LSTD} to $[\Phi'\Pi(I - \gamma P)\Phi]^{-1} [\Phi'\Pi\bar{R}]$ as a preliminary result. Thus, θ_{LSTD} converges for both absorbing and ergodic chains as long the assumptions of Lemma (5) are satisfied.

Suppose, on the other hand, that rank$(\Phi) = n < m$. This means that the state representations are linearly independent but contain extraneous features. Therefore, there are more adjustable parameters than there are constraints, and an infinite number of parameter vectors θ satisfy (16). The stochastic approximation algorithms TD(λ) and NTD(λ)

converge to some θ that satisfies (16). Which one they find depends on the order in which the states are visited. LS TD does not converge, since $\left[\frac{1}{t} \sum_{k=1}^{t} \phi_k (\phi_k - \gamma \phi_{k+1})' \right]$ is not invertible in this case. However, RLS TD converges to some θ that satisfies (16). In this case, too, the θ to which the algorithm converges depends on the order in which the states are visited.

6. Choosing an Algorithm

When TD(λ) and NTD(λ) algorithms are used with function approximators that are linear in the parameters, they involve $O(m)$ costs at each time step when measured either in terms of the number of basic computer operations, or in terms of memory requirements, where m is the length of the feature vectors representing the states. Algorithm LS TD's costs are $O(m^3)$ in time and $O(m^2)$ in space at each time step, while RLS TD's are $O(m^2)$ in both time and space. TD(λ) and NTD(λ) are clearly superior in terms of cost per time step. However, LS TD and RLS TD are more efficient estimators in the statistical sense. They extract more information from each additional observation. Therefore, we would expect LS TD and RLS TD to converge more rapidly than do TD(λ) and NTD(λ). The use of LS TD and RLS TD is justified, then, if the increased costs per time step are offset by increased convergence rate.

The performance of TD(λ) is sensitive to a number of interrelated factors that do not affect the performance of either LS TD or RLS TD. Convergence of TD(λ) can be dramatically slowed by a poor choice of the step–size (α) and trace (λ) parameters. The algorithm can become unstable if α is too large, causing θ_t to diverge. TD(λ) is also sensitive to the norms of the feature vectors representing the states. Judicious choice of α and λ can prevent instability, but at the price of decreased learning rate. The performance of TD(λ) is also sensitive to $\|\theta_0 - \theta^*\|$, the distance between θ^* and the initial estimate for θ^*. NTD(λ) is sensitive to these same factors, but normalization reduces the sensitivity. In contrast, algorithms LS TD and RLS TD are insensitive to all of these factors. Use of LS TD and RLS TD eliminates the possibility of poor performance due to unlucky choice of parameters.

The transient behavior of a learning algorithm is also important. TD(λ) and NTD(λ) remain stable (assuming that the step–size parameter is small enough) no matter what sequence of states is visited. This is not true for LS TD and RLS TD. If $C_t^{-1} = \left[C_0^{-1} + \sum_{k=1}^{t} \rho_k \hat{\omega}_k' \right]$ is ill–conditioned or singular for some time t, then the estimate θ_t can very far from θ^*. LS TD will recover from this transient event, and is assured of converging eventually to θ^*. The version of RLS TD described in Section 5.4 will not recover if C_t^{-1} is singular. It may or may not recover from an ill–conditioned C_t^{-1}, depending on the machine arithmetic. However, there are well–known techniques for protecting RLS algorithms from transient instability (Goodwin & Sin, 1984).

TD(λ), NTD(λ), and RLS TD have an advantage over LS TD in the case of extraneous features, as discussed in Section 5.5. TD(λ), NTD(λ), and RLS TD converge to the correct value function in this situation, while LS TD does not.

None of the factors discussed above makes a definitive case for one algorithm over another in all situations. The choice depends finally on the computational cost structure imposed on the user of these algorithms.

7. The TD Error Variance

One of the interesting characteristics of the TD error term,

$$e_{TD}(\theta_{t-1}) = R_t + \gamma\phi'_{t+1}\theta_{t-1} - \phi'_t\theta_{t-1},$$

is that it does not go to zero as θ_t converges to θ^*, except in the trivial case of a deterministic Markov chain. This is readily verified by inspection of (10). We define the *TD error variance*, σ_{TD}, of a Markov chain as follows:

$$
\begin{aligned}
\sigma_{TD} &= E\left\{e_{TD}(\theta^*)^2\right\}\\
&= E\left\{\left[R_t + \gamma\phi'_{t+1}\theta^* - \phi'_t\theta^*\right]^2\right\}\\
&= \sum_{x\in X}\pi_x\sum_{y\in X}P(x.y)\left[R(x,y) + \gamma\phi'_y\theta^* - \phi'_x\theta^*\right]^2.
\end{aligned}
$$

σ_{TD} is the variance of the TD error term under the assumptions that θ_t has converged to θ^*, and that the states (and the corresponding TD errors) are sampled on–line by following a sample path of the Markov chain. σ_{TD} is a measure of the noise that cannot be removed from any of the TD learning rules (TD(λ), NTD(λ), LS TD, or RLS TD), even after parameter convergence. It seems reasonable to expect that experimental convergence rates depend on σ_{TD}.

8. Experiments

This section describes two experiments designed to demonstrate the advantage in convergence speed that can be gained through using least–squares techniques. Both experiments compare the performance of NTD(λ) with that of RLS TD in the on–line estimation of the value function of a randomly generated ergodic Markov chain, the first with five states and the second with fifty states (see Appendix B for the specification of the smaller of the Markov chains). The conditions of Theorem 2 are satisfied in these experiments, so that the lengths of the state representation vectors equal five and fifty respectively in the two experiments. In a preliminary series of experiments, not reported here, NTD(λ) always performed at least as well as TD(λ), while showing less sensitivity to the choice of parameters, such as initial step size. Appendix C describes the algorithm we used to set the step size parameters for NTD(λ). Figures 4, 5, and 6 show the experimental results.

The x-axis of Figure 4 measures the TD error variance of the test Markov chain, which was varied over five distinct values from $\sigma_{TD} = 10^{-1}$ through $\sigma_{TD} = 10^3$ by scaling the cost function R. The state transition probability function, P, and the state representations,

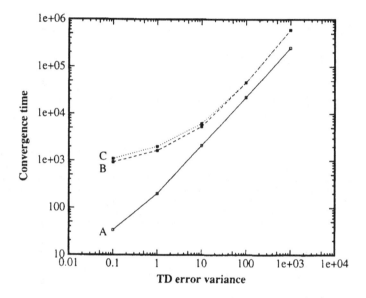

Figure 4. Comparison of RLS TD and NTD(λ) on a randomly generated 5 state ergodic Markov chain. The x-axis measures the TD error variance of the test Markov chain, which was varied over five distinct values from $\sigma_{TD} = 10^{-1}$ through $\sigma_{TD} = 10^3$ by scaling the cost function R. The y-axis measures the average convergence time over 100 training runs of on-line learning. There was one time step counted for each interaction with the environment. The parameter vector was considered to have converged when the average of the error $\|\theta_t - \theta^*\|_\infty$ fell below 10^{-2} and stayed below this value thereafter. Graph A shows the performance of RLS TD. Graph B shows the performance of NTD(λ) where $\|\theta_0 - \theta^*\|_2 = 1$. Graph C shows the performance of NTD(λ) where $\|\theta_0 - \theta^*\|_2 = 2$.

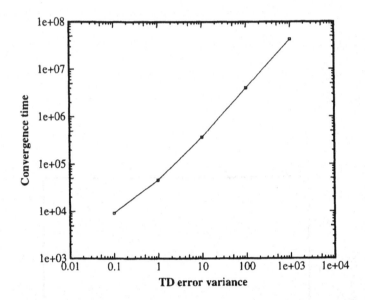

Figure 5. Performance of RLS TD on a randomly generated 50 state ergodic Markov chain. The x-axis measures the TD error variance of the test Markov chain, which was varied over five distinct values from $\sigma_{TD} = 10^{-1}$ through $\sigma_{TD} = 10^3$ by scaling the cost function R. The y-axis measures the average convergence time over 100 training runs of on–line learning. The parameter vector was considered to have converged when the average of the error $\|\theta_t - \theta^*\|_\infty$ fell below 10^{-2} and stayed below this value thereafter.

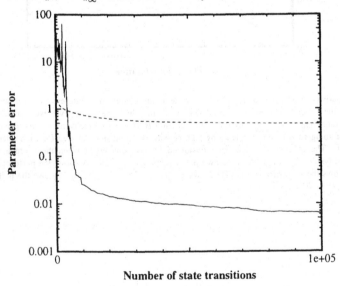

Figure 6. Learning curves for RLS TD (solid) and NTD(λ) (dashed) on a randomly generated 50 state ergodic Markov chain, for $\sigma_{TD} = 1$. The x-axis measures the number of state transitions, or the number of time steps. The y-axis measures the average parameter error over 100 training runs of on–line learning. The parameter error was defined as $\|\theta_t - \theta^*\|_\infty$.

Φ, were left unchanged. The y-axis of Figure 4 measures the average convergence time over 100 training runs of on–line learning. This was computed as follows. For each of the 100 training runs, $\|\theta_t - \theta^*\|_\infty$ was recorded at each time step (where $\|\cdot\|_\infty$ denotes the l_∞, or max, norm). These 100 error curves were averaged to produce the mean error curve. Finally, the mean error curve was inspected to find the time step t at which the average error fell below 10^{-2} and stayed below 10^{-2} for all the simulated times steps thereafter.

Graph A of Figure 4 shows the performance of RLS TD. The other two graphs show the performance of NTD(λ) given different initial values for θ_0. Graph B shows the performance of NTD(λ) when θ_0 was chosen so that $\|\theta_0 - \theta^*\|_2 = 1$ (where $\|\cdot\|_2$ denotes the l_2, or Euclidean, norm). Graph C shows the performance of NTD(λ) when θ_0 was chosen so that $\|\theta_0 - \theta^*\|_2 = 2$. One can see that the performance of NTD(λ) is sensitive to the distance of the initial parameter vector from θ^*. In contrast, the performance of RLS TD is not sensitive to this distance (θ_0 for Graph A was the same as that for Graph B). The performance of NTD(λ) is also sensitive to the settings of four control parameters: λ, α_0, c, and τ. The parameters c and τ govern the evolution of the sequence of step–size parameters (see Appendix C). A search for the best set of control parameters for NTD(λ) was performed for each experiment in an attempt to present NTD(λ) in the best light.[3] The control parameter ϵ (see Equation 4) was held constant at 1.0 for all experiments.

Figure 4 shows a number of things. First, RLS TD outperformed NTD(λ) at every level of σ_{TD}. RLS TD always converged at least twice as fast as NTD(λ), and did much better than that at lower levels of σ_{TD}. Next, we see that, at least for RLS TD, convergence time is a linear function of σ_{TD}: increase σ_{TD} by a factor of 10, and the convergence time can be expected to increase by a factor of 10. This relationship is less clear for NTD(λ), although the curves seem to follow the same rule for larger σ_{TD}. It appears that the effect of the initial distance from θ to θ^*, $\|\theta_0 - \theta^*\|_2$, is significant when σ_{TD} is small but becomes less important, and is finally eliminated, as σ_{TD} increases.

Figures 5 and 6 present the results of repeating the experiment described above for a randomly generated ergodic Markov chain with fifty states. Each state of this larger Markov chain has a possible transition to five other states, on average. Figure 5 shows that the convergence rates for RLS TD follow the same pattern seen in Figure 4: the convergence time rises at the same rate σ_{TD} rises. We attempted to experimentally test the convergence times of NTD(λ) on this problem as we did on the smaller problem. However, we were unable to achieve convergence to the criterion ($\|\theta_t - \theta^*\|_\infty < 10^{-2}$) for any value of σ_{TD} or any selection of the parameters λ, α_0, c, and τ. Figure 6 compares the learning curves generated by RLS TD and NTD(λ) for $\sigma_{TD} = 1$. The parameters governing the behavior of NTD(λ) were the best we could find. After some initial transients, RLS TD settles very rapidly toward convergence, while NTD(λ) settles very slowly toward convergence, making almost no progress for tens of thousands of time steps. These results indicate that the relative advantage of using the RLS TD algorithm may actually improve as the size of the problem grows, despite the order $O(m^2)$ cost required by RLS TD at each time step.

The results shown in Figures 4, 5, and 6 suggest that the use of RLS TD instead of TD(λ) or NTD(λ) is easily justified. RLS TD's costs per time step are an order of $m = |X|$ more expensive in both time and space than the costs for TD(λ) or NTD(λ). However, in the example problems, RLS TD always converged significantly faster than TD(λ) or NTD(λ), and was at least an order of m faster for smaller σ_{TD}. RLS TD has the significant additional advantage that it has no control parameters that have to be adjusted. In contrast, it required a very extensive search to select settings of the control parameters α_0, c, and τ of NTD(λ) to show this algorithm in a good light.

9. Conclusion

We presented three new TD learning algorithms, NTD(λ), LS TD, and RLS TD, and we proved probability 1 convergence for these algorithms under appropriate conditions. These algorithms have a number of advantages over previously proposed TD learning algorithms. NTD(λ) is a normalized version of TD(λ) used with a linear–in–the–parameters function approximator. The normalization serves to reduce the algorithm's sensitivity to the choice of control parameters. LS TD is a Least–Squares algorithm for finding the value function of a Markov chain. Although LS TD is more expensive per time step than the algorithms TD(λ) and NTD(λ), it converges more rapidly and has no control parameters that need to be set, reducing the chances for poor performance. RLS TD is a recursive version of LS TD.

We also defined the TD error variance of a Markov chain. σ_{TD}. σ_{TD} is a measure of the noise that is inherent in any TD learning algorithm, even after the parameters have converged to θ^*. Based on our experiments, we conjecture that the convergence rate of a TD algorithm depends linearly on σ_{TD}(Figure 4). This relationship is very clear for RLS TD, but also seems to hold for NTD(λ) for larger σ_{TD}.

The theorems concerning convergence of LS TD (and RLS TD) can be generalized in at least two ways. First, the immediate rewards can be random variables instead of constants. $R(x, y)$ would then designate the *expected* reward of a transition from state x to state y. The second change involves the way the states (and state transitions) are sampled. Throughout this chapter we have assumed that the states are visited along sample paths of the Markov chain. This need not be the case. All that is necessary is that there is some limiting distribution, π, of the states selected for update, such that $\pi_x > 0$ for all states x.

One of the goals of using a parameterized function approximator (of which the linear–in–the–parameters approximators considered in this article are the simplest examples) is to store the value function more compactly than it could be stored in a lookup table. Function approximators that are linear in the parameters do not achieve this goal if the feature vectors representing the states are linearly independent, since in this case they use the same amount of memory as a lookup table. However, we believe that the results presented here and elsewhere on the performance of TD algorithms with function approximators that are linear in the parameters are first steps toward understanding the performance of TD algorithms using more compact representations.

Appendix A

Proofs of Lemmas

In preparation for the proofs of Lemmas 3 and 4, we first examine the sum $\sum_{y \in X} P(x, y)(R(x, y) - \bar{r}_x)$ for an arbitrary state x:

$$
\begin{aligned}
\sum_{y \in X} P(x, y)(R(x, y) - \bar{r}_x) &= \sum_{y \in X} P(x, y)R(x, y) - \sum_{y \in X} P(x, y)\bar{r}_x \\
&= \sum_{y \in X} P(x, y)R(x, y) - \bar{r}_x \\
&= \bar{r}_x - \bar{r}_x \\
&= 0.
\end{aligned}
$$

Proof of Lemma 3: The result in the preceding paragraph leads directly to a proof that $E\{\eta\} = 0$:

$$
\begin{aligned}
E\{\eta\} &= E\{R(x, y) - \bar{r}_x\} \\
&= \sum_{x \in X} \pi_x \sum_{y \in X} P(x, y)(R(x, y) - \bar{r}_x) \\
&= \sum_{x \in X} \pi_x \cdot 0 \\
&= 0,
\end{aligned}
$$

and to a proof that $\mathrm{Cor}(\omega, \eta) = 0$:

$$
\begin{aligned}
\mathrm{Cor}(\omega, \eta) &= E\{\omega\eta\} \\
&= \sum_{x \in X} \pi_x \sum_{y \in X} P(x, y)[\omega_x \eta_{xy}] \\
&= \sum_{x \in X} \pi_x \sum_{y \in X} P(x, y)\omega_x(R(x, y) - \bar{r}_x) \\
&= \sum_{x \in X} \pi_x \omega_x \sum_{y \in X} P(x, y)(R(x, y) - \bar{r}_x) \\
&= \sum_{x \in X} \pi_x \omega_x \cdot 0 \\
&= 0.
\end{aligned}
$$

■

Proof of Lemma 4: First we consider $\mathrm{Cor}(\rho, \eta)$:

$$
\mathrm{Cor}(\rho, \eta) = E\{\rho\eta'\}
$$

$$
\begin{aligned}
&= \sum_{x \in X} \pi_x \sum_{y \in X} P(x,y) \left[\rho_x \eta'_{xy} \right] \\
&= \sum_{x \in X} \pi_x \sum_{y \in X} P(x,y) \phi_x (R(x,y) - \bar{r}_x)' \\
&= \sum_{x \in X} \pi_x \phi_x \sum_{y \in X} P(x,y) (R(x,y) - \bar{r}_x)' \\
&= \sum_{x \in X} \pi_x \phi_x \cdot 0 \\
&= 0.
\end{aligned}
$$

Now for $\text{Cor}(\rho, \zeta)$:

$$
\begin{aligned}
\text{Cor}(\rho, \zeta) &= E\{\rho \zeta'\} \\
&= \sum_{x \in X} \pi_x \sum_{y \in X} P(x,y) \left[\rho_x \zeta'_{xy} \right] \\
&= \sum_{x \in X} \pi_x \sum_{y \in X} P(x,y) \phi_x (\gamma \sum_{z \in X} P(x,z) \phi_z - \gamma \phi_y)' \\
&= \sum_{x \in X} \pi_x \phi_x \sum_{y \in X} P(x,y) (\gamma \sum_{z \in X} P(x,z) \phi'_z) - \\
&\quad\;\; \sum_{x \in X} \pi_x \phi_x \sum_{y \in X} P(x,y) \gamma \phi'_y \\
&= \sum_{x \in X} \pi_x \phi_x \gamma \sum_{z \in X} P(x,z) \phi'_z - \sum_{x \in X} \pi_x \phi_x \gamma \sum_{y \in X} P(x,y) \phi'_y \\
&= 0.
\end{aligned}
$$

∎

Proof of Lemma 5: Equation 11 (repeated here) gives us the tth estimate found by algorithm LS TD for θ^*:

$$
\theta_t = \left[\frac{1}{t} \sum_{k=1}^{t} \phi_k (\phi_k - \gamma \phi_{k+1})' \right]^{-1} \left[\frac{1}{t} \sum_{k=1}^{t} \phi_k R_k \right].
$$

As t grows we have by condition (2) that the *sampled* transition probabilities between each pair of states approaches the *true* transition probabilities, P, with probability 1. We also have by condition (3) that each state $x \in X$ is visited in the proportion π_x with probability 1. Therefore, given condition (4) we can express the limiting estimate found by algorithm LS TD, θ_{LSTD}, as

$$
\theta_{\text{LSTD}} = \lim_{t \to \infty} \theta_t
$$

$$= \lim_{t \to \infty} \left[\frac{1}{t} \sum_{k=1}^{t} \phi_k (\phi_k - \gamma \phi_{k+1})' \right]^{-1} \left[\frac{1}{t} \sum_{k=1}^{t} \phi_k R_k \right]$$

$$= \left[\lim_{t \to \infty} \frac{1}{t} \sum_{k=1}^{t} \phi_k (\phi_k - \gamma \phi_{k+1})' \right]^{-1} \left[\lim_{t \to \infty} \frac{1}{t} \sum_{k=1}^{t} \phi_k R_k \right]$$

$$= \left[\sum_x \pi_x \sum_y P(x,y) \phi_x (\phi_x - \gamma \phi_y)' \right]^{-1} \left[\sum_x \pi_x \phi_x \sum_y P(x,y) R(x,y) \right]$$

$$= \left[\sum_x \pi_x \sum_y P(x,y) \phi_x (\phi_x - \gamma \phi_y)' \right]^{-1} \left[\sum_x \pi_x \bar{R}_x \phi_x \right]$$

$$= [\Phi' \Pi (I - \gamma P) \Phi]^{-1} [\Phi' \Pi \bar{R}] .$$

∎

Appendix B

The Five–State Markov Chain Example

The transition probability function of the five–state Markov chain used in the experiments appears in matrix form as follows, where the entry in row i, column j is the probability of a transition from state i to state j (rounded to two decimal places):

$$\begin{bmatrix} 0.42 & 0.13 & 0.14 & 0.03 & 0.28 \\ 0.25 & 0.08 & 0.16 & 0.35 & 0.15 \\ 0.08 & 0.20 & 0.33 & 0.17 & 0.22 \\ 0.36 & 0.05 & 0.00 & 0.51 & 0.07 \\ 0.17 & 0.24 & 0.19 & 0.18 & 0.22 \end{bmatrix}$$

The feature vectors representing the states (rounded to two decimal places) are listed as the rows of the following matrix Φ:

$$\begin{bmatrix} 74.29 & 34.61 & 73.48 & 53.29 & 7.79 \\ 61.60 & 48.07 & 34.68 & 36.19 & 82.02 \\ 97.00 & 4.88 & 8.51 & 87.89 & 5.17 \\ 41.10 & 40.13 & 64.63 & 92.67 & 31.09 \\ 7.76 & 79.82 & 43.78 & 8.56 & 61.11 \end{bmatrix}$$

The immediate rewards were specified by the following matrix (which has been rounded to two decimal places), where the entry in row i, column j determines the immediate reward for a transition from state i to state j. The matrix was scaled to produce the different TD error variance values used in the experiments. The relative sizes of the immediate rewards remained the same.

$$R = \begin{bmatrix} 104.66 & 29.69 & 82.36 & 37.49 & 68.82 \\ 75.86 & 29.24 & 100.37 & 0.31 & 35.99 \\ 57.68 & 65.66 & 56.95 & 100.44 & 47.63 \\ 96.23 & 14.01 & 0.88 & 89.77 & 66.77 \\ 70.35 & 23.69 & 73.41 & 70.70 & 85.41 \end{bmatrix}$$

Appendix C

Selecting Step–Size Parameters

The convergence theorem for NTD(0) (Bradtke, 1994) requires a separate step–size parameter, $\alpha(x)$, for each state x, that satisfies the Robbins and Monro (Robbins & Monro, 1951) criteria

$$\sum_{k=1}^{\infty} \alpha_k(x) = \infty \quad \text{and} \quad \sum_{k=1}^{\infty} \alpha_k(x)^2 < \infty$$

with probability 1, where $\alpha_k(x)$ is the step–size parameter for the k–th visitation of state x. Instead of a separate step–size parameter for each state, we used a single parameter α_t, which we decreased at every time step. For each state x there is a corresponding subsequence $\{\alpha_t\}_x$ that is used to update the value function when x is visited. We conjecture that if the original sequence $\{\alpha_t\}$ satisfies the Robbins and Monro criteria, then these subsequences also satisfy the criteria, with probability 1. The overall convergence rate may be decreased by use of a single step–size parameter since each subsequence will contain fewer large step sizes.

The step–size parameter sequence $\{\alpha_t\}$ was generated using the "search then converge" algorithm described by Darken, Chang, and Moody (Darken, et al., 1992):

$$\alpha_t = \alpha_0 \frac{1 + \frac{c}{\alpha_0} \frac{t}{\tau}}{1 + \frac{c}{\alpha_0} \frac{t}{\tau} + \tau \frac{t^2}{\tau^2}}.$$

The choice of parameters α_0, c, and τ determines the transition of learning from "search mode" to "converge mode". Search mode describes the time during which $t \ll \tau$. Converge mode describes the time during which $t \gg \tau$. α_t is nearly constant in search mode, while $\alpha_t \approx \frac{c}{t}$ in converge mode. The ideal choice of step–size parameters moves θ_t as quickly as possible into the vicinity of θ^* during search mode, and then settles into converge mode.

Notes

1. If the set of feature vectors is linearly independent, then there exist parameter values such that *any* real-valued function of X can be approximated with zero error by a function approximator linear in the parameters. Using terminology from adaptive control (Goodwin & Sin, 1984), this situation is said to satisfy the *exact matching* condition for arbitrary real–valued functions of X.

2. m can not be less than rank(Φ). If $m >$ rank(Φ), then $[\Phi'\Pi(I - \gamma P)\Phi]$ is an $(m \times m)$ matrix with rank less than m. It is therefore not invertible.

3. The search for the best settings for λ, α_0, c, and τ was the limiting factor on the size of the state space for this experiment.

References

Anderson, C. W., (1988). Strategy learning with multilayer connectionist representations. Technical Report 87-509.3, GTE Laboratories Incorporated, Computer and Intelligent Systems Laboratory, 40 Sylvan Road, Waltham, MA 02254.

Barto, A. G. , Sutton, R. S. & Anderson, C. W. (1983). Neuronlike elements that can solve difficult learning control problems. *IEEE Transactions on Systems, Man, and Cybernetics*, 13:835–846.

Bradtke, S. J., (1994). *Incremental Dynamic Programming for On-Line Adaptive Optimal Control*. PhD thesis, University of Massachusetts, Computer Science Dept. Technical Report 94-62.

Darken, C. Chang, J. & Moody, J., (1992) Learning rate schedules for faster stochastic gradient search. In *Neural Networks for Signal Processing 2 — Proceedings of the 1992 IEEE Workshop*. IEEE Press.

Dayan, P., (1992). The convergence of TD(λ) for general λ. *Machine Learning*, 8:341–362.

Dayan, P. & Sejnowski, T.J., (1994). TD(λ): Convergence with probability 1. *Machine Learning*.

Goodwin, G.C. & Sin, K.S., (1984). *Adaptive Filtering Prediction and Control*. Prentice-Hall, Englewood Cliffs, N.J.

Jaakkola, T, Jordan, M.I. & Singh, S.P., (1994). On the convergence of stochastic iterative dynamic programming algorithms. *Neural Computation*, 6(6).

Kemeny, J.G. & Snell, J.L., (1976). *Finite Markov Chains*. Springer-Verlag, New York.

Ljung, L. & Söderström, T., (1983). *Theory and Practice of Recursive Identification*. MIT Press, Cambridge, MA.

Lukes, G., Thompson, B. & Werbos, P., (1990). Expectation driven learning with an associative memory. In *Proceedings of the International Joint Conference on Neural Networks*, pages I:521–524.

Robbins, H & Monro, S., (1951). A stochastic approximation method. *Annals of Mathematical Statistics*, 22:400–407.

Söderström, T. & Stoica, P.G., (1983). *Instrumental Variable Methods for System Identification*. Springer-Verlag, Berlin.

Sutton. A.S., (1984). *Temporal Credit Assignment in Reinforcement Learning*. PhD thesis, Department of Computer and Information Science, University of Massachusetts at Amherst, Amherst, MA 01003.

Sutton, R.S., (1988). Learning to predict by the method of temporal differences. *Machine Learning*, 3:9–44.

Tesauro, G.J., (1992). Practical issues in temporal difference learning. *Machine Learning*, 8(3/4):257–277.

Tsitsiklis, J.N., (1993). Asynchronous stochastic approximation and Q-learning. Technical Report LIDS-P-2172, Laboratory for Information and Decision Systems, MIT, Cambridge, MA.

Watkins, C. J. C. H., (1989). *Learning from Delayed Rewards*. PhD thesis, Cambridge University, Cambridge, England.

Watkins, C. J. C. H. & Dayan, P., (1992). Q-learning. *Machine Learning*, 8(3/4):257–277, May 1992.

Werbos, P.J., (1987). Building and understanding adaptive systems: A statistical/numerical approach to factory automation and brain research. *IEEE Transactions on Systems. Man, and Cybernetics*, 17(1):7–20.

Werbos, P.J., (1988). Generalization of backpropagation with application to a recurrent gas market model. *Neural Networks*, 1(4):339–356, 1988.

Werbos, P.J., (1990). Consistency of HDP applied to a simple reinforcement learning problem. *Neural Networks*, 3(2):179–190.

Werbos, P.J., (1992). Approximate dynamic programming for real-time control and neural modeling. In D. A. White and D. A. Sofge, editors, *Handbook of Intelligent Control: Neural, Fuzzy, and Adaptive Approaches*, pages 493–525. Van Nostrand Reinhold, New York.

Young, P., (1984). *Recursive Estimation and Time-series Analysis*. Springer–Verlag.

Received November 10, 1994
Accepted March 10, 1995
Final Manuscript October 4, 1995

References

Machine Learning, 22, 59–94 (1996)

Feature-Based Methods
for Large Scale Dynamic Programming

JOHN N. TSITSIKLIS AND BENJAMIN VAN ROY   jnt@mit.edu, bvr@mit.edu
Laboratory for Information and Decision Systems
Massachusetts Institute of Technology, Cambridge, MA 02139

Editor: Leslie Pack Kaelbling

Abstract. We develop a methodological framework and present a few different ways in which dynamic programming and compact representations can be combined to solve large scale stochastic control problems. In particular, we develop algorithms that employ two types of feature-based compact representations; that is, representations that involve feature extraction and a relatively simple approximation architecture. We prove the convergence of these algorithms and provide bounds on the approximation error. As an example, one of these algorithms is used to generate a strategy for the game of Tetris. Furthermore, we provide a counter-example illustrating the difficulties of integrating compact representations with dynamic programming, which exemplifies the shortcomings of certain simple approaches.

Keywords: Compact representation, curse of dimensionality, dynamic programming, features, function approximation, neuro-dynamic programming, reinforcement learning.

1. Introduction

Problems of sequential decision making under uncertainty (stochastic control) have been studied extensively in the operations research and control theory literature for a long time, using the methodology of dynamic programming (Bertsekas, 1995). The "planning problems" studied by the artificial intelligence community are of a related nature although, until recently, this was mostly done in a deterministic setting leading to search or shortest path problems in graphs (Korf, 1987). In either context, realistic problems have usually proved to be very difficult mostly due to the large size of the underlying state space or of the graph to be searched. In artificial intelligence, this issue is usually addressed by using heuristic *position evaluation functions*; chess playing programs are a prime example (Korf, 1987). Such functions provide a rough evaluation of the quality of a given state (or board configuration in the context of chess) and are used in order to rank alternative actions.

In the context of dynamic programming and stochastic control, the most important object is the *cost–to–go function*, which evaluates the expected future cost to be incurred, as a function of the current state. Similarly with the artificial intelligence context, cost–to–go functions are used to assess the consequences of any given action at any particular state. Dynamic programming provides a variety of methods for computing cost-to-go functions. Due to the curse of dimensionality, however, the practical applications of dynamic programming are somewhat limited; they involve certain problems in which the cost–to–go function has a simple analytical form (e.g., controlling a linear system subject to a quadratic cost) or to problems with a manageably small state space.

In most of the stochastic control problems that arise in practice (control of nonlinear systems, queueing and scheduling, logistics, etc.) the state space is huge. For example, every possible configuration of a queueing system is a different state, and the number of states increases exponentially with the number of queues involved. For this reason, it is essentially impossible to compute (or even store) the value of the cost–to–go function at every possible state. The most sensible way of dealing with this difficulty is to generate a compact parametric representation (compact representation, for brevity), such as an artificial neural network, that approximates the cost–to–go function and can guide future actions, much the same as the position evaluators are used in chess. Since a compact representation with a relatively small number of parameters may approximate a cost-to-go function, we are required to compute only a few parameter values rather than as many values as there are states.

There are two important preconditions for the development of an effective approximation. First, we need to choose a compact representation that can closely approximate the desired cost-to-go function. In this respect, the choice of a suitable compact representation requires some practical experience or theoretical analysis that provides some rough information on the shape of the function to be approximated. Second, we need effective algorithms for tuning the parameters of the compact representation. These two objectives are often conflicting. Having a compact representation that can approximate a rich set of functions usually means that there is a large number of parameters to be tuned and/or that the dependence on the parameters is nonlinear, and in either case, there is an increase in the computational complexity involved.

It is important to note that methods of selecting suitable parameters for standard function approximation are inadequate for approximation of cost-to-go functions. In function approximation, we are given training data pairs $\{(x_1, y_1), \ldots, (x_K, y_K)\}$ and must construct a function $y = f(x)$ that "explains" these data pairs. In dynamic programming, we are interested in approximating a cost-to-go function $y = V(x)$ mapping states to optimal expected future costs. An ideal set of training data would consist of pairs $\{(x_1, y_1), \ldots, (x_K, y_K)\}$, where each x_i is a state and each y_i is a sample of the future cost incurred starting at state x_i when the system is optimally controlled. However, since we do not know how to control the system at the outset (in fact, our objective is to figure out how to control the system), we have no way of obtaining such data pairs. An alternative way of making the same point is to note that the desirability of a particular state depends on how the system is controlled, so observing a poorly controlled system does not help us tell how desirable a state will be when the system is well controlled. To approximate a cost-to-go function, we need variations of the algorithms of dynamic programming that work with compact representations.

The concept of approximating cost-to-go functions with compact representations is not new. Bellman and Dreyfus (1959) explored the use of polynomials as compact representations for accelerating dynamic programming. Whitt (1978) and Reetz (1977) analyzed approaches of reducing state space sizes, which lead to compact representations. Schweitzer and Seidmann (1985) developed several techniques for approximating cost-to-go functions using linear combinations of fixed sets of basis functions. More recently, reinforcement learning researchers have developed a number of approaches, including temporal-difference learning (Sutton, 1988) and Q–learning (Watkins and Dayan, 1992), which have been used for dynamic programming with many types of compact representation, especially artificial neural networks.

Aside from the work of Whitt (1988) and Reetz (1977), the techniques that have been developed largely rely on heuristics. In particular, there is a lack of formal proofs guaranteeing sound results. As one might expect from this, the methods have generated a mixture of success stories and failures. Nevertheless, the success stories – most notably the world-class backgammon player of Tesauro (1992) – inspire great expectations in the potential of compact representations and dynamic programming.

The main aim of this paper is to provide a methodological foundation and a rigorous assessment of a few different ways that dynamic programming and compact representations can be combined to form the basis of a rational approach to difficult stochastic control problems. Although heuristics have to be involved at some point, especially in the selection of a particular compact representation, it is desirable to retain as much as possible of the non-heuristic aspects of the dynamic programming methodology. A related objective is to provide results that can help us assess the efficacy of alternative compact representations.

Cost-to-go functions are generally nonlinear, but often demonstrate regularities similar to those found in the problems tackled by traditional function approximation. There are several types of compact representations that one can use to approximate a cost–to–go function. (a) Artificial neural networks (e.g., multi-layer perceptrons) present one possibility. This approach has led to some successes, such as Tesauro's backgammon player which was mentioned earlier. Unfortunately, it is very hard to quantify or analyze the performance of neural-network-based techniques. (b) A second form of compact representation is based on the use of feature extraction to map the set of states onto a much smaller set of feature vectors. By storing a value of the cost–to–go function for each possible feature vector, the number of values that need to be computed and stored can be drastically reduced and, if meaningful features are chosen, there is a chance of obtaining a good approximation of the true cost-to-go function. (c) A third approach is to choose a parametric form that maps the feature space to cost–to–go values and then try to compute suitable values for the parameters. If the chosen parametric representation is simple and structured, this approach may be amenable to mathematical analysis. One such approach, employing linear approximations, will be studied here.

In this paper, we focus on dynamic programming methods that employ the latter two types of compact representations, i.e., the feature–based compact representations. We provide a general framework within which one can reason about such methods. We also suggest variants of the value iteration algorithm of dynamic programming that can be used in conjunction with the representations we propose. We prove convergence results for our algorithms and then proceed to derive bounds on the difference between optimal performance and the performance obtained using our methods. As an example, one of the techniques presented is used to generate a strategy for Tetris, the arcade game.

This paper is organized as follows. In Section 2, we introduce the Markov decision problem (MDP), which provides a mathematical setting for stochastic control problems, and we also summarize the value iteration algorithm and its properties. In Section 3, we propose a conceptual framework according to which different approximation methodologies can be studied. To illustrate some of the difficulties involved with employing compact representations for dynamic programming, in Section 4, we describe a "natural" approach for dynamic programming with compact representations and then present a counter-example demonstrating the shortcomings of such an approach. In Section 5, we propose a variant of the value iteration algorithm that employs a look-up table in feature space rather than in state space. We also discuss a theorem that ensures its convergence and provides bounds on the accuracy of resulting approximations. Section 6 discusses an application of the algorithm from Section 5 to the game of Tetris. In Section 7, we present our second approximation methodology, which employs feature extraction and linear approximations. Again, we provide a convergence theorem as well as bounds on the performance it delivers. This general methodology encompasses many types of compact representations, and in Sections 8 and 9 we provide two subclasses: interpolative representations and localized basis function architectures. Two technical results that are central to our convergence theorems are presented in the Appendices A and B. In particular, Appendix A proves a theorem involving transformations that preserve contraction properties of an operator, and Appendix B reviews a result on stochastic approximation algorithms involving maximum norm contractions. Appendices C and D provide proofs of the convergence theorems of Sections 5 and 7, respectively.

2. Markov Decision Problems

In this section, we introduce Markov decision problems, which provide a model for sequential decision making problems under uncertainty (Bertsekas, 1995).

We consider infinite horizon, discounted Markov decision problems defined on a finite state space S. Throughout the paper, we let n denote the cardinality of S and, for simplicity, assume that $S = \{1, \ldots, n\}$. For every state $i \in S$, there is a finite set $U(i)$ of possible control actions and a set of nonnegative scalars $p_{ij}(u)$, $u \in U(i)$, $j \in S$, such that $\sum_{j \in S} p_{ij}(u) = 1$ for all $u \in U(i)$. The scalar $p_{ij}(u)$ is interpreted as the probability of a transition to state j, given that the current state is i and the control u is applied. Furthermore, for every state i and control u, there is a random variable c_{iu} which represents the one-stage cost if action u is applied at state i. We assume that the variance of c_{iu} is finite for every $i \in S$ and $u \in U(i)$. In this paper, we treat only Markov decision problems for which transition probabilities $p_{ij}(u)$ and expected immediate costs $E[c_{iu}]$ are known. However, the ideas presented generalize to the context of algorithms such as Q–learning, which assume no knowledge of transition probabilities and costs.

A *stationary policy* is a function π defined on S such that $\pi(i) \in U(i)$ for all $i \in S$. Given a stationary policy, we obtain a discrete-time Markov chain $s^\pi(t)$ with transition probabilities

$$\Pr(s^\pi(t+1) = j \mid s^\pi(t) = i) = p_{ij}(\pi(i)).$$

Let $\beta \in [0, 1)$ be a discount factor. For any stationary policy π and initial state i, the cost-to-go V_i^π is defined by

$$V_i^\pi = E\left[\sum_{t=0}^{\infty} \beta^t c(t) \middle| s^\pi(0) = i \right],$$

where $c(t) = c_{s^\pi(t), \pi(s^\pi(t))}$. In much of the dynamic programming literature, the mapping from states to cost-to-go values is referred to as the cost-to-go function. However, since the state spaces we consider in this paper are finite, we choose to think of the mapping in terms of a cost-to-go vector whose components are the cost-to-go values of various states. Hence, given the cost-to-go vector V^π of policy π, the cost-to-go value of policy π at state i is the ith component of V^π. The optimal cost-to-go vector V^* is defined by

$$V_i^* = \min_\pi V_i^\pi, \qquad i \in S.$$

It is well known that the optimal cost-to-go vector V^* is the unique solution to Bellman's equation:

$$V_i^* = \min_{u \in U(i)} \left(E[c_{iu}] + \beta \sum_{j \in S} p_{ij}(u) V_j^* \right), \qquad \forall i \in S. \tag{1}$$

This equation simply states that the optimal cost–to–go starting from a state i is equal to the minimum, over all actions u that can be taken, of the immediate expected cost $E[c_{iu}]$ plus the suitably discounted expected cost–to–go V_j^* from the next state j, assuming that an optimal policy will be followed in the future.

The Markov decision problem is to find a policy π^* such that

$$V_i^{\pi^*} = V_i^*, \qquad \forall i \in S.$$

This is usually done by computing V_i^*, and then choosing π^* as a function which satisfies

$$\pi^*(i) = \arg \min_{u \in U(i)} \left(E[c_{iu}] + \beta \sum_{j \in S} p_{ij}(u) V_j^* \right), \qquad \forall i \in S.$$

If we can not compute V^* but can obtain an approximation V to V^*, we might generate a reasonable control policy π_V satisfying

$$\pi_V(i) = \arg \min_{u \in U(i)} \left(E[c_{iu}] + \beta \sum_{j \in S} p_{ij}(u) V_j \right), \qquad \forall i \in S.$$

Intuitively, this policy considers actual immediate costs and uses V to judge future consequences of control actions. Such a policy is sometimes called a *greedy policy* with respect to the cost-to-go vector V, and as V approaches V^*, the performance of a greedy policy π_V approaches that of an optimal policy π^*.

There are several algorithms for computing V^* but we only discuss the value iteration algorithm which forms the basis of the algorithms to be considered later on. We start with some notation. We define $T_i : \Re^n \longmapsto \Re$ by

$$T_i(V) = \min_{u \in U(i)} \left(E[c_{iu}] + \beta \sum_{j \in S} p_{ij}(u)V_j \right). \qquad \forall i \in S. \tag{2}$$

We then define the *dynamic programming operator* $T : \Re^n \mapsto \Re^n$ by

$$T(V) = (T_1(V), \ldots, T_n(V)).$$

In terms of this notation, Bellman's equation simply asserts that $V^* = T(V^*)$ and V^* is the unique fixed point of T. The value iteration algorithm is described by

$$V(t+1) = T(V(t)),$$

where $V(0)$ is an arbitrary vector in $\Re^n$ used to initialize the algorithm. Intuitively, each $V(t)$ is an estimate (though not necessarily a good one) of the true cost–to–go function V^*, which gets replaced by the hopefully better estimate $T(V(t))$.

Let $\| \cdot \|_\infty$ be the maximum norm defined for every vector $x = (x_1, \ldots, x_n) \in \Re^n$ by $\|x\|_\infty = \max_i |x_i|$. It is well known (Bertsekas, 1995) and easy to check that T is a contraction with respect to the maximum norm, that is, for all $V, V' \in \Re^n$,

$$\|T(V) - T(V')\|_\infty \leq \beta \|V - V'\|_\infty.$$

For this reason, the sequence $V(t)$ produced by the value iteration algorithm converges to V^*, at the rate of a geometric progression. Unfortunately, this algorithm requires that we maintain and update a vector V of dimension n and this is essentially impossible when n is extremely large.

For notational convenience, it is useful to define for each policy π the operator $T_i^\pi : \Re^n \mapsto \Re$:

$$T_i^\pi(V) = E[c_{i\pi(i)}] + \beta \sum_{j \in S} p_{ij}(\pi(i))V_j.$$

for each $i \in S$. The operator T^π is defined by

$$T^\pi(V) = (T_1^\pi(V), \ldots, T_n^\pi(V)).$$

It is well known that T^π is also a contraction of the maximum norm and that V^π is its unique fixed point (Bertsekas, 1995). Note that, for any vector $V \in \Re^n$ we have

$$T(V) = T^{\pi^V}(V),$$

since the cost-minimizing control action in Equation (2) is given by the greedy policy.

3.   Compact Representations and Features

As mentioned in the introduction, the size of state spaces typically grows exponentially with the number of variables involved. Because of this, it is often impractical to compute and store every component of a cost-to-go vector. We set out to overcome this limitation by using compact representations to approximate cost-to-go vectors. In this section, we develop a formal framework for reasoning about compact representations and features as groundwork for subsequent sections, where we will discuss ways of using compact representations for dynamic programming. The setting is in many respects similar to that in (Schweitzer and Seidman, 1985).

A compact representation can be thought of as a scheme for recording a high-dimensional cost-to-go vector $V \in \Re^n$ using a lower-dimensional parameter vector $W \in \Re^m$ ($m \ll n$). Such a scheme can be described by a mapping $\tilde{V} : \Re^m \mapsto \Re^n$ which to any given parameter vector $W \in \Re^m$ associates a cost-to-vector $\tilde{V}(W)$. In particular, each component $\tilde{V}_i(W)$ of the mapping is the ith component of a cost-to-go vector represented by the parameter vector W. Note that, although we may wish to represent an arbitrary vector V in $\Re^n$, such a scheme allows for exact representation only of those vectors V which happen to lie in the range of $\tilde{V}$.

Let us define a *feature* f as a function from the state space S into a finite set Q of feature values. For example, if the state i represents the number of customers in a queueing system, a possible and often interesting feature f is defined by $f(0) = 0$ and $f(i) = 1$ if $i > 0$. Such a feature focuses on whether a queue is empty or not.

Given a Markov decision problem, one may wish to use several features $f_1, \ldots, f_K$, each one being a function from the state space S to a finite set Q_k, $k = 1, \ldots, K$. Then, to each state $i \in S$, we associate the feature vector $F(i) = (f_1(i), \ldots, f_K(i))$. Such a feature vector is meant to represent the most salient properties of a given state. Note that the resulting set of all possible feature vectors is the Cartesian product of the sets Q_k and its cardinality increases exponentially with the number of features.

In a feature-based compact representation, each component $\tilde{V}_i$ of the mapping $\tilde{V}$ is a function of the corresponding feature vector $F(i)$ and the parameter vector W (but not an explicit function of the state value i). Hence, for some function $g : (\prod_{k=1}^{K} Q_k) \times \Re^m \mapsto \Re$,

$$\tilde{V}_i(W) = g(F(i), W). \tag{3}$$

If each feature takes on real values, we have $Q_k \subset \Re$ for all k, in which case it may be natural to define the function g over all possible real feature values, $g : \Re^K \times \Re^m \mapsto \Re$, even though g will only ever be computed over a finite domain. Figure 1 illustrates the structure of a feature-based compact representation.

In most problems of interest, V_i^* is a highly complicated function of i. A representation like the one in Equation (3) attempts to break the complexity of V^* into less complicated mappings g and F. There is usually a trade-off between the complexity of g and F and different choices lead to drastically different structures. As a general principle, the feature extraction function F is usually hand crafted and relies on whatever human experience or intelligence is available. The function g represents the choice of an *architecture* used for approximation and the vector W are the free parameters (or weights) of the chosen architecture. When a compact representation is used for static function approximation, the values for the parameters W are chosen using some optimization algorithm, which could range from linear regression to backpropagation in neural networks. In this paper, however, we will develop parameter selection techniques for dynamic programming (rather than function approximation). Let us first discuss some alternative architectures.

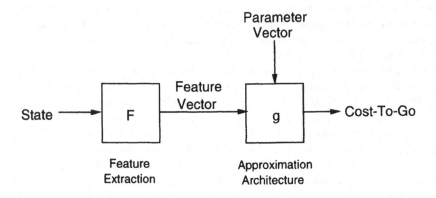

Figure 1. Block structure of a feature-based compact representation.

Look-Up Tables

One possible compact representation can be obtained by employing a look-up table in feature space, that is, by assigning one value to each point in the feature space. In this case, the parameter vector W contains one component for each possible feature vector. The function g acts as a hashing function, selecting the component of W corresponding to a given feature vector. In one extreme case, each feature vector corresponds to a single state, there are as many parameters as states, and $\tilde{V}$ becomes the identity function. On the other hand, effective feature extraction may associate many states with each feature vector so that the optimal cost-to-go values of states associated to any particular feature vector are close. In this scenario, the feature space may be much smaller that the state space, reducing the number of required parameters. Note, however, that the number of possible feature vectors increases exponentially with the number of features. For this reason, look-up tables are only practical when there are very few features.

Using a look-up table in feature space is equivalent to partitioning the state space and then using a common value for the cost-to-go from all the states in any given partition. In this context, the set of states which map to a particular feature vector forms one partition. By identifying one such partition per possible feature vector, the feature extraction mapping F defines a partitioning of the state space. The function g assigns each component of the parameter vector to a partition. For conceptual purposes, we choose to view this type of representation in terms of state aggregation, rather than feature-based look-up tables. As we will see in our formulation for Tetris, however, the feature-based look-up table interpretation is often more natural in applications.

We now develop a mathematical description of state aggregation. Suppose that the state space $S = \{1, ..., n\}$ has been partitioned into m disjoint subsets $S_1, \ldots, S_m$, where m is the same as the dimension of the parameter vector W. The compact representations we consider take on the following form:

$$\tilde{V}_i(W) = W_j,$$

for any $i \in S_j$.

There are no inherent limitations to the representational capability of such an architecture. Whatever limitations this approach may have are actually connected with the availability of useful features. To amplify this point, let us fix some $\epsilon > 0$ and let us define, for all j,

$$S_j = \{i \mid j\epsilon \leq V_i^* < (j+1)\epsilon\}.$$

Using this particular partition, the function V^* can be approximated with an accuracy of ϵ. The catch is of course that since V^* is unknown, we are unable to form the sets S_j. A different way of making the same point is to note that the most useful feature of a state is its optimal cost–to–go but, unfortunately, this is what we are trying to compute in the first place.

Linear Architectures

With a look-up table, we need to store one parameter for every possible value of the feature vector $F(i)$, and, as already noted, the number of possible values increases exponentially with the number of features. As more features are deemed important, look-up tables must be abandoned at some point and a different kind of parametric representation is now called for. For instance, a representation of the following form can be used:

$$\tilde{V}_i(W) = \sum_{k=1}^{K} W_k f_k(i). \tag{4}$$

This representation approximates a cost-to-go function using a linear combination of features. This simplicity makes it amenable to rigorous analysis, and we will develop an algorithm for dynamic programming with such a representation. Note that the number of parameters only grows linearly with the number of features. Hence, unlike the case of look-up tables, the number of features need not be small. However, it is important to choose features that facilitate the linear approximation.

Many popular function approximation architectures fall in the class captured by Equation (4). Among these are radial basis functions, wavelet networks, polynomials, and more generally all approximation methods that involve a fixed set of basis functions. In this paper, we will discuss two types of these compact representations that are compatible with our algorithm – a method based on linear interpolation and localized basis functions.

Nonlinear Architectures

The architecture, as described by g, could be a nonlinear mapping such as a feedforward neural network (multi-layer perceptron) with parameters W. The feature extraction mapping F could be either entirely absent or it could be included to facilitate the job of the neural network. Both of these options were used in the backgammon player of Tesauro and, as expected, the inclusion of features led to improved performance. Unfortunately, as was mentioned in the introduction, there is not much that can be said analytically in this context.

4. Least-Squares Value Iteration: A Counter-Example

Given a set of k samples $\{(i_1, V_{i_1}^*), (i_2, V_{i_2}^*), ..., (i_K, V_{i_K}^*)\}$ of an optimal cost-to-go vector V^*, we could approximate the vector with a compact representation $\tilde{V}$ by choosing parameters W to minimize an error function such as

$$\sum_{k=1}^{K} \left(\tilde{V}_{i_k}(W) - V_{i_k}^*\right)^2.$$

i.e., by finding the "least-squares fit." Such an approximation conforms to the spirit of traditional function approximation. However, as discussed in the introduction, we do not have access to such samples of the optimal cost-to-go vector. To approximate an optimal cost-to-go vector, we must adapt dynamic programming algorithms such as the value iteration algorithm so that they manipulate parameters of compact representations.

For instance, we could start with a parameter vector $W(0)$ corresponding to an initial cost-to-go vector $\tilde{V}(W(0))$, and then generate a sequence $\{W(t)|t = 1, 2, ...\}$ of parameter vectors such that $\tilde{V}(W(t + 1))$ approximates $T(\tilde{V}(W(t)))$. Hence, each iteration approximates a traditional value iteration. The hope is that, by approximating individual value iterations in such a way, the sequence of approximations converges to an accurate approximation of the optimal cost-to-go vector, which is what value iteration converges to.

It may seem as though any reasonable approximation scheme could be used to generate each approximate value iteration. For instance, the "least-squares fit" is an obvious candidate. This involves selecting $W(t + 1)$ by setting

$$W(t + 1) = \arg\min_{W} \sum_{i=1}^{n} \left(\tilde{V}(W) - T(\tilde{V}(W(t)))\right)^2. \tag{5}$$

However, in this section we will identify subtleties that make the choice of criterion for parameter selection crucial. Furthermore, an approximation method that is compatible with one type of compact representation may generate poor results when a different compact representation is employed.

We will now develop a counter-example that illustrates the shortcomings of such a combination of value iteration and least-squares approximation. This analysis is particularly interesting, since the algorithm is closely related to Q-learning (Watkins and Dayan, 1992) and temporal-difference learning (TD(λ)) (Sutton, 1988), with λ set to 0. The counter-example discussed demonstrates the short-comings of some (but not all) variants of Q-learning and temporal-difference learning that are employed in practice. [1]

Bertsekas (1994) described a counter-example to methods like the one defined by Equation (5). His counter-example involves a Markov decision problem and a compact representation that could generate a close approximation (in terms of Euclidean distance) of the optimal cost-to-go vector, but fails to do so when algorithms like the one we have described are used. In particular, the parameter vector does converge to some $W^* \in \Re^m$, but, unfortunately, this parameter vector generates a poor estimate of the

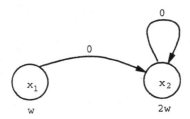

Figure 2. A counter-example.

optimal cost-to-go vector (in terms of Euclidean distance), that is,

$$\|\tilde{V}(W^*) - V^*\|_2 \gg \min_{W \in \mathbb{R}^n} \|\tilde{V}(W) - V^*\|_2,$$

where $\| \cdot \|_2$ denotes the Euclidean norm. With our upcoming counter-example, we show that much worse behavior is possible: even when the compact representation can generate a perfect approximation of the optimal cost-to-go function (i.e., $\min_W \|\tilde{V}(W) - V^*\|_2 = 0$), the algorithm may diverge.

Consider the simple Markov decision problem depicted in Figure 2. The state space consists of two states, x_1 and x_2, and at state x_1 a transition is always made to x_2, which is an absorbing state. There are no control decisions involved. All transitions incur 0 cost. Hence, the optimal cost-to-go function assigns 0 to both states.

Suppose a feature f is defined over the state space so that $f(x_1) = 1$ and $f(x_2) = 2$, and a compact representation of the form

$$\tilde{V}_i(w) = wf(i), \qquad i \in \{x_1. x_2\}.$$

is employed, where w is scalar. When we set w to 0, we get $\tilde{V}(w) = V^*$. so a perfect representation of the optimal cost-to-go vector is possible.

Let us investigate the behavior of the least-squares value iteration algorithm with the Markov decision problem and compact representation we have described. The parameter w evolves as follows:

$$
\begin{aligned}
w(t+1) &= \arg \min_{w \in \mathbb{R}^m} \sum_{i \in S} \left(\tilde{V}_i(w) - T_i(\tilde{V}(w(t))) \right)^2 \\
&= \arg \min_{w \in \mathbb{R}} \left((w - 32w(t))^2 + (2w - 32w(t))^2 \right).
\end{aligned}
$$

and we obtain

$$w(t+1) = \frac{6}{5} \beta w(t). \tag{6}$$

Hence, if $\beta > \frac{5}{6}$ and $w(0) \neq 0$. the sequence diverges. Counter-examples involving Markov decision problems that allow several control actions at each state can also be produced. In that case, the least-squares approach to value iteration can generate poor control strategies even when the optimal cost-to-go vector can be represented.

The shortcomings of straightforward procedures such as least-squares value iteration characterize the challenges involved with combining compact representations and dynamic programming. The remainder of this paper is dedicated to the development of approaches that guarantee more graceful behavior.

5. Value Iteration with Look-Up Tables

As a starting point, let us consider what is perhaps the simplest possible type of compact representation. This is the feature-based look-up table representation described in Section 3. In this section, we discuss a variant of the value iteration algorithm that has sound convergence properties when used in conjunction with such representations. We provide a convergence theorem, which we formally prove in Appendix C. We also point out relationships between the presented algorithm and previous work in the fields of dynamic programming and reinforcement learning.

5.1. Algorithmic Model

As mentioned earlier, the use of a look-up table in feature space is equivalent to state aggregation. We choose this latter viewpoint in our analysis. We consider a partition of the state space $S = \{1, \ldots, n\}$ into subsets $S_1, S_2, \ldots, S_m$; in particular, $S = S_1 \cup S_2 \cup \cdots \cup S_m$ and $S_i \cap S_j = \emptyset$ if $i \neq j$. Let $\tilde{V} : \Re^m \mapsto \Re^n$, the function which maps a parameter vector W to a cost-to-go vector V, be defined by:

$$\tilde{V}_i(W) = W_j, \qquad \forall i \in S_j.$$

Let $\mathcal{N}$ be the set of nonnegative integers. We employ a discrete variable t, taking on values in $\mathcal{N}$, which is used to index successive updates of the parameter vector W. Let $W(t)$ be the parameter vector at time t. Let Γ^j be an infinite subset of $\mathcal{N}$ indicating the set of times at which an update of the jth component of the parameter vector is performed. For each set S_j, $j = 1, \ldots, m$, let $p^j(\cdot)$ be a probability distribution over the set S_j. In particular, for every $i \in S_j$, $p^j(i)$ is the probability that a random sample from S_j is equal to i. Naturally, we have $p^j(i) \geq 0$ and $\sum_{i \in S_j} p^j(i) = 1$.

At each time t, let $X(t)$ be an m–dimensional vector whose jth component is a random representative of the set S_j, sampled according to the probability distribution $p^j(\cdot)$. We assume that each such sample is generated independently from everything else that takes place in the course of the algorithm.[2]

The value iteration algorithm applied at state $X_j(t)$ would update the value $V_{X_j(t)}$, which is represented by W_j, by setting it equal to $T_{X_j(t)}(V)$. Given the compact representation that we are using and given the current parameter vector $W(t)$, we actually need to set W_j to $T_{X_j(t)}(\tilde{V}(W(t)))$. However, in order to reduce the sensitivity of the algorithm to the randomness caused by the random sampling, W_j is updated in that direction with a small stepsize. We therefore end up with the following update formula:

$$W_j(t+1) = (1 - \alpha_j(t))W_j(t) + \alpha_j(t)T_{X_j(t)}(\tilde{V}(W(t))), \qquad t \in \Gamma^j, \qquad (7)$$

$$W_i(t+1) = W_i(t), \qquad t \notin \Gamma^i. \qquad (8)$$

Here, $\alpha_j(t)$ is a stepsize parameter between 0 and 1. In order to bring Equations (7) and (8) into a common format, it is convenient to assume that $\alpha_j(t)$ is defined for every j and t, but that $\alpha_j(t) = 0$ for $t \notin \Gamma^j$.

In a simpler version of this algorithm, we could define a single probability distribution $p(\cdot)$ over the entire state space S such that for each subset S_j, we have $\sum_{i \in S_j} p_X(i) > 0$. Then, defining $x(t)$ as a state sampled according to the $p(\cdot)$, updates of the form

$$W_j(t+1) = (1 - \alpha_j(t))W_j(t) + \alpha_j(t)T_{x(t)}(\tilde{V}(W(t))), \qquad \text{if } x(t) \in S_j, \qquad (9)$$

$$W_j(t+1) = W_j(t), \qquad \text{if } x(t) \notin S_j, \qquad (10)$$

can be used. The simplicity of this version – primarily the fact that samples are taken from only one distribution rather than many – makes it attractive for implementation. This version has a potential shortcoming, though. It does not involve any adaptive exploration of the feature space; that is, the choice of the subset S_j to be sampled does not depend on past observations. This rules out the possibility of adapting the distribution to concentrate on a region of the feature space that appears increasingly significant as approximation of the cost–to–go function ensues. Regardless, this simple version is the one chosen for application to the Tetris playing problem which is reported in Section 5.

We view all of the variables introduced so far, namely, $\alpha_j(t)$, $X_j(t)$, and $W(t)$, as random variables defined on a common probability space. The reason for $\alpha_j(t)$ being a random variable is that the decision whether W_j will be updated at time t (and, hence, whether $\alpha_j(t)$ will be zero or not) may depend on past observations. Let $\mathcal{F}(t)$ be the set of all random variables that have been realized up to and including the point at which the stepsize $\alpha_j(t)$ is fixed but just before $X_j(t)$ is generated.

5.2. Convergence Theorem

Before stating our convergence theorem, we must introduce the following standard assumption concerning the stepsize sequence:

Assumption 1 a) For all i, the stepsize sequence satisfies

$$\sum_{t=0}^{\infty} \alpha_i(t) = \infty. \qquad \text{w.p.1.} \qquad (11)$$

b) There exists some (deterministic) constant C such that

$$\sum_{t=0}^{\infty} \alpha_i^2(t) \leq C. \qquad \text{w.p.1.} \qquad (12)$$

Following is the convergence theorem:

THEOREM 1 *Let Assumption 1 hold.*
(a) With probability 1, the sequence $W(t)$ converges to W^*, the unique vector whose components solve the following system of equations:

$$W_j^* = \sum_{i \in S_j} p^j(i) T_i(\tilde{V}(W^*)), \qquad \forall j. \tag{13}$$

Define V^* as the optimal cost-to-go vector and $e \in \Re^m$ by

$$e_i = \max_{j,l \in S_i} |V_j^* - V_l^*|, \qquad \forall i \in \{1, ..., m\}.$$

Recall that $\pi_{\tilde{V}(W^*)}$ denotes a greedy policy with respect to cost-to-go vector $\tilde{V}(W^*)$, i.e.,

$$\pi_{\tilde{V}(W^*)}(i) = \arg \min_{u \in U(i)} \left(E[c_{iu}] + \beta \sum_{j \in S} p_{ij}(u) \tilde{V}_j(W^*) \right).$$

The following hold:
(b)

$$\|\tilde{V}(W^*) - V^*\|_\infty \leq \frac{\|e\|_\infty}{1 - \beta},$$

(c)

$$\|V^{\pi_{\tilde{V}(W^*)}} - V^*\|_\infty \leq \frac{2\beta \|e\|_\infty}{(1 - \beta)^2}.$$

(d) there exists an example for which the bounds in (b) and (c) both hold with equality.

A proof of Theorem 1 is provided in Appendix C. We prove the theorem by showing that the algorithm corresponds to a stochastic approximation involving a maximum norm contraction, and then appeal to a theorem concerning asynchronous stochastic approximation due to Tsitsiklis (1994) (see also (Jaakola, Jordan, and Singh, 1994)), which is discussed in Appendix B, and a theorem concerning multi-representation contractions presented and proven in Appendix A.

5.3. The Quality of Approximations

Theorem 1 establishes that the quality of approximations is determined by the quality of the chosen features. If the true cost–to–go function V^* can be accurately represented in the form $\tilde{V}(W)$, then the computed parameter values deliver near optimal performance. This is a desirable property.

The distressing aspect of Theorem 1 is the wide margin allowed by the worst-case bound. As the discount factor approaches unity, the $\frac{1}{1-\beta}$ term explodes. Since discount factors close to one are most common in practice, this is a severe weakness. However, achieving or nearly achieving the worst-case bound in real world applications may be a rare event. These weak bounds are to be viewed as the minimum desired properties for a method to be sound. As we have seen in Section 4, even this is not guaranteed by some other methods in current practice.

5.4. *Role of the Sampling Distributions*

The worst-case bounds provided by Theorem 1 are satisfied for any set of state–sampling distributions. The distribution of probability among states within a particular partition may be arbitrary. Sampling only a single state per partition constitutes a special case which satisfies the requirement. For this special case, a decaying stepsize is unnecessary. If a constant stepsize of one is used in such a setting, the algorithm becomes an asynchronous version of the standard value iteration algorithm applied to a reduced Markov decision problem that has one state per partition of the original state space; the convergence of such an algorithm is well known (Bertsekas, 1982; Bertsekas and Tsitsiklis, 1989). Such a state space reduction is analogous to that brought about by state space discretization, which is commonly applied to problems with continuous state spaces. Whitt (1978) considered this method of discretization and derived the bounds of Theorem 1, for the case where a single state is sampled in each partition. Our result can be viewed as a generalization of Whitt's, allowing the use of arbitrary sampling distributions.

When the state aggregation is perfect in that the true optimal cost-to-go values for all states in any particular partition are equal, the choice of sampling function is insignificant. This is because, independent of the distribution, the error bound is zero when there is no fluctuation of optimal cost-to-go values within any partition. In contrast, when V^* fluctuates within partitions, the error achieved by a feature-based approximation can depend on the sampling distribution. Though the derived bound limits the error achieved using any set of state distributions, the choice of distributions may play an important role in attaining errors significantly lower than this worst case bound. It often appears desirable to distribute the probability among many representative states in each partition. If only a few states are sampled, the error can be magnified if these states do not happen to be representative of the whole partition. On the other hand, if many states are chosen, and their cost-to-go values are in some sense averaged, a cost-to-go value representative of the entire partition may be generated. It is possible to develop heuristics to aid in choosing suitable distributions, but the relationship between sampling distributions and approximation error is not yet clearly understood or quantified.

5.5. *Related Work*

As was mentioned earlier, Theorem 1 can be viewed as an extension to the work of Whitt (1978). However, our philosophy is much different. Whitt was concerned with discretizing a continuous state space. Our concern here is to exploit human intuition concerning useful features and heuristic state sampling distributions to drastically reduce the dimensionality of a dynamic programming problem.

Several other researchers have considered ways of aggregating states to facilitate dynamic programming. Bertsekas and Castañon (1989) developed an adaptive aggregation scheme for use with the policy iteration algorithm. Rather than relying on feature extraction, this approach automatically and adaptively aggregates states during the course of an algorithm based on probability transition matrices under greedy policies.

The algorithm we have presented in this section is closely related to Q-learning and temporal-difference learning (TD(λ)) in the case where λ is set to 0. In fact, Theorem 1 can easily be extended so that it applies to TD(0) or Q-learning when used in conjunction with feature-based look-up tables. Since the convergence and efficacy of TD(0) and Q-learning in this setting have not been theoretically established in the past, our theorem sheds new light on these algorithms.

In considering what happens when applying the Q-learning algorithm to partially observable Markov decision problems, Jaakola, Singh and Jordan (1995) prove a convergence theorem similar to part (a) of Theorem 1. Their analysis involves a scenario where the state aggregation is inherent because of incomplete state information – i.e., a policy must choose the same action within a group of states because there is no way a controller can distinguish between different states within the group – and is not geared towards accelerating dynamic programming in general.

6. Example: Playing Tetris

As an example, we used the algorithm from the previous section to generate a strategy for the game of Tetris. In this section we discuss the process of formulating Tetris as a Markov decision problem, choosing features, and finally, generating and assessing a game strategy. The objective of this exercise was to verify that feature-based value iteration can deliver reasonable performance for a rather complicated problem. Our objective was not to construct the best possible Tetris player, and for this reason, no effort was made to construct and use sophisticated features.

6.1. Problem Formulation

We formulated the game of Tetris as a Markov decision problem, much in the same spirit as the Tetris playing programs of Lippman, Kukolich and Singer (1993). Each state of the Markov decision problem is recorded using a two–hundred–dimensional binary vector (the *wall vector*) which represents the configuration of the current wall of bricks and a seven–dimensional binary vector which identifies the current falling piece.[3] The Tetris screen is twenty squares high and ten squares wide, and each square is associated with a component of the wall vector. The component corresponding to a particular square is assigned 1 if the square is occupied by a brick and 0 otherwise. All components of the seven–dimensional vector are assigned 0 except for the one associated with the piece which is currently falling (there are seven types of pieces).

At any time, the set of possible decisions includes the locations and orientations at which we can place the falling piece on the current wall of bricks. The subsequent state is determined by the resulting wall configuration and the next random piece that appears. Since the resulting wall configuration is deterministic and there are seven possible pieces, there are seven potential subsequent states for any action, each of which occurs with equal probability. An exception is when the wall is higher than sixteen rows. In this circumstance, the game ends, and the state is absorbing.

Each time an entire row of squares is filled with bricks, the row vanishes, and the portion of the wall previously supported falls by one row. The goal in our version of Tetris is to maximize the expected number of rows eliminated during the course of a game. Though we generally formulate Markov decision problems in terms of minimizing costs, we can think of Tetris as a problem of maximizing rewards, where rewards are negative costs. The reward of a transition is the immediate number of rows eliminated. To ensure that the optimal cost–to–go from each state is finite, we chose a discount factor of $\beta = 0.9999$.

In the vast majority of states, there is no scoring opportunity. In other words, given a random wall configuration and piece, chances are that no decision will lead to an immediate reward. When a human being plays Tetris, it is crucial that she makes decisions in anticipation of long–term rewards. Because of this, simple policies that play Tetris such as those that make random decisions or even those that make greedy decisions (i.e., decisions that maximize immediate rewards with no concern for the future) rarely score any points in the course of a game. Decisions that deliver reasonable performance reflect a degree of "foresight."

6.2. Some Simple Features

Since each combination of wall configuration and current piece constitute a separate state, the state space of Tetris is huge. As a result, classical dynamic programming algorithms are inapplicable. Feature-based value iteration, on the other hand, *can* be used. In order to demonstrate this, we chose some simple features and applied the algorithm.

The two features employed in our experiments were the height of the current wall and the number of holes (empty squares with bricks both above and below) in the wall. Let us denote the set of possible heights by $H = \{0, ..., 20\}$, and the set of possible numbers of holes by $L = \{0, ..., 200\}$. We can then think of the feature extraction process as the application of a function $F : S \mapsto H \times L$.

Note that the chosen features do not take into account the shape of the current falling piece. This may initially seem odd, since the decision of where to place a piece relies on knowledge of its shape. However, the cost–to–go function actually only needs to enable the assessment of alternative decisions. This would entail assigning a value to each possible placement of the current piece on the current wall. The cost–to–go function thus needs only to evaluate the desirability of each resulting wall configuration. Hence, features that capture salient characteristics of a wall configuration are sufficient.

6.3. A Heuristic Evaluation Function

As a baseline Tetris–playing program, we produced a simple Tetris player that bases state assessments on the two features. The player consists of a quadratic function $g : H \times L \mapsto \Re$ which incorporates some heuristics developed by the authors. Then, although the composition of feature extraction and the rule based system's evaluation function,

$g \circ F$, is not necessarily an estimate of the optimal cost-to-go vector, the expert player follows a greedy policy based on the composite function.

The average score of this Tetris player on a hundred games was 31 (rows eliminated). This may seem low since arcade versions of Tetris drastically inflate scores. To gain perspective, though, we should take into account the fact that an experienced human Tetris player would take about three minutes to eliminate thirty rows.

6.4. Value Iteration with a Feature-Based Look-Up Table

We synthesized two Tetris playing programs by applying the feature-based value iteration algorithm. These two players differed in that each relied on different state–sampling distributions.

The first Tetris player used the states visited by the heuristic player as sample states for value iterations. After convergence, the average score of this player on a hundred games was 32. The fact that this player does not do much better than the heuristic player is not surprising given the simplicity of the features on which both players base position evaluations. This example reassures us, nevertheless, that feature-based value iteration converges to a reasonable solution.

We may consider the way in which the first player was constructed unrealistic, since it relied on a pre-existing heuristic player for state sampling. The second Tetris player eliminates this requirement by uses an *ad hoc* state sampling algorithm. In sampling a state, the sampling algorithm begins by sampling a maximum height for the wall of bricks from a uniform distribution. Then, for each square below this height, a brick is placed in the square with probability $\frac{3}{4}$. Each unsupported row of bricks is then allowed to fall until every row is supported. The player based on this sampling function gave an average score of 11 (equivalent to a human game lasting about one and a half minutes).

The experiments performed with Tetris provide some assurance that feature-based value iteration produces reasonable control policies. In some sense, Tetris is a worst-case scenario for the evaluation of automatic control algorithms, since humans excel at Tetris. The goal of algorithms that approximate dynamic programming is to generate reasonable control policies for large scale stochastic control problems that we have no other reasonable way of addressing. Such problems would not be natural to humans, and any reasonable policy generated by feature-based value iteration would be valuable. Furthermore, the features chosen for this study were very crude; perhaps with the introduction of more sophisticated features, feature-based value iteration would excel in Tetris. As a parting note, an additional lesson can be drawn from the fact that two strategies generated by feature–based value iteration were of such disparate quality. This is that the sampling distribution plays an important role.

7. Value Iteration with Linear Architectures

We have discussed the use of feature-based look-up tables with value iteration, and found that their use can significantly accelerate dynamic programming. However, employing a look-up table with one entry per feature vector is viable only when the number of feature vectors is reasonably small. Unfortunately, the number of possible feature vectors grows exponentially with the dimension of the feature space. When the number of features is fairly large, alternative compact representations, requiring fewer parameters, must be used. In this section, we explore one possibility which involves a linear approximation architecture. More formally, we consider compact representations of the form

$$\tilde{V}_i(W) = \sum_{k=1}^{K} W_k f_k(i) = W^T F(i), \qquad \forall i \in S. \tag{14}$$

where $W \in \Re^K$ is the parameter vector, $F(i) = (f_1(i), ..., f_K(i)) \in \Re^K$ is the feature vector associated with state i, and the superscript T denotes transpose. This type of compact representation is very attractive since the number of parameters is equal to the number of dimensions of, rather than the number of elements in, the feature space.

We will describe a variant of the value iteration algorithm that, under certain assumptions on the feature mapping, is compatible with compact representations of this form, and we will provide a convergence result and bounds on the quality of approximations. Formal proofs are presented in Appendix D.

7.1. Algorithmic Model

The iterative algorithm we propose is an extension to the standard value iteration algorithm. At the outset, K representative states $i_1, ..., i_K$ are chosen, where K is the dimension of the parameter vector. Each iteration generates an improved parameter vector $W(t+1)$ from a parameter vector $W(t)$ by evaluating $T_i(\tilde{V}(W(t)))$ at states $i_1, ..., i_K$ and then computing $W(t+1)$ so that $\tilde{V}_i(W(t+1)) = T_i(\tilde{V}(W(t)))$ for $i \in \{i_1, ..., i_K\}$. In other words, the new cost-to-go estimate is constructed by fitting the compact representation to $T(V)$, where V is the previous cost-to-go estimate, by fixing the compact representation at $i_1, ..., i_K$. If suitable features and representative states are chosen, $\tilde{V}(W(t))$ may converge to a reasonable approximation of the optimal cost-to-go vector V^*. Such an algorithm has been considered in the literature (Bellman (1959), Reetz (1977), Morin (1979)). Of these references, only (Reetz (1977)), establishes convergence and error bounds. However, Reetz's analysis is very different from what we will present and is limited to problems with one-dimensional state spaces.

If we apply an algorithm of this type to the counter-example of Section 4, with $K = 1$ and $i_1 = x_1$, we obtain $w(t+1) = 2\beta w(t)$, and if $\beta > \frac{1}{2}$, the algorithm diverges. Thus, an algorithm of this type is only guaranteed to converge for a subclass of the compact representations described by Equation (14). To characterize this subclass, we introduce the following assumption which restricts the types of features that may be employed:

Assumption 2 Let $i_1, ..., i_K \in S$ be the pre-selected states used by the algorithm.
(a) The vectors $F(i_1), ..., F(i_K)$ are linearly independent.
(b) There exists a value $\beta' \in [\beta, 1)$ such that for any state $i \in S$ there exist $\theta_1(i), ...,$

$\theta_K(i) \in \Re$ *with*

$$\sum_{k=1}^{K} |\theta_k(i)| \leq 1,$$

and

$$F(i) = \frac{\beta'}{\beta} \sum_{k=1}^{K} \theta_k(i) F(i_k).$$

In order to understand the meaning of this condition, it is useful to think about the feature space defined by $\{F(i)|i \in S\}$ and its convex hull. In the special case where $\beta = \beta'$ and under the additional restrictions $\sum_{k=1}^{K} \theta_k(i) = 1$ for all i, and $\theta_k(i) \geq 0$, the feature space is contained in the $(K - 1)$-dimensional simplex with vertices $F(i_1), \ldots, F(i_K)$. Allowing β' to be strictly greater than β introduces some slack and allows the feature space to extend a bit beyond that simplex. Finally, if we only have the condition $\sum_{k=1}^{K} |\theta_k(i)| \leq 1$, the feature space is contained in the convex hull of the vectors $\pm \frac{\beta'}{\beta} F(i_1), \pm \frac{\beta'}{\beta} F(i_2), \ldots, \pm \frac{\beta'}{\beta} F(i_K)$.

The significance of the geometric interpretation lies in the fact that the extrema of a linear function within a convex polyhedron must be located at the corners. Formally, Assumption 2 ensures that

$$\|\tilde{V}(W)\|_\infty \leq \frac{\beta'}{\beta} \max_k |\tilde{V}_{i_k}(W)|.$$

The upcoming convergence proof capitalizes on this property.

To formally define our algorithm, we need to define a few preliminary notions. First, the representation described by Equation (14) can be rewritten as

$$\tilde{V}(W) = MW, \tag{15}$$

where $M \in \Re^{n \times K}$ is a matrix with the ith row equal to $F(i)^T$. Let $L \in \Re^{K \times K}$ be a matrix with the kth row being $F(i_k)^T$. Since the rows of L are linearly independent, there exists a unique matrix inverse $L^{-1} \in \Re^{K \times K}$. We define $M^\dagger \in \Re^{K \times n}$ as follows. For $k \in \{1, \ldots, K\}$, the i_kth column is the same as the kth column of L^{-1}; all other entries are zero. Assuming, without loss of generality, that $i_1 = 1, \ldots, i_k = K$, we have

$$M^\dagger M = [L^{-1} \ 0] \begin{bmatrix} L \\ G \end{bmatrix} = L^{-1}L = I,$$

where $I \in \Re^{K \times K}$ is the identity matrix and G represents the remaining rows of M. Hence, $M^\dagger$ is a left inverse of M.

Our algorithm proceeds as follows. We start by selecting a set of K states, $i_1, \ldots, i_K$, and an initial parameter vector $W(0)$. Then, defining T' as $M^\dagger \circ T \circ M$, successive parameter vectors are generated using the following update rule:

$$W(t + 1) = T'(W(t)). \tag{16}$$

7.2. *Computational Considerations*

We will prove shortly that the operation T' applied during each iteration of our algorithm is a contraction in the parameter space. Thus, the difference between an intermediate parameter vector $W(t)$ and the limit W^* decays exponentially with the time index t. Hence, in practice, the number of iterations required should be reasonable.[4]

The reason for using a compact representation is to alleviate the computational time and space requirements of dynamic programming, which traditionally employs an exhaustive look-up table, storing one value per state. Even when the parameter vector is small and the approximate value iteration algorithm requires few iterations, the algorithm would be impractical if the computation of T' required time or memory proportional to the number of states. Let us determine the conditions under which T' can be computed in time polynomial in the number of parameters K rather than the number of states n.

The operator T' is defined by

$$T'(W) = M^\dagger T(MW).$$

Since $M^\dagger$ only has K nonzero columns, only K components of $T(MW)$ must be computed: we only need to compute $T_i(MW)$ for $i = i_1, ..., i_k$. Each iteration of our algorithm thus takes time $O(K^2 t_T)$ where t_T is the time taken to compute $T_i(MW)$ for a given state i. For any state i, $T_i(MW)$ takes on the form

$$T_i(MW) = \min_{u \in U(i)} \left(E[c_{iu}] + \sum_{j \in S} p_{ij}(u) W^T F(i) \right).$$

The amount of time required to compute $\sum_{j \in S} p_{ij}(u) W^T F(i)$ is $O(N_s K)$, where N_s is the maximum number of possible successor states under any control action (i.e., states j such that $p_{ij}(u) > 0$). By considering all possible actions $u \in U(i)$ in order to perform the required minimization, $T_i(MW)$ can be computed in time $O(N_u N_s K)$ where N_u is maximum number of control actions allowed at any state. The computation of T' thus takes time $O(N_u N_s K^3)$.

Note that for many control problems of practical interest, the number of control actions allowed at a state and the number of possible successor states grow exponentially with the number of state variables. For problems in which the number of possible successor states grows exponentially, methods involving Monte-Carlo simulations may be coupled with our algorithm to reduce the computational complexity to a manageable level. We do not discuss such methods in this paper since we choose to concentrate on the issue of compact representations. For problems in which the number of control actions grows exponentially, on the other hand, there is no satisfactory solution, except to limit choices to a small subset of allowed actions (perhaps by disregarding actions that seem "bad" *a priori*). In summary, our algorithm is suitable for problems with large state spaces and can be modified to handle cases where an action taken at a state can potentially lead to any of a large number of successor states, but the algorithm is not geared to solve problems where an extremely large number of control actions is allowed.

7.3. Convergence Theorem

Let us now proceed with our convergence result for value iteration with linear architectures.[5]

THEOREM 2 *Let Assumption 2 hold.*
(a) There exists a vector $W^* \in \Re^K$ such that $W(t)$ converges to W^*.
(b) T' is a contraction, with contraction coefficient β', with respect to a norm $\|\cdot\|$ on $\Re^K$ defined by

$$\|W\| = \|MW\|_\infty.$$

Let V^* be the optimal cost-to-go vector, and define ϵ by letting

$$\epsilon = \inf_{W \in \Re^K} \|V^* - \tilde{V}(W)\|_\infty,$$

where V^* is the optimal cost-to-go vector. Recall that $\pi_{\tilde{V}(W^*)}$ denotes a greedy policy with respect to cost-to-go vector $\tilde{V}(W^*)$, i.e.,

$$\pi_{\tilde{V}(W^*)}(i) = \arg\min_{u \in U(i)} \left(E[c_{iu}] + \beta \sum_{j \in S} p_{ij}(u) \tilde{V}_j(W^*) \right).$$

The following hold:
(c)

$$\|V^* - \tilde{V}(W^*)\|_\infty \le \frac{\beta + \beta'}{\beta(1 - \beta')}\epsilon,$$

(d)

$$\|V^{\pi_{\tilde{V}(W^*)}} - V^*\|_\infty \le \frac{2(\beta + \beta')}{(1 - \beta)(1 - \beta')}\epsilon,$$

(e) there exists an example for which the bounds of (c) and (d) hold with equality.

This result is analogous to Theorem 1. The algorithm is guaranteed to converge and, when the compact representation can perfectly represent the optimal cost-to-go vector, the algorithm converges to it. Furthermore, the accuracy of approximations generated by the algorithm decays gracefully as the propriety of the compact representation diminishes. The proof of this Theorem involves a straightforward application of Theorem 3 concerning multi-representation contractions, which is presented in Appendix D.

Theorem 2 provides some assurance of reasonable behavior when feature-based linear architectures are used for dynamic programming. However, the theorem requires that the chosen representation satisfies Assumption 2, which seems very restrictive. In the next two sections, we discuss two types of compact representations that satisfy Assumption 2 and may be of practical use.

8. Example: Interpolative Representations

One possible compact representation can be produced by specifying values of K states in the state space, and taking weighted averages of these K values to obtain values of other states. This approach is most natural when the state space is a grid of points in a Euclidean space. Then, if cost-to-go values at states sparsely distributed in the grid are computed, values at other points can be generated via interpolation. Other than the case where the states occupy a Euclidean space, interpolation-based representations may be used in settings where there seems to be a small number of "prototypical" states that capture key features. Then, if cost-to-go values are computed for these states, cost-to-go values at other states can be generated as weighted averages of cost-to-go values at the "prototypical" states.

For a more formal presentation of interpolation-based representations, let $S = \{1, \ldots, n\}$ be the states in the original state space and let $i_1, \ldots, i_K \in S$ be the states for which values are specified. The kth component of the parameter vector $W \in \Re^K$ stores the cost-to-go value of state i_k. We are then dealing with the representation

$$\tilde{V}_i(W) = \begin{cases} W_i, & \text{if } i \in \{i_1, \ldots, i_K\}, \\ W^T F(i), & \text{otherwise,} \end{cases} \tag{17}$$

where $F(i) \in \Re^K$ is a vector used to interpolate at state i. For any $i \in S$, the vector $F(i)$ is fixed; it is a part of the interpolation architecture, as opposed to the parameters W which are to be adjusted by an algorithm. The choice of the components $f_k(i)$ of $F(i)$ is generally based on problem-specific considerations. For the representations we consider in this section, we require that each component $f_k(i)$ of $F(i)$ be nonnegative and $\sum_{k=1}^K f_k(i) = 1$ for any state i.

In relation to feature–based methods, we could view the vector $F(i)$ as the feature vector associated with state i. To bring Equation (17) into a uniform format, let us define vectors $\{F(i_1), \ldots, F(i_K)\}$ as the usual basis vectors of $\Re^m$ so that we have

$$\tilde{V}_i(W) = W^T F(i), \quad \forall i \in S.$$

To apply the algorithm from Section 6, we should show that Assumption 2 of Theorem 2 is satisfied. Assumption 2(a) is satisfied by the fact that $F(i_1), \ldots, F(i_K)$ are the basis vectors of $\Re^m$. This fact also implies that $F(i) = \sum_{k=1}^K \theta_k(i) F(i_k)$ for $\theta_k(i) = f_k(i)$. Since the components of $F(i)$ sum to one, Assumption 2(b) is satisfied with $\beta' = \beta$. Hence, this interpolative representation is compatible with the algorithm of Section 6.

9. Example: Localized Basis Functions

Compact representations consisting of linear combinations of localized basis functions have attracted considerable interest as general architectures for function approximation. Two examples are radial basis function (Poggio and Girosi, 1990) and wavelet networks (Bakshi and Stephanopoulos, 1993). With these representations, states are typically contained in a Euclidean space $\Re^d$ (typically forming a finite grid). Let us continue to view the state space as $S = \{1, \ldots, n\}$. Each state index is associated with a point $x^i \in \Re^d$. With a localized basis function architecture, the cost-to-go value of state $i \in S$ takes on the following form:

$$\tilde{V}_i(W) = \sum_{k=1}^{K} W_k \phi(x^i, \mu_k, \sigma_k), \tag{18}$$

where $W \in \Re^d$ is the parameter vector, and the function $\phi : \Re^d \times \Re^d \times \Re \mapsto \Re$ is the chosen basis function. In the case of radial basis functions, for instance, ϕ is a Gaussian, and the second and third arguments, $\mu_k \in \Re^d$ and $\sigma_k \in \Re$, specify the center and dilation, respectively. More formally,

$$\phi(x, \mu, \sigma) = a e^{-\frac{\|x - \mu\|_2^2}{2\sigma^2}}, \qquad \forall x, \mu \in \Re^d, \sigma \in \Re,$$

where $\|\cdot\|_2$ denotes the Euclidean norm and $a \in \Re$ is a normalization factor. Without loss of generality, we assume that the height at the center of each basis function is normalized to one. In the case of radial basis functions, this means $a = 1$. For convenience, we will assume that $\mu_k = x^{i_k}$ for $k \in \{1, \ldots, K\}$, where $i_1, \ldots, i_K$ are preselected states in S. In other words, each basis function is centered at a point that corresponds to some state.

Architectures employing localized basis functions are set apart from other compact representations by the tendency for individual basis functions to capture only local characteristics of the function to be approximated. This is a consequence of the fact that $\phi(x, \mu, \sigma)$ generates a significant value only when x is close to μ. Otherwise, $\phi(x, \mu, \sigma)$ is extremely small. Intuitively, each basis function captures a feature that is local in Euclidean space. More formally, we use locality to imply that a basis function, ϕ, has maximum magnitude at the center, so $\phi(\mu, \mu, \sigma) = 1$ while $|\phi(x, \mu, \sigma)| < 1$ for $x \neq \mu$. Furthermore, $|\phi(x, \mu, \sigma)|$ generally decreases as $\|x - \mu\|_2$ increases, and the dilation parameter controls the rate of this decrease. Hence, as the dilation parameter is decreased, $|\phi(x, c, \sigma)|$ becomes increasingly localized, and formally, for all $x \neq \mu$, we have

$$\lim_{\sigma \to 0} \phi(x, \mu, \sigma) = 0.$$

In general, when a localized basis function architecture is used for function approximation, the centers and dilations are determined via some heuristic method which employs data and any understanding about the problem at hand. Then, the parameter vector W is determined, usually via solving a least-squares problem. In this section, we explore the use of localized basis functions to solve dynamic programming, rather than function approximation, problems. In particular, we show that, under certain assumptions, the algorithm of Section 6 may be used to generate parameters for approximation of a cost-to-go function.

To bring localized basis functions into our feature-based representation framework, we can view an individual basis function, with specified center and dilation parameter, as a feature. Then, given a basis function architecture which linearly combines K basis functions, we can define

$$f_k(i) = \phi(x^i, \mu_k, \sigma_k). \qquad \forall i \in S,$$

and a feature mapping $F(i) = (f_1(i), \ldots, f_K(i))$. The architecture becomes a special case of the familiar feature-based representation from Section 6:

$$\tilde{V}(W) = g(F(i), W) = \sum_{k=1}^{K} W_k f_k(i), \qquad \forall i \in S. \tag{19}$$

We now move on to show how the algorithm introduced in Section 6 may be applied in conjunction with localized basis function architectures. To do this, we will provide conditions on the architecture that are sufficient to ensure satisfaction of Assumption 2. The following formal assumption summarizes the sufficient condition we present.

Assumption 3 (a) For all $k \in \{1, \dots, K\}$,

$$f_k(i_k) = 1.$$

(b) For all $j \in \{1, \dots, K\}$,

$$\sum_{k \neq j} |f_k(i_j)| < 1.$$

(c) With δ defined by

$$\delta \equiv \max_{j \in \{1, \dots, K\}} \sum_{k \neq j} |f_k(i_j)|.$$

there exists a $\beta' \in [\beta, 1)$ such that, for all $i \in S$,

$$\sum_{k=1}^{K} |f_k(i)| \leq \frac{\beta'}{\beta}(1 - \delta).$$

Intuitively, δ is a bound on the influence of other basis functions on the cost-to-go value at the center of a particular basis function. By decreasing the dilation parameters of the basis functions, we can make δ arbitrarily small. Combined with the fact that $\max_{i \in S} \sum_{k=1}^{K} F(i)$ approaches unity as the dilation parameter diminishes, this implies that we can ensure satisfaction of Assumption 3 by choosing sufficiently small dilation parameters. In practice, a reasonable size dilation parameter may be desirable, and Assumption 3 may often be overly restrictive.

We will show that Assumption 3 guarantees satisfaction of Assumption 2 of Theorem 2. This will imply that, under the given restrictions, localized basis function architectures are compatible with the algorithm of Section 6. We start by choosing the states $\{i_1, \dots, i_K\}$ to be those corresponding to node centers. Hence, we have $x^{i_k} = \mu_k$ for all k.

Define $B \in \Re^{K \times K}$ as a matrix whose kth column is the feature vector $F(i_k)$. Define $\| \cdot \|_1 : \Re^K \mapsto \Re$ as the l_1 norm on $\Re^K$. Suppose we choose a vector $\theta \in \Re^K$ with $\|\theta\|_1 = 1$. Using Assumptions 3(a) and 3(b), we obtain

$$\|B\theta\|_1 = \left\| \sum_{j=1}^{K} \theta_j F(i_j) \right\|_1$$

$$= \sum_{k=1}^{K} \left| \sum_{j=1}^{K} \theta_j f_k(i_j) \right|$$

$$\geq \sum_{k=1}^{K} \left(|\theta_k| - \sum_{j \neq k} |\theta_j| |f_k(i_j)| \right)$$

$$= 1 - \sum_{j=1}^{K} |\theta_j| \sum_{k \neq j} |f_k(i_j)|$$

$$\geq 1 - \sum_{j=1}^{K} |\theta_j| \delta$$

$$= 1 - \delta$$

$$> 0.$$

Hence, B is nonsingular. It follows that the columns of B, which are the vectors $F(i_1), \ldots, F(i_k)$, are linearly independent. Thus, Assumption 2(a) is satisfied.

We now place an upper bound on $\|B^{-1}\|_1$, the l_1–induced norm on B^{-1}:

$$\|B^{-1}\|_1 = \max_{x \in \Re^K} \frac{\|B^{-1}x\|_1}{\|x\|_1}$$

$$= \min_{\theta \in \Re^K} \frac{\|\theta\|_1}{\|B\theta\|_1}$$

$$\leq \frac{1}{1 - \delta}$$

Let us define $\theta(i) = B^{-1}F(i)$ so that

$$F(i) = \sum_{k=1}^{K} \theta_k(i) F(i_k).$$

For any i, we can put a bound on $\|\theta(i)\|_1$ as follows:

$$\|\theta(i)\|_1 = \|B^{-1}F(i)\|_1$$

$$\leq \|B^{-1}\|_1 \|F(i)\|_1$$

$$\leq \frac{\|F(i)\|_1}{1 - \delta}$$

$$\leq \frac{\beta'}{\beta}.$$

Hence, Assumption 2(b) is satisfied. It follows that the algorithm of Section 6 may be applied to localized basis function architectures that satisfy Assumption 3.

10. Conclusion

We have proved convergence and derived error bounds for two algorithms that employ feature–based compact representations to approximate cost–to–go functions. The use of compact representations can potentially lead to the solution of many stochastic control problems that are computationally intractable to classical dynamic programming.

The algorithms described in this paper rely on the use of features that summarize the most salient characteristics of a state. These features are typically hand-crafted using available experience and intuition about the underlying Markov decision problem. If appropriate features are chosen, the algorithms lead to good solutions. When it is not clear what features are appropriate, several choices may be tried in order to arrive at a set of features that enables satisfactory performance. However, there is always a possibility that a far superior choice of features exists but has not been considered.

The approximation architectures we have considered are particularly simple. More complex architectures such as polynomials or artificial neural networks may lead to better approximations. Unfortunately, the algorithms discussed are not compatible with such architectures. The development of algorithms that guarantee sound behavior when used with more complex architectures is an area of open research.

Acknowledgments

The use of feature extraction to aggregate states for dynamic programming was inspired by Dimitri Bertsekas. We thank Rich Sutton for clarifying the relationship of the counter-example of Section 4 with TD(0). The choice of Tetris as a test-bed was motivated by earlier developments of Tetris learning algorithms by Richard Lippman and Linda Kukolich at Lincoln Laboratories. Early versions of this paper benefited from the proofreading of Michael Branicky and Peter Marbach. This research was supported by the NSF under grant ECS 9216531 and by EPRI under contract 8030-10.

Appendix A

Multi-Representation Contractions

Many problems requiring numerical computation can be cast in the abstract framework of fixed point computation. Such computation aims at finding a fixed point $V^* \in \Re^n$ of a mapping $T : \Re^n \mapsto \Re^n$; that is, solving the equation $V = T(V)$. One typical approach involves generating a sequence $\{V(t)|t = 0, 1, 2, ...\}$ using the update rule $V(t+1) = T(V(t))$ with the hope that the sequence will converge to V^*. In the context of dynamic programming, the function T could be the value iteration operator, and the fixed point is the optimal cost-to-go vector.

In this appendix, we deal with a simple scenario where the function T is a contraction mapping – that is, for some vector norm $\| \cdot \|$, we have $\|T(V) - T(V')\| \leq \beta\|V - V'\|$ for all $V, V' \in \Re^n$ and some $\beta \in [0, 1)$. Under this assumption, the fixed point of T is unique, and a proof of convergence for the iterative method is trivial.

When the number of components n is extremely large (n often grows exponentially with the number of variables involved in a problem), the computation of T is inherently slow. One potential way to accelerate the computation is to map the problem onto a smaller space $\Re^m$ ($m \ll n$), which can be thought of as a parameter space. This can be done by defining a mapping $\tilde{V} : \Re^m \mapsto \Re^n$ and a pseudo-inverse $\tilde{V}^\dagger : \Re^n \mapsto \Re^m$. The mapping $\tilde{V}$ can be thought of as a compact representation. A solution can be approximated by finding the fixed point of a mapping $T' : \Re^m \mapsto \Re^m$ defined by $T' = \tilde{V}^\dagger \circ T \circ \tilde{V}$. The hope is that $\tilde{V}(W^*)$ is close to a fixed point of T if W^* is a fixed point of T'. Ideally, if the compact representation can exactly represent a fixed point $V^* \in \Re^n$ of T – that is, if there exists a $W \in \Re^m$ such that $\tilde{V}(W) = V^*$ – then W should be a fixed point of T'. Furthermore, if the compact representation cannot exactly, but can closely, represent the fixed point $V^* \in \Re^n$ of T then W^* should be close to V^*. Clearly, choosing a mapping $\tilde{V}$ for which $\tilde{V}(W)$ may closely approximate fixed points of T requires some intuition about where fixed points should generally be found in $\Re^n$.

A.1. Formal Framework

Though the theorem we will prove generalizes to arbitrary metric spaces, to promote readability, we only treat normed vector spaces. We are given the mappings $T : \Re^n \mapsto \Re^n$, $\tilde{V} : \Re^m \mapsto \Re^n$, and $\tilde{V}^\dagger : \Re^n \mapsto \Re^m$. We employ a vector norm on $\Re^n$ and a vector norm on $\Re^m$, denoting both by $\| \cdot \|$. We have $m < n$, so the norm being used in a particular expression can be determined by the dimension of the argument. Define a mapping $T' : \Re^m \mapsto \Re^m$ by $T' = \tilde{V}^\dagger \circ T \circ \tilde{V}$. We make two sets of assumptions. The first concerns the mapping T.

Assumption 4 *The mapping T is a contraction with contraction coefficient $\beta \in [0, 1)$ with respect to $\| \cdot \|$. Hence, for all $V, V' \in \Re^n$,*

$$\|T(V) - T(V')\| \leq \beta \|V - V'\|.$$

Our second assumption defines the relationships between $\tilde{V}$ and $\tilde{V}^\dagger$.

Assumption 5 *The following hold for the mappings $\tilde{V}$ and $\tilde{V}^\dagger$:*
(a) For all $W \in \Re^m$,

$$W = \tilde{V}^\dagger(\tilde{V}(W)).$$

(b) There exists a $\beta' \in [\beta, 1)$ such that, for all $W, W' \in \Re^m$,

$$\|\tilde{V}(W) - \tilde{V}(W')\| \leq \frac{\beta'}{\beta} \|W - W'\|.$$

(c) For all $V, V' \in \Re^n$,

$$\|\tilde{V}^\dagger(V) - \tilde{V}^\dagger(V')\| \leq \|V - V'\|.$$

Intuitively, part (a) ensures that $\tilde{V}^\dagger$ is a pseudo-inverse of $\tilde{V}$. Part (b) forces points that are close in $\Re^m$ to map to points that are close in $\Re^n$, and part (c) ensures the converse, nearby points in $\Re^n$ must project onto points in $\Re^m$ that are close.

A.2. Theorem and Proof

Since T is a contraction mapping, it has a unique fixed point V^*. Let

$$\epsilon = \inf_{W \in \Re^m} \|V^* - \tilde{V}(W)\|.$$

THEOREM 3 *Let Assumptions 4 and 5 hold.*
(a) We have

$$\|T'(W) - T'(W')\| \leq \beta'\|W - W'\|,$$

for all $W, W' \in \Re^m$.
(b) If W^* is the fixed point of T', then

$$\|V^* - \tilde{V}(W^*)\| \leq \frac{\beta + \beta'}{\beta(1 - \beta')}\epsilon.$$

This theorem basically shows that T' is a contraction mapping, and if V^* can be closely approximated by the compact representation then W^* provides a close representation of V^*.

Proof of Theorem 3 (a) Take arbitrary $W, W' \in \Re^m$. Then,

$$
\begin{aligned}
\|T'(W) - T'(W')\| &= \|\tilde{V}^\dagger(T(\tilde{V}(W))) - \tilde{V}^\dagger(T(\tilde{V}(W')))\| \\
&\leq \|T(\tilde{V}(W)) - T(\tilde{V}(W'))\| \\
&\leq \beta\|\tilde{V}(W) - \tilde{V}(W')\| \\
&\leq \beta'\|W - W'\|.
\end{aligned}
$$

Hence, T' is a contraction mapping with contraction coefficient β'.
(b) Let $\epsilon' = \epsilon + \delta$ for some $\delta > 0$. Choose $W_{opt} \in \Re^m$ such that $\|V^* - \tilde{V}(W_{opt})\| < \epsilon'$. Then,

$$
\begin{aligned}
\|W_{opt} - T'(W_{opt})\| &= \|\tilde{V}^\dagger(\tilde{V}(W_{opt})) - \tilde{V}^\dagger(T(\tilde{V}(W_{opt})))\| \\
&\leq \|\tilde{V}(W_{opt}) - T(\tilde{V}(W_{opt}))\| \\
&\leq \|\tilde{V}(W_{opt}) - V^*\| + \|T(\tilde{V}(W_{opt})) - V^*\| \\
&< \epsilon' + \beta\epsilon' \\
&= (1 + \beta)\epsilon'.
\end{aligned}
$$

Now we can place a bound on $\|W^* - W_{opt}\|$:

$$
\begin{aligned}
\|W^* - W_{opt}\| &\leq \|W^* - T'(W_{opt})\| + \|T'(W_{opt}) - W_{opt}\| \\
&< \beta'\|W^* - W_{opt}\| + (1 + \beta)\epsilon',
\end{aligned}
$$

and it follows that

$$\|W^* - W_{opt}\| < \frac{1 + \beta}{1 - \beta'}\epsilon'.$$

Next, a bound can be put on $\|V^* - \tilde{V}(W^*)\|$:

$$\|V^* - \tilde{V}(W^*)\| \leq \|V^* - \tilde{V}(W_{opt})\| + \|\tilde{V}(W_{opt}) - \tilde{V}(W^*)\|$$

$$< \epsilon' + \frac{\beta'}{\beta}\|W_{opt} - W^*\|$$

$$< \epsilon' + \frac{\beta'}{\beta}\left(\frac{1+\beta}{1-\beta'}\right)\epsilon'$$

$$= \frac{\beta + \beta'}{\beta(1-\beta')}\epsilon'.$$

Since δ can be arbitrarily small, the proof is complete.   □

Appendix B

Asynchronous Stochastic Approximation

Consider an algorithm that performs noisy updates of a vector $V \in \Re^n$, for the purpose of solving a system of equations of the form $T(V) = V$. Here T is a mapping from $\Re^n$ into itself. Let $T_1, \ldots, T_n : \Re^n \mapsto \Re$ be the corresponding component mappings; that is, $T(V) = (T_1(V), \ldots, T_n(V))$ for all $V \in \Re^n$.

Let $\mathcal{N}$ be the set of nonnegative integers, let $V(t)$ be the value of the vector V at time t, and let $V_i(t)$ denote its ith component. Let Γ^i be an infinite subset of $\mathcal{N}$ indicating the set of times at which an update of V_i is performed. We assume that

$$V_i(t+1) = V_i(t), \qquad t \notin \Gamma^i. \tag{B.1}$$

and

$$V_i(t+1) = V_i(t) + \alpha_i(t)\Big(T_i(V(t)) - V_i(t) + \eta_i(t)\Big), \qquad t \in \Gamma^i. \tag{B.2}$$

Here, $\alpha_i(t)$ is a stepsize parameter between 0 and 1, and $\eta_i(t)$ is a noise term. In order to bring Equations (B.1) and (B.2) into a unified form, it is convenient to assume that $\alpha_i(t)$ and $\eta_i(t)$ are defined for every i and t, but that $\alpha_i(t) = 0$ for $t \notin \Gamma^i$.

Let $\mathcal{F}(t)$ be the set of all random variables that have been realized up to and including the point at which the stepsizes $\alpha_i(t)$ for the tth iteration are selected, but just before the noise term $\eta_i(t)$ is generated. As in Section 2, $\|\cdot\|_\infty$ denotes the maximum norm. The following assumption concerns the statistics of the noise.

Assumption 6 (a) For every i and t, we have $E[\eta_i(t) \mid \mathcal{F}(t)] = 0$.
(b) There exist (deterministic) constants A and B such that

$$E[\eta_i^2(t) \mid \mathcal{F}(t)] \leq A + B\|V(t)\|_\infty^2. \qquad \forall i, t.$$

We then have the following result (Tsitsiklis, 1994) (related results are obtained in (Jaakola, Jordan, and Singh, 1994)):

THEOREM 4 *Let Assumption 6 and Assumption 1 of Section 5 on the stepsizes $\alpha_i(t)$ hold and suppose that the mapping T is a contraction with respect to the maximum norm. Then, $V(t)$ converges to the unique fixed point V^* of T, with probability 1.*

Appendix C
Proof of Theorem 1

(a) To prove this result, we will bring the aggregated state value iteration algorithm into a form to which Theorems 3 and 4 (from the Appendices A and B) can be applied and we will then verify that the assumptions of these theorems are satisfied.

Let us begin by defining a function $T' : \Re^m \mapsto \Re^m$, which in some sense is a noise-free version of our update procedure on $W(t)$. In particular, the jth component T'_j of T' is defined by

$$T'_j(W) = E\Big[T_{X_j}(\tilde{V}(W))\Big] = \sum_{i \in S_j} p^j(i) T_i(\tilde{V}(W)). \tag{C.1}$$

The update equations (7) and (8) can be then rewritten as

$$W_j(t+1) = (1 - \alpha_j(t))W_j(t) + \alpha_j(t)(T'_j(W) + \eta_j(t)). \tag{C.2}$$

where the random variable $\eta_j(t)$ is defined by

$$\eta_j(t) = T_{X_j(t)}(\tilde{V}(W(t))) - E\Big[T_{X_j(t)}(\tilde{V}(W(t)))\Big].$$

Given that each $X_j(t)$ is a random sample from S_j whose distribution is independent of $\mathcal{F}(t)$ we obtain

$$E[\eta_j(t) \mid \mathcal{F}(t)] = E[\eta_j(t)] = 0.$$

Our proof consists of two parts. First, we use Theorem 3 to establish that T' is a maximum norm contraction. Once this is done, the desired convergence result follows from Theorem 4.

Let us verify the assumptions required by Theorem 3. First, let us define a function $\tilde{V}^\dagger : \Re^n \mapsto \Re^m$ as

$$\tilde{V}^\dagger_j(V) = \sum_{i \in S_j} p^j(i) V_i.$$

This function is a pseudo-inverse of $\tilde{V}$ since, for any $W \in \Re^M$,

$$\tilde{V}^\dagger_j(\tilde{V}(W)) = \sum_{i \in S_j} p^j(i) \tilde{V}_i(W) = W_j.$$

We can express T' as $T' = \tilde{V}^\dagger \circ T \circ \tilde{V}$, to bring it into the form of Theorem 3. In this context, the vector norm we have in mind for both $\Re^n$ and $\Re^m$ is the maximum norm, $\| \cdot \|_\infty$.

Assumption 4 is satisfied since T is a contraction mapping. We will now show that $\tilde{V}^\dagger$, $\tilde{V}$, and T, satisfy Assumption 5. Assumption 5(a) is satisfied since $\tilde{V}^\dagger$ is a pseudo-inverse of $\tilde{V}$. Assumption 5(b) is satisfied with $\beta' = \beta$ since

$$
\begin{aligned}
\|\tilde{V}(W) - \tilde{V}(W')\|_\infty &= \max_{i \in S} \left| \tilde{V}_i(W) - \tilde{V}_i(W') \right| \\
&= \max_{j \in \{1,\dots,m\}} |W_j - W'_j| \\
&= \|W - W'\|_\infty .
\end{aligned}
$$

Assumption 5(c) is satisfied because

$$
\begin{aligned}
\|\tilde{V}^\dagger(V) - \tilde{V}^\dagger(V')\|_\infty &= \max_{j \in \{1,\dots,m\}} \left| \sum_{i \in S_j} p^j(i)(V_i - V'_i) \right| \\
&\le \max_{j \in \{1,\dots,m\}} \max_{i \in S_j} |V_i - V'_i| \\
&= \|V - V'\|_\infty .
\end{aligned}
$$

Hence, Theorem 3 applies, and T' must be a maximum norm contraction with contraction coefficient β.

Since T' is a maximum norm contraction, Theorem 4 now applies as long as its assumptions hold. We have already shown that

$$
E[\eta_j(t) \mid \mathcal{F}(t)] = 0,
$$

so Assumption 6(a) is satisfied. As for Assumption 6(b) on the variance of $\eta_j(t)$, the conditional variance of our noise term satisfies

$$
\begin{aligned}
E[\eta_j^2(t) \mid \mathcal{F}(t)] &= E\left[\left(T_{X_j(t)}(\tilde{V}(W(t))) - E[T_{X_j(t)}(\tilde{V}(W(t)))] \right)^2 \right] \\
&\le 4 \left(\max_{i \in S} T_i(\tilde{V}(W(t))) \right)^2 \\
&\le 8 \max_{i,u}(E[c_{iu}])^2 + 8\|W(t)\|_\infty^2 .
\end{aligned}
$$

Hence, Theorem 4 applies and our proof is complete.

(b) If the maximum fluctuation of V^* within a particular partition is e_i then the minimum error that can be attained using a single constant to approximate the cost-to-go of every state within the partition is $\frac{e_i}{2}$. This implies that $\min_W \|\tilde{V}(W) - V^*\|_\infty$, the minimum error that can be attained by an aggregated state representation, is $\frac{\|e\|_\infty}{2}$. Hence, by substituting ϵ with $\frac{\|e\|_\infty}{2}$, and recalling that we have $\beta' = \beta$, the result follows from Theorem 3(b).

(c) Now that we have a bound on the maximum norm between the optimal cost-to-go estimate and the true optimal cost-to-go vector, we can place a bound on the maximum norm between the cost of a greedy policy with respect to $\tilde{V}(W^*)$ and the optimal policy as follows:

$$
\|V^{\pi_{\tilde{V}(W^*)}} - V^*\|_\infty \le \|V^{\pi_{\tilde{V}(W^*)}} - T(\tilde{V}(W^*))\|_\infty + \|T(\tilde{V}(W^*)) - V^*\|_\infty .
$$

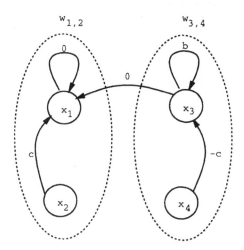

Figure C.1. An example for which the bound holds with equality.

Since $T(V) = T^{\pi V}(V)$ for all $V \in \Re^n$, we have

$$\|V^{\pi \tilde{V}(w^*)} - V^*\|_\infty \leq \|V^{\pi \tilde{V}(w^*)} - T^{\pi \tilde{V}(w^*)}(\tilde{V}(W^*))\|_\infty + \|T(\tilde{V}(W^*)) - V^*\|_\infty$$
$$\leq \beta \|V^{\pi \tilde{V}(w^*)} - \tilde{V}(W^*)\|_\infty + \beta \|\tilde{V}(W^*) - V^*\|_\infty$$
$$\leq \beta \|V^{\pi \tilde{V}(w^*)} - V^*\|_\infty (1 + \beta) \|V^* - \tilde{V}(W^*)\|_\infty.$$

It follows that

$$\|V^{\pi \tilde{V}(w^*)} - V^*\|_\infty \leq \frac{2\beta}{1 - \beta} \|\tilde{V}(W^*) - V^*\|_\infty$$
$$\leq \frac{2\beta}{(1 - \beta)^2} \|e\|_\infty.$$

(d) Consider the four–state Markov decision problem shown in Figure C.1. The states are x_1, x_2, x_3, and x_4, and we form two partitions, the first consisting of x_1 and x_2, and the second containing the remaining two states. All transition probabilities are one. No control decisions are made at states x_1, x_2, or x_4. State x_1 is a zero-cost absorbing state. In state x_2 a transition to state x_1 is inevitable, and, likewise, when in state x_4, a transition to state x_3 always occurs. In state x_3 two actions are allowed: *move* and *stay*. The transition cost for each state–action pair is deterministic, and the arc labels in Figure C.1 represent the values. Let c be an arbitrary positive constant and, let b, the cost of staying in state x_3, be defined as $b = \frac{2\beta c - \delta}{1 - \beta}$, with $\delta < 2\beta c$. Clearly, the optimal cost-to-go values at x_1, x_2, x_3, and x_4 are 0, c, 0, $-c$, respectively, and $\|e\| = c$.

Now, let us define sampling distributions, within each partition, that will be used with the algorithm. In the first partition, we always sample x_2 and in the second partition, we always sample x_4. Consequently, the algorithm will converge to partition values $w_{1,2}^*$ and $w_{3,4}^*$ satisfying the following equations:

$$w_{1,2}^* = c + \beta w_{1,2}^*$$
$$w_{3,4}^* = -c + \beta w_{3,4}^*.$$

It is not hard to see that the unique solution is

$$w_{1,2}^* = \frac{c}{1 - \beta}$$

$$w_{3,4}^* = \frac{-c}{1 - \beta}.$$

The bound of part (b) is therefore satisfied with equality.

Consider the greedy policy with respect to w. For $\delta > 0$, the *stay* action is chosen at state x_3, and the total discounted cost incurred starting at state x_3 is $\frac{2\beta c - \delta}{(1-\beta)^2}$. When $\delta = 0$, both actions, *stay* and *move*, are legitimate choices. If *stay* is chosen, the bound of part (c) holds with equality.   □

Appendix D

Proof of Theorem 2

(a) By defining $\tilde{V}^\dagger(V) = M^\dagger V$, we have $T' = \tilde{V}^\dagger \circ T \circ \tilde{V}$, which fits into the framework of multi-representation contractions. Our proof consists of a straightforward application of Theorem 3 from Appendix A (on multi-representation contractions). We must show that the technical assumptions of Theorem 3 are satisfied. To complete the multi-representation contraction framework, we must define a norm in our space of cost-to-go vectors and a norm in the parameter space. In this context, as a metric for parameter vectors, let us define a norm $\| \cdot \|$ by

$$\|W\| = \frac{\beta}{\beta'} \|MW\|_\infty.$$

Since M has full column rank, $\| \cdot \|$ has the standard properties of a vector norm. For cost-to-go vectors, we employ the maximum norm in $\Re^n$ as our metric.

We know that T is a maximum norm contraction, so Assumption 4 is satisfied. Assumption 5(a) is satisfied since, for all $W \in \Re^K$,

$$\tilde{V}_i^\dagger(\tilde{V}(W)) = M^\dagger MW$$
$$= W.$$

Assumption 5(b) follows from our definition of $\| \cdot \|$ and the fact that $\beta' \in [\beta, 1)$:

$$\|W - W'\| = \frac{\beta}{\beta'} \|MW - MW'\|_\infty$$

$$= \frac{\beta}{\beta'} \|\tilde{V}(W) - \tilde{V}(W')\|_\infty.$$

Showing that Assumption 2 implies Assumption 5(c) is the heart of this proof. To do this, we must show that, for arbitrary cost-to-go vectors V and V',

$$\|V - V'\|_\infty \geq \|V^\dagger(V) - V^\dagger(V')\|. \tag{D.1}$$

Define $D = \frac{\beta}{\beta'}M(V^\dagger(V) - V^\dagger(V'))$. Then, for arbitrary $i \in S$ we have

$$|D_i| = \frac{\beta}{\beta'}|F^T(i)(V^\dagger(V) - V^\dagger(V'))|.$$

Under Assumption 2 there exist positive constants $\theta_1(i), ..., \theta_K(i) \in \Re$, with $\sum_{k=1}^K |\theta_k(i)| \leq 1$, such that $F(i) = \frac{\beta'}{\beta}\sum_{k=1}^K \theta_k(i)F(i_k)$. It follows that, for such $\theta_1(i), ..., \theta_K(i) \in \Re$,

$$
\begin{aligned}
|D_i| &\leq \frac{\beta}{\beta'}|(\frac{\beta'}{\beta}\sum_{k=1}^K \theta_k(i)F^T(i_k))(V^\dagger(V) - V^\dagger(V'))| \\
&\leq \max_k |F^T(i_k)(V_i^\dagger(V) - V_i^\dagger(V'))| \\
&\leq |D_{i_k}| \\
&= |V_{i_k} - V'_{i_k}| \\
&\leq \|V - V'\|_\infty.
\end{aligned}
$$

Inequality (D.1) follows. Hence, Theorem 3 applies, implying parts (a), (b), and (c), of Theorem 2.

Part (d) can be proven using the same argument as in the proof of Theorem 1(c). For part (e), we can use the same example as that used to prove part (d) of Theorem 1.

□

Notes

1. To those familiar with Q-learning or temporal-difference learning: the counter-example applies to cases where temporal-difference or Q-learning updates are performed at states that are sampled uniformly from the entire state space. Often times, however, temporal-difference methods assume that sample states are generated by following a randomly produced complete trajectory. In our example, this would correspond to starting at state x_1, moving to state x_2, and then doing an infinite number of self-transitions from state x_2 to itself. If this mechanism is used, our example is no longer divergent, in agreement with results of Dayan (1992).

2. We take the point of view that each of these samples is independently generated from the same probability distribution. If the samples were generated by a simulation experiment, as Monte-Carlo simulation under some fixed policy, independence would fail to hold. This would complicate somewhat the convergence analysis, but can be handled as in (Jaakola, Singh and Jordan, 1995).

3. The way in which state is recorded is inconsequential, so we have made no effort to minimize the number of vector components required.

4. To really ensure a reasonable order of growth for the number of required iterations, we would have to characterize a probability distribution for the difference between the initial parameter vector $W(0)$ and the goal W^* as well as how close to the goal W^* the parameter vector $W(t)$ must be in order for the error bounds to hold.

5. Related results have been obtained independently by Gordon (1995).

References

Bakshi, B. R. & Stephanopoulos G., (1993)."Wave-Net: A Multiresolution, Hierarchical Neural Network with Localized Learning," AIChE Journal, vol. 39, no. 1, pp. 57-81.

Barto, A. G., Bradtke, S. J., & Singh, S. P., (1995). "Real–time Learning and Control Using Asynchronous Dynamic Programming," Aritificial Intelligence, vol. 72, pp. 81-138.

Bellman, R. E. & Dreyfus, S. E., (1959)."Functional Approximation and Dynamic Programming," Math. Tables and Other Aids Comp., Vol. 13, pp. 247-251.

Bertsekas, D. P., (1995).Dynamic Programming and Optimal Control, Athena Scientific, Bellmont, MA.

Bertsekas, D. P., (1994)."A Counter-Example to Temporal Differences Learning," Neural Computation, vol. 7, pp. 270-279.

Bertsekas D. P. & Castañon, D. A., (1989). "Adaptive Aggregation for Infinite Horizon Dynamic Programming," IEEE Transactions on Automatic Control, Vol. 34, No. 6, pp. 589-598.

Bertsekas, D. P. & Tsitsiklis, J. N., (1989). Parallel and Distributed Computation: Numerical Methods, Prentice Hall, Englewood Cliffs, NJ.

Dayan, P. D., (1992). "The Convergence of TD(λ) for General λ," Machine Learning, vol. 8, pp. 341-362.

Gordon, G. J., (1995)."Stable Function Approximation in Dynamic Programming," Technical Report: CMU-CS-95-103, Carnegie Mellon University.

Jaakola, T., Jordan M. I., & Singh, S. P., (1994)."On the Convergence of Stochastic Iterative Dynamic Programming Algorithms," Neural Computation, Vol. 6, No. 6.

Jaakola T., Singh, S. P., & Jordan, M. I., (1995). "Reinforcement Learning Algorithms for Partially Observable Markovian Decision Processes," in Advances in Neural Information Processing Systems 7, J. D. Cowan, G. Tesauro, and D. Touretzky, editors, Morgan Kaufmann.

Korf, R. E., (1987). "Planning as Search: A Quantitative Approach." Artificial Intelligence, vol. 33, pp. 65-88.

Lippman, R. P., Kukolich, L. & Singer, E., (1993). "LNKnet: Neural Network, Machine-Learning, and Statistical Software for Pattern Classification," The Lincoln Laboratory Journal, vol. 6, no. 2, pp. 249-268.

Morin, T. L., (1987). "Computational Advances in Dynamic Programming." in Dynamic Programming and Its Applications, edited by Puterman, M.L., pp. 53-90.

Poggio, T. & Girosi, F., (1990). "Networks for Approximation and Learning," Proceedings of the IEEE, vol. 78, no. 9, pp. 1481-1497.

Reetz, D., (1977). "Approximate Solutions of a Discounted Markovian Decision Process," Bonner Mathematische Schriften, vol. 98: Dynamische Optimierung, pp. 77-92.

Schweitzer, P. J., & Seidmann, A., (1985). "Generalized Polynomial Approximations in Markovian Decision Processes," Journal of Mathematical Analysis and Applications, vol. 110, pp. 568-582.

Sutton, R. S., (1988). "Learning to Predict by the Method of Temporal Differences," Machine Learning, vol. 3, pp. 9-44.

Tesauro, G., (1992). "Practical Issues in Temporal Difference Learning," Machine Learning, vol. 8, pp. 257-277.

Tsitsiklis, J. N., (1994). "Asynchronous Stochastic Approximation and Q-Learning," Machine Learning, vol. 16, pp. 185-202.

Watkins, C. J. C. H., Dayan, P., (1992). "Q–learning," Machine Learning, vol. 8, pp. 279-292.

Whitt, W., (1978). Approximations of Dynamic Programs I. Mathematics of Operations Research, vol. 3, pp. 231-243.

Received Decmeber 2, 1994
Accepted March 29, 1995
Final Manuscript October 13, 1995

Machnine Learning, , 95–121 (1996)

On the Worst-Case Analysis of Temporal-Difference Learning Algorithms

ROBERT E. SCHAPIRE   schapire@research.att.com
AT&T Bell Laboratories, 600 Mountain Avenue, Room 2A-424, Murray Hill, NJ 07974

MANFRED K. WARMUTH   manfred@cse.ucsc.edu
Computer and Information Sciences, University of California, Santa Cruz, CA 95064

Editor: Leslie Pack Kaelbling

Abstract. We study the behavior of a family of learning algorithms based on Sutton's method of temporal differences. In our on-line learning framework, learning takes place in a sequence of trials, and the goal of the learning algorithm is to estimate a discounted sum of all the reinforcements that will be received in the future. In this setting, we are able to prove general upper bounds on the performance of a slightly modified version of Sutton's so-called TD(λ) algorithm. These bounds are stated in terms of the performance of the best linear predictor on the given training sequence, and are proved without making any statistical assumptions of any kind about the process producing the learner's observed training sequence. We also prove lower bounds on the performance of any algorithm for this learning problem, and give a similar analysis of the closely related problem of learning to predict in a model in which the learner must produce predictions for a whole batch of observations before receiving reinforcement.

Keywords: machine learning, temporal-difference learning, on-line learning, worst-case analysis

1. Introduction

As an example, consider the problem of estimating the present value of a company. At the end of each quarter t, a company returns a profit r_t. In terms of its future profits, what is the company worth today? One possible answer is simply the sum total of all future profits $\sum_{k=0}^{\infty} r_{t+k}$, but this is clearly an unsatisfactory measure of present worth since a dollar earned today is certainly worth more than a dollar earned ten years from now. Indeed, taking into account inflation and the exponential growth rate of money that is invested, it can be argued that future profits drop in value exponentially with time.

For this reason, it is common to discount profits r_{t+k} earned k time steps in the future by γ^k, where $\gamma < 1$ is a parameter that estimates the rate at which future profits diminish in value. This leads to a definition of the present value of the company as the discounted sum

$$y_t := \sum_{k=0}^{\infty} \gamma^k r_{t+k}. \tag{1}$$

Suppose now that we want to predict or estimate the present value y_t as defined in Eq. (1). Obviously, if we know all the future profits $r_t, r_{t+1}, \ldots$, then we can compute y_t directly, but it would be absurd to assume that the future is known in the present.

Instead, we consider the problem of estimating y_t based on current observations that can be made about the world and the company. We summarize these observations abstractly by a vector $\mathbf{x}_t \in \mathbb{R}^N$. This vector might include, for instance, the company's profits in recent quarters, current sales figures, the state of the economy as measured by gross national product, etc.

Thus, at the beginning of each quarter t, the vector $\mathbf{x}_t$ is observed and an estimate $\hat{y}_t \in \mathbb{R}$ is formulated of the company's present value y_t. At the end of the quarter, the company returns profit r_t. The goal is to make the estimates $\hat{y}_t$ as close as possible to y_t.

We study this prediction problem more abstractly as follows: At each point in time $t = 1, 2, \ldots$, a learning agent makes an observation about the current state of its environment, which is summarized by a real vector $\mathbf{x}_t \in \mathbb{R}^N$. After having made this observation, the learning agent receives some kind of feedback from its environment, which is summarized by a real number r_t. The goal of the learning agent is to learn to predict the discounted sum y_t given in Eq. (1) where $\gamma \in [0, 1)$ is some fixed constant called the *discount rate parameter*.

At each time step t, after receiving the instance vector $\mathbf{x}_t$ and prior to receiving the reinforcement signal r_t, we ask that the learning algorithm make a prediction $\hat{y}_t$ of the value of y_t. We measure the performance of the learning algorithm in terms of the discrepancy between $\hat{y}_t$ and y_t. There are many ways of measuring this discrepancy; in this paper, we use the quadratic loss function. That is, we define the *loss* of the learning algorithm at time t to be $(\hat{y}_t - y_t)^2$, and the loss for an entire sequence of predictions is just the sum of the losses at each trial. Thus, the goal of the learning algorithm is to minimize its loss over a sequence of observation/feedback trials.

We study the worst-case behavior of a family of learning algorithms based on Sutton's (1988) *method of temporal differences* . Specifically, we analyze a slightly modified version of Sutton's so-called TD(λ) algorithm in a worst-case framework that makes no statistical assumptions of any kind. All previous analyses of TD(λ) have relied heavily on stochastic assumptions about the nature of the environment that is generating the data observed by the learner (Dayan, 1992; Dayan & Sejnowski, 1994; Jaakkola, Jordan & Singh, 1993; Sutton, 1988; Watkins, 1989). For instance, the learner's environment is often modeled by a Markov process. We apply some of our results to Markov processes later in the paper.

The primary contribution of our paper is to introduce a method of worst-case analysis to the area of temporal-difference learning. We present upper bounds on the loss incurred by our temporal-difference learning algorithm (denoted by TD*(λ)) which hold even when the sequence of observations $\mathbf{x}_t$ and reinforcement signals r_t is arbitrary.

To make our bounds meaningful in such an adversarial setting, we compare the performance of the learning algorithm to the loss that would be incurred by the best prediction function among a family of prediction functions; in this paper, this class will always be the set of linear prediction functions. More precisely, for any vector $\mathbf{u} \in \mathbb{R}^N$, let

$$L^\ell(\mathbf{u}, S) := \sum_{t=1}^{\ell} (\mathbf{u} \cdot \mathbf{x}_t - y_t)^2$$

denote the loss of vector $\mathbf{u}$ on the first ℓ trials of training sequence S. That is, $L^\ell(\mathbf{u}, S)$ is the loss that would be incurred by a prediction function that predicts $\mathbf{u} \cdot \mathbf{x}_t$ on each observation vector $\mathbf{x}_t$.

We compare the performance of our learning algorithms to the performance of the best vector $\mathbf{u}$ (of bounded norm) that minimizes the loss on the given sequence. For example, we prove below that, for any training sequence S, the loss on the first ℓ trials of TD$^*(1)$ is at most[1]

$$\min_{\substack{||\mathbf{u}|| \leq U \\ L^\ell(\mathbf{u},S) \leq K}} \left(L^\ell(\mathbf{u}, S) + 2\sqrt{K} U X_0 c_\gamma + ||\mathbf{u}||^2 X_0^2 c_\gamma^2 \right) \tag{2}$$

where $c_\gamma = (1 + \gamma)/(1 - \gamma)$. (Here, U, X_0 and K are parameters that are used to "tune" the algorithm's "learning rate:" specifically, it is assumed that $||\mathbf{x}_t|| \leq X_0$, and that $\min\{L^\ell(\mathbf{u}, S) : ||\mathbf{u}|| \leq U\} \leq K$. Various methods are known for guessing these parameters when they are unknown; see, for instance, Cesa-Bianchi, Long and Warmuth's paper (1993).) Thus, TD$^*(1)$ will perform reasonably well, provided that there exists some linear predictor $\mathbf{u}$ that gives a good fit to the training sequence.

To better understand bounds such as those given in Eq. (2), it is often helpful to consider the average per-trial loss that is guaranteed by the bound. Suppose for the moment, as is likely to be the case in practice, that U, X_0 and γ are fixed, and that K grows linearly with the number of trials ℓ, so that $K = O(\ell)$. Then Eq. (2) implies that the average per-trial loss of TD$^*(1)$ (i.e., the total cumulative loss of TD$^*(1)$ divided by the number of trials ℓ) is at most

$$\min_{\substack{||\mathbf{u}|| \leq U \\ L^\ell(\mathbf{u},S) \leq K}} \left(\frac{L^\ell(\mathbf{u}, S)}{\ell} + O\left(\frac{1}{\sqrt{\ell}} \right) \right).$$

In other words, as the number of trials ℓ becomes large, the average per-trial loss of TD$^*(1)$ rapidly approaches the average loss of the best vector $\mathbf{u}$. Furthermore, the rate of convergence is given explicitly as $O(1/\sqrt{\ell})$.

Note that the above result, like all the others presented in this paper, provides a characterization of the learner's performance after only a *finite* number of time steps. In contrast, most previous work on TD(λ) has focused on its asymptotic performance. Moreover, previous researchers have focused on the convergence of the learner's hypothesis to a "true" or "optimal" model of the world. We, on the other hand, take the view that the learner's one and only goal is to make good predictions, and we therefore measure the learner's performance entirely by the quality of its predictions.

The upper bound given in Eq. (2) on the performance of TD$^*(1)$ is derived from a more general result we prove on the worst-case performance of TD$^*(\lambda)$ for general λ. Our bounds for the special case when $\lambda = 0$ or $\lambda = 1$ can be stated in closed form. The proof techniques used in this paper are similar but more general than those used by Cesa-Bianchi, Long and Warmuth (1993) in their analysis of the Widrow-Hoff algorithm (corresponding to the case that $\gamma = 0$).

Note that $\min\{L^\ell(\mathbf{u}, S) : \mathbf{u} \in \mathbb{R}^N\}$ is the best an arbitrary linear model can do that knows all $y_1 \cdots y_\ell$ ahead of time. If the on-line learner were given y_t at the end of trial t (i.e., if $\gamma = 0$), then the Widrow-Hoff algorithm would achieve a worst case bound of

$$\min_{\substack{\|\mathbf{u}\| \leq U \\ L^\ell(\mathbf{u},S) \leq K}} \left(L^\ell(\mathbf{u}, S) + 2\sqrt{K}UX_0 + \|\mathbf{u}\|^2 X_0{}^2 \right)$$

(matching the bound in Eq. (2) with γ set to 0). However, in our model, the learner is given only the reinforcements r_t, even though its goal is to accurately estimate the infinite sum y_t given in Eq. (1). Intuitively, as γ gets larger, this task becomes more difficult since the learner must make predictions about events farther and farther into the future. All of our worst-case loss bounds depend explicitly on γ and, not surprisingly, these bounds typically tend to infinity or become vacuous as γ approaches 1. Thus, our bounds quantify the price one has to pay for giving the learner successively less information.

In addition to these upper bounds, we prove a general lower bound on the loss of *any* algorithm for this prediction problem. Such a lower bound may be helpful in determining what kind of worst-case bounds can feasibly be proved. None of our upper bounds match the lower bound; it is an open question whether this remaining gap can be closed (this is possible in certain special cases, such as when $\gamma = 0$).

Finally, we consider a slightly different, but closely related learning model in which the learner is given a whole batch of instances at once and the task is to give a prediction for all instances before an outcome is received for each instance in the batch. The loss in a trial t is $\|\hat{\mathbf{y}}_t - \mathbf{y}_t\|^2$, where $\hat{\mathbf{y}}_t$ is the vector of predictions and $\mathbf{y}_t$ the vector of outcomes. Again, the goal is to minimize the additional total loss summed over all trials in excess of the total loss of the best linear predictor (of bounded norm).

In this batch model all instances count equally and the exact outcome for each instance is received at the end of each batch. A special case of this model is when the algorithm has to make predictions on a whole batch of instances before receiving the *same* outcome for all of them (a case studied by Sutton (1988)).

We again prove worst-case bounds for this model (extending Cesa-Bianchi, Long and Warmuth's (1993) previous analysis for the noise-free case). We also prove matching lower bounds for this very general model, thus proving that our upper bounds are the optimal worst-case bounds.

The paper is outlined as follows. Section 2 describes the on-line model for temporal difference learning. Section 3 gives Sutton's original temporal difference learning algorithm TD(λ) and introduces our new algorithm TD*(λ). Section 4 contains the worst-case loss bounds for the new algorithm, followed by Section 5 containing a lower bound for the on-line model. In Section 6, we illustrate our results with an application of TD*(1) to obtain a kind of convergence result in a Markov-process setting. We present our results for the batch model in Section 7. Finally, we discuss the merits of the method of worst-case analysis in Section 8.

2. The prediction model

In this section, we describe our on-line learning model. Throughout the paper, N denotes the dimension of the learning problem. Each *trial t* $(t = 1, 2, \ldots)$ proceeds as follows:

1. The learner receives instance vector $\mathbf{x}_t \in \mathbb{R}^N$.

2. The learner is required to compute a prediction $\hat{y}_t \in \mathbb{R}$.

3. The learner receives a *reinforcement signal* $r_t \in \mathbb{R}$.

The goal of the learner is to predict not merely the next reinforcement signal, but rather a discounted sum of all of the reinforcements that will be received in the future. Specifically, the learner is trying to make its prediction $\hat{y}_t$ as close as possible to

$$y_t := \sum_{k=0}^{\infty} \gamma^k r_{t+k}$$

where $\gamma \in [0, 1)$ is a fixed parameter of the problem. (We will always assume that this infinite sum converges absolutely for all t.)

Note that if we multiply y_t by the constant $1 - \gamma$, we obtain a weighted average of all the future r_t's; that is, $(1 - \gamma)y_t$ is a weighted average of $r_t, r_{t+1}, \ldots$. Thus it might be more natural to use the variables $y'_t = y_t(1 - \gamma)$. (For instance, if all r_t equal r, then the modified variables y'_t all equal r as well.) However, for the sake of notational simplicity, we use the variables y_t instead (as was done by Sutton (1988) and others).

The infinite sequence of pairs of instances $\mathbf{x}_t$ and reinforcement signals r_t is called a *training sequence* (usually denoted by S). The loss of the learner at trial t is $(y_t - \hat{y}_t)^2$, and the total loss of an algorithm A on the first ℓ trials is

$$L^\ell(A, S) := \sum_{t=1}^{\ell} (y_t - \hat{y}_t)^2.$$

Similarly, the total loss of a weight vector $\mathbf{u} \in \mathbb{R}^N$ on the first ℓ trials is defined to be

$$L^\ell(\mathbf{u}, S) := \sum_{t=1}^{\ell} (y_t - \mathbf{u} \cdot \mathbf{x}_t)^2.$$

The purpose of this paper is to exhibit algorithms whose loss is guaranteed to be "not too much worse" than the loss of the *best* weight vector for the entire sequence. Thus, we would like to show that if there exists a weight vector $\mathbf{u}$ that fits the training sequence well, then the learner's predictions will also be reasonably good.

3. Temporal-difference algorithms

We focus now on a family of learning algorithms that are only a slight modification of those considered by Sutton (1988). Each of these algorithms is parameterized by a real number $\lambda \in [0, 1]$. For any sequence S and $t = 1, 2, \cdots$, let

$$\mathbf{X}_t^\lambda := \sum_{k=1}^{t} (\gamma\lambda)^{t-k}\mathbf{x}_k \tag{3}$$

be a weighted sum of all previously observed instances $\mathbf{x}_k$. The parameter λ controls how strong an influence past instances have. For instance, when $\lambda = 0$, $\mathbf{X}_t^0 = \mathbf{x}_t$ so only the most recent instance is considered.

The learning algorithm $TD(\lambda)$ works by maintaining a weight vector $\mathbf{w}_t \in \mathbb{R}^N$. The initial weight vector $\mathbf{w}_1$ may be arbitrary, although in the simplest case $\mathbf{w}_1 = \mathbf{0}$. The weight vector $\mathbf{w}_t$ is then updated to the new weight vector $\mathbf{w}_{t+1}$ using the following update rule:

$$\mathbf{w}_{t+1} := \mathbf{w}_t + \eta_t(r_t + \gamma\hat{y}_{t+1} - \hat{y}_t)\mathbf{X}_t^\lambda. \tag{4}$$

As suggested by Sutton (1988), the weight vectors are updated using $\mathbf{X}_t^\lambda$ rather than $\mathbf{x}_t$, allowing instances prior to $\mathbf{x}_t$ to have a diminishing influence on the update.

The constant η_t appearing in Eq. (4) is called the *learning rate* on trial t. We will discuss later how to set the learning rates using prior knowledge about the training sequence.

In Sutton's original presentation of $TD(\lambda)$, and in most of the subsequent work on the algorithm, the prediction at each step is simply $\hat{y}_t = \mathbf{w}_t \cdot \mathbf{x}_t$. We, however, have found that a variant on this prediction rule leads to a simpler analysis, and, moreover, we were unable to obtain worst-case loss bounds for the original algorithm TD(λ) as strong as the bounds we prove for the new algorithm.

Our variant of TD(λ) uses the same update (4) for the weight vector as the original algorithm, but predicts as follows:

$$\begin{aligned}
\hat{y}_t &:= \mathbf{w}_t \cdot \mathbf{x}_t + \sum_{k=1}^{t-1} (\gamma\lambda)^{t-k}(\mathbf{w}_t \cdot \mathbf{x}_k - \hat{y}_k) \\
&= \mathbf{w}_t \cdot \mathbf{X}_t^\lambda - \sum_{k=1}^{t-1} (\gamma\lambda)^{t-k}\hat{y}_k.
\end{aligned} \tag{5}$$

This new algorithm, which we call TD*(λ), is summarized in Fig. 1.

The rule (4) for updating $\mathbf{w}_{t+1}$ has $\mathbf{w}_{t+1}$ implicit in $\hat{y}_{t+1}$, so at first it seems impossible to do this update rule.[2] However, by multiplying Eq. (4) by $\mathbf{X}_{t+1}^\lambda$, one can first solve for $\hat{y}_{t+1}$ and then compute $\mathbf{w}_{t+1}$. Specifically, this gives a solution for $\hat{y}_{t+1}$ of

$$\frac{(\mathbf{w}_t + \eta_t(r_t - \hat{y}_t)\mathbf{X}_t^\lambda) \cdot \mathbf{X}_{t+1}^\lambda - \sum_{k=1}^{t}(\gamma\lambda)^{t+1-k}\hat{y}_k}{1 - \eta_t\gamma\mathbf{X}_t^\lambda \cdot \mathbf{X}_{t+1}^\lambda}$$

Algorithm TD$^*(\lambda)$
    Parameters: discount rate $\gamma \in [0, 1)$
                 $\lambda \in [0, 1]$
                 start vector $\mathbf{w}_1 \in \mathbb{R}^N$
                 method of computing learning rate η_t
    Given: training sequence $\mathbf{x}_1, r_1, \mathbf{x}_2, r_2, \ldots$
    Predict: $\hat{y}_1, \hat{y}_2, \ldots$
    Procedure:

        get $\mathbf{x}_1$

        $\mathbf{X}_1^\lambda \leftarrow \mathbf{x}_1$

        $\hat{y}_1 \leftarrow \mathbf{w}_1 \cdot \mathbf{X}_1^\lambda$

        for $t = 1, 2, \ldots$

            predict $\hat{y}_t$ $(*\ \hat{y}_t = \mathbf{w}_t \cdot \mathbf{X}_t^\lambda - \sum_{k=1}^{t-1} (\gamma\lambda)^{t-k} \hat{y}_k\ *)$

            get r_t

            get $\mathbf{x}_{t+1}$

            $\mathbf{X}_{t+1}^\lambda \leftarrow \mathbf{x}_{t+1} + (\gamma\lambda)\mathbf{X}_t^\lambda$

            compute η_t

$$\hat{y}_{t+1} \leftarrow \frac{\mathbf{w}_t \cdot \mathbf{x}_{t+1} + \eta_t(r_t - \hat{y}_t)\mathbf{X}_t^\lambda \cdot \mathbf{X}_{t+1}^\lambda}{1 - \eta_t \gamma \mathbf{X}_t^\lambda \cdot \mathbf{X}_{t+1}^\lambda}$$

$$\mathbf{w}_{t+1} \leftarrow \mathbf{w}_t + \eta_t(r_t + \gamma\hat{y}_{t+1} - \hat{y}_t)\mathbf{X}_t^\lambda$$

        end

Figure 1. Pseudocode for TD$^*(\lambda)$.

$$= \frac{(\mathbf{w}_t + \eta_t(r_t - \hat{y}_t)\mathbf{X}_t^\lambda) \cdot \mathbf{X}_{t+1}^\lambda - (\gamma\lambda)\mathbf{w}_t \cdot \mathbf{X}_t^\lambda}{1 - \eta_t\gamma\mathbf{X}_t^\lambda \cdot \mathbf{X}_{t+1}^\lambda}$$

$$= \frac{\mathbf{w}_t \cdot \mathbf{x}_{t+1} + \eta_t(r_t - \hat{y}_t)\mathbf{X}_t^\lambda \cdot \mathbf{X}_{t+1}^\lambda}{1 - \eta_t\gamma\mathbf{X}_t^\lambda \cdot \mathbf{X}_{t+1}^\lambda}$$

where, in the first equality, we assume inductively that Eq. (5) holds at trial t. Thus, we can solve successfully for $\hat{y}_{t+1}$ provided that $\gamma\eta_t\mathbf{X}_t^\lambda \cdot \mathbf{X}_{t+1}^\lambda \neq 1$, as will be the case for all the values of η_t we consider. Also, note that $\hat{y}_{t+1}$ is computed after the instance $\mathbf{x}_{t+1}$ is received but before the reinforcement r_{t+1} is available (see Fig. 1 for details).

Note that for the prediction $\hat{y}_t = \mathbf{w}_t \cdot \mathbf{x}_t$ of TD(λ),

$$\begin{aligned}
\nabla_{\mathbf{w}_t}(y_t - \hat{y}_t)^2 &= -2\eta_t(y_t - \hat{y}_t)\mathbf{x}_t \\
&\approx -2\eta_t(r_t + \gamma\hat{y}_{t+1} - \hat{y}_t)\mathbf{x}_t.
\end{aligned}$$

(Since $y_t = r_t + \gamma y_{t+1}$ is not available to the learner, it is approximated by $r_t + \gamma\hat{y}_{t+1}$.) Thus with the prediction rule of TD(λ) the update rule (4) *is not* gradient descent for all choices of λ. In contrast, with the new prediction rule (5) of $\text{TD}^*(\lambda)$, the update rule (4) used by both algorithms *is* gradient descent,[3] since if $\hat{y}_t$ is set according to the new prediction rule then

$$\nabla_{\mathbf{w}_t}(y_t - \hat{y}_t)^2 = -2\eta_t(y_t - \hat{y}_t)\mathbf{X}_t^\lambda$$
$$\approx -2\eta_t(r_t + \gamma\hat{y}_{t+1} - \hat{y}_t)\mathbf{X}_t^\lambda.$$

We can also motivate the term $-\sum_{k=1}^{t-1}(\gamma\lambda)^{t-k}\hat{y}_k$ in the prediction rule of $\text{TD}^*(\lambda)$ given in Eq. (5): In this paper, we are comparing the total loss of the algorithm with the total loss of the best linear predictor, so both algorithms $\text{TD}(\lambda)$ and $\text{TD}^*(\lambda)$ try to match the y_t's with an on-line linear model. In particular, if $y_t = \mathbf{w} \cdot \mathbf{x}_t$ (that is, the y_t's are a linear function of the $\mathbf{x}_t$'s) and the initial weight vector is the "correct" weight vector $\mathbf{w}$, then the algorithms should always predict correctly (so that $\hat{y}_t = y_t$) and the weight vector $\mathbf{w}_t$ of the algorithms should remain unchanged. It is easy to prove by induction that both algorithms have this property.

Thus, in sum, the prediction rule $\hat{y}_t = \mathbf{w}_t \cdot \mathbf{X}_t^\lambda + c_t$ is motivated by gradient descent, where c_t is any term that does not depend on the weight vector $\mathbf{w}_t$. The exact value for c_t is derived using the fact that, in the case described above, we want $\hat{y}_t = y_t$ for all t.

4. Upper bounds for $\text{TD}^*(\lambda)$

In proving our upper bounds, we begin with a very general lemma concerning the performance of $\text{TD}^*(\lambda)$. We then apply the lemma to derive an analysis of some special cases of interest.

LEMMA 1 *Let* $\gamma \in [0, 1)$, $\lambda \in [0, 1]$, *and let* S *be an arbitrary training sequence such that* $\|\mathbf{X}_t^\lambda\| \leq X_\lambda$ *for all trials t. Let* $\mathbf{u}$ *be any weight vector, and let* $\ell > 0$.
If we execute $\text{TD}^*(\lambda)$ *on* S *with initial vector* $\mathbf{w}_1$ *and learning rates* $\eta_t = \eta$ *where* $0 < \eta X_\lambda^2 \gamma < 1$, *then*

$$L^\ell(\text{TD}^*(\lambda), S) \leq \inf\left\{\frac{bL^\ell(\mathbf{u}, S) + \|\mathbf{u} - \mathbf{w}_1\|^2}{C_b} : b > 0, C_b > 0\right\}$$

where C_b *equals*

$$2\eta - \eta^2 X_\lambda^2(1+\gamma^2) - \frac{\eta^2}{b}\left(1 + \left(\frac{\gamma - \gamma\lambda}{1 - \gamma\lambda}\right)^2\right)$$
$$-2\left(\left|\eta - \frac{\eta^2}{b}\right|\right)\left(\frac{\gamma - \gamma\lambda}{1 - \gamma\lambda}\right)\gamma\lambda - 2\left(\left|\left(\eta - \frac{\eta^2}{b}\right)(\gamma - \gamma\lambda) - \eta^2 X_\lambda^2\gamma\right|\right).$$

Proof: For $1 \leq t \leq \ell$, we let $e_t = y_t - \hat{y}_t$, and $e_{\mathbf{u}.t} = y_t - \mathbf{u} \cdot \mathbf{x}_t$. We further define $e_{\ell+1} = 0$. Note that the loss of the algorithm at trial t is e_t^2 and the loss of $\mathbf{u}$ is $e_{\mathbf{u}.t}^2$. Since, for $t < \ell$,

$$r_t + \gamma\hat{y}_{t+1} - \hat{y}_t = r_t + \gamma\hat{y}_{t+1} - (r_t + \gamma y_{t+1}) + y_t - \hat{y}_t$$
$$= e_t - \gamma e_{t+1}.$$

we can write Eq. (4), the update of our algorithm, conveniently as

$$\mathbf{w}_{t+1} = \mathbf{w}_t + \eta(e_t - \gamma e_{t+1})\mathbf{X}_t^\lambda. \tag{6}$$

To simplify the proof, we also define

$$\mathbf{w}_{\ell+1} = \mathbf{w}_\ell + \eta e_\ell \mathbf{X}_\ell^\lambda$$

so that Eq. (6) holds for all $t \leq \ell$. (In fact, this definition of $\mathbf{w}_{\ell+1}$ differs from the vector that would actually be computed by TD$^*(\lambda)$. This is not a problem, however, since we are here only interested in the behavior of the algorithm on the first ℓ trials which are unaffected by this change.)

We use the function progr$_t$ to measure how much "closer" the algorithm gets to u during trial t as measured by the distance function $|| \cdot ||^2$:

$$\text{progr}_t = ||\mathbf{u} - \mathbf{w}_t||^2 - ||\mathbf{u} - \mathbf{w}_{t+1}||^2.$$

Let $\Delta\mathbf{w}_t = \mathbf{w}_{t+1} - \mathbf{w}_t$ for $t \leq \ell$. We have that

$$\begin{aligned}
||\Delta\mathbf{w}_t||^2 &= \eta^2(e_t - \gamma e_{t+1})^2||\mathbf{X}_t^\lambda||^2 \\
&\leq \eta^2 X_\lambda{}^2(e_t - \gamma e_{t+1})^2
\end{aligned}$$

and that

$$\begin{aligned}
\Delta\mathbf{w}_t \cdot (\mathbf{w}_t - \mathbf{u}) &= \eta(e_t - \gamma e_{t+1})(\mathbf{w}_t \cdot \mathbf{X}_t^\lambda - \mathbf{u} \cdot \mathbf{X}_t^\lambda) \\
&= \eta(e_t - \gamma e_{t+1}) \sum_{k=1}^{t} (\gamma\lambda)^{t-k}(\hat{y}_k - \mathbf{u} \cdot \mathbf{x}_k) \\
&= \eta(e_t - \gamma e_{t+1}) \sum_{k=1}^{t} (\gamma\lambda)^{t-k}(e_{\mathbf{u},k} - e_k)
\end{aligned}$$

where the second equality follows from Eqs. (3) and (5), and the last equality from the fact that

$$\begin{aligned}
\hat{y}_k - \mathbf{u} \cdot \mathbf{x}_k &= \hat{y}_k - y_k + y_k - \mathbf{u} \cdot \mathbf{x}_k \\
&= e_{\mathbf{u},k} - e_k.
\end{aligned}$$

Since $-\text{progr}_t = 2\Delta\mathbf{w}_t \cdot (\mathbf{w}_t - \mathbf{u}) + ||\Delta\mathbf{w}_t||^2$, we have that

$$\begin{aligned}
-||\mathbf{w}_1 - \mathbf{u}||^2 &\leq ||\mathbf{w}_{\ell+1} - \mathbf{u}||^2 - ||\mathbf{w}_1 - \mathbf{u}||^2 \\
&= -\sum_{t=1}^{\ell} \text{progr}_t \\
&\leq 2\eta \sum_{t=1}^{\ell} \left[(e_t - \gamma e_{t+1}) \sum_{k=1}^{t} (\gamma\lambda)^{t-k}(e_{\mathbf{u},k} - e_k) \right] \\
&\quad + \eta^2 X_\lambda{}^2 \sum_{t=1}^{\ell} (e_t - \gamma e_{t+1})^2. \tag{7}
\end{aligned}$$

This can be written more concisely using matrix notation as follows: Let $\mathbf{Z}_k$ be the $\ell \times \ell$ matrix whose entry (i, j) is defined to be 1 if $j = i + k$ and 0 otherwise. (For instance, $\mathbf{Z}_0$ is the identity matrix.) Let $\mathbf{D} = \mathbf{Z}_0 - \gamma \mathbf{Z}_1$, and let

$$\mathbf{V} = \sum_{t=0}^{\ell-1} (\gamma \lambda)^t \mathbf{Z}_t.$$

Finally, let $\mathbf{e}$ (respectively, $\mathbf{e_u}$) be the length ℓ vector whose t^{th} element is e_t (respectively, $e_{u,t}$). Then the last expression in Eq. (7) is equal to

$$\eta^2 X_\lambda{}^2 \mathbf{e}^T \mathbf{D}^T \mathbf{De} + 2\eta \mathbf{e}^T \mathbf{D}^T \mathbf{V}^T (\mathbf{e_u} - \mathbf{e}). \tag{8}$$

This can be seen by noting that, by a straightforward computation, the tth element of $\mathbf{De}$ is $e_t - \gamma e_{t+1}$, and the tth element of $\mathbf{V}^T(\mathbf{e_u} - \mathbf{e})$ is

$$\sum_{k=1}^{t} (\gamma \lambda)^{t-k} (e_{u,k} - e_k).$$

We also used the identity $(\mathbf{De})^T = \mathbf{e}^T \mathbf{D}^T$.

Using the fact that $2\mathbf{p}^T \mathbf{q} \le ||\mathbf{p}||^2 + ||\mathbf{q}||^2$ for any pair of vectors $\mathbf{p}, \mathbf{q} \in \mathbb{R}^\ell$, we can upper bound Eq. (8), for any $b > 0$, by

$$\eta^2 X_\lambda{}^2 \mathbf{e}^T \mathbf{D}^T \mathbf{De} - 2\eta \mathbf{e}^T \mathbf{D}^T \mathbf{V}^T \mathbf{e} + \frac{\eta^2}{b} \mathbf{e}^T \mathbf{D}^T \mathbf{V}^T \mathbf{VDe} + b \mathbf{e_u}{}^T \mathbf{e_u} \tag{9}$$

(where we use $\mathbf{p} = (\eta/\sqrt{b})\, \mathbf{VDe}$ and $\mathbf{q} = \sqrt{b}\, \mathbf{e_u}$). Defining

$$\mathbf{M} = \eta^2 X_\lambda{}^2 \mathbf{D}^T \mathbf{D} - \eta(\mathbf{VD} + \mathbf{D}^T \mathbf{V}^T) + \frac{\eta^2}{b} \mathbf{D}^T \mathbf{V}^T \mathbf{VD},$$

and noting that $\mathbf{e}^T \mathbf{D}^T \mathbf{V}^T \mathbf{e} = \mathbf{e}^T \mathbf{VDe}$, we can write Eq. (9) simply as

$$\mathbf{e}^T \mathbf{Me} + b \mathbf{e_u}{}^T \mathbf{e_u}.$$

Note that $\mathbf{M}$ is symmetric. It is known that in this case

$$\max_{\mathbf{e} \ne 0} \frac{\mathbf{e}^T \mathbf{Me}}{\mathbf{e}^T \mathbf{e}} = \rho(\mathbf{M}) \tag{10}$$

where $\rho(\mathbf{M})$ is the largest eigenvalue of $\mathbf{M}$. (See, for instance, Horn and Johnson (1985) for background on matrix theory.) Thus, for all vectors $\mathbf{e}$, $\mathbf{e}^T \mathbf{Me} \le \rho(\mathbf{M})\mathbf{e}^T \mathbf{e}$. It follows from Eq. (7) that

$$-||\mathbf{w}_1 - \mathbf{u}||^2 \le b \sum_{t=1}^{\ell} e_{u,t}{}^2 + \rho(\mathbf{M}) \sum_{t=1}^{\ell} e_t{}^2$$

so

$$-\rho(\mathbf{M}) \sum_{t=1}^{\ell} e_t^2 \leq ||\mathbf{w}_1 - \mathbf{u}||^2 + b \sum_{t=1}^{\ell} e_{\mathbf{u},t}^2.$$

In the appendix, we complete the proof by arguing that $\rho(\mathbf{M}) \leq -C_b$.   ∎

Having proved Lemma 1 in gory generality, we are now ready to apply it to some special cases to obtain bounds that are far more palatable. We begin by considering the case that $\lambda = 0$. Note that TD(0) and TD*(0) are identical.

THEOREM 1 *Let* $0 \leq \gamma < 1$, *and let* S *be any sequence of instances/reinforcements. Assume that we know a bound* X_0 *for which* $||\mathbf{x}_t|| \leq X_0$.

If TD*(0) *uses any start vector* $\mathbf{w}_1$ *and learning rates* $\eta_t = \eta = 1/(X_0^2 + 1)$, *we have for all* $\ell > 0$ *and for all* $\mathbf{u} \in \mathbb{R}^N$:

$$L^\ell(\mathrm{TD}^*(0), S) \leq \frac{(1 + X_0^2)(L^\ell(\mathbf{u}, S) + ||\mathbf{w}_1 - \mathbf{u}||^2)}{1 - \gamma^2}. \tag{11}$$

Assume further that we know bounds K *and* U *such that for some* $\mathbf{u}$ *we have* $L^\ell(\mathbf{u}, S) \leq K$ *and* $||\mathbf{w}_1 - \mathbf{u}|| \leq U$. *Then for the learning rate*

$$\eta_t = \eta = \frac{U}{X_0 \sqrt{K} + X_0^2 U}$$

we have that

$$L^\ell(\mathrm{TD}^*(0), S) \leq \frac{L^\ell(\mathbf{u}, S) + 2U X_0 \sqrt{K} + X_0^2 ||\mathbf{w}_1 - \mathbf{u}||^2}{1 - \gamma^2}. \tag{12}$$

Proof: When $\lambda = 0$, C_b simplifies to

$$2\eta - \eta^2 \left(X_0^2 + \frac{1}{b} \right)(1 + \gamma^2) - 2\gamma \left| \eta - \eta^2 \left(X_0^2 + \frac{1}{b} \right) \right|.$$

To minimize the loss bound given in Lemma 1, we need to maximize C_b with respect to η. It can be shown that, in this case, C_b is maximized, for fixed b, when

$$\eta = \frac{1}{X_0^2 + 1/b}. \tag{13}$$

The first bound (Eq. (11)) is then obtained by choosing $b = 1$.

If bounds K and U are known as stated in the theorem, an optimal choice for b can be derived by plugging the choice for η given in Eq. (13) into the bound in Lemma 1, and replacing $L^\ell(\mathbf{u}, S)$ by K and $||\mathbf{u} - \mathbf{w}_1||^2$ by U^2. This gives

$$\frac{(bK + U^2)(X_0^2 + 1/b)}{1 - \gamma^2}$$

which is minimized when $b = U/(X_0\sqrt{K})$. Plugging this choice of b into the bound of Lemma 1 (and setting η as in Eq. (13)) gives the bound

$$\frac{\left(UL^\ell(\mathbf{u}, S)/X_0\sqrt{K} + ||\mathbf{u} - \mathbf{w}_1||^2\right)\left(X_0{}^2 + X_0\sqrt{K}/U\right)}{1 - \gamma^2}$$

$$= \frac{L^\ell(\mathbf{u}, S) + L^\ell(\mathbf{u}, S)X_0U/\sqrt{K} + ||\mathbf{u} - \mathbf{w}_1||^2X_0\sqrt{K}/U + ||\mathbf{u} - \mathbf{w}_1||^2X_0{}^2}{1 - \gamma^2}$$

$$\leq \frac{L^\ell(\mathbf{u}, S) + 2X_0U\sqrt{K} + ||\mathbf{u} - \mathbf{w}_1||^2X_0{}^2}{1 - \gamma^2}.$$

∎

Next, we consider the case that $\lambda = 1$.

THEOREM 2 Let $0 \leq \gamma < 1$, $\ell > 0$ and let S be any sequence of instances/reinforcements. Assume that we know a bound X_1 for which $||\mathbf{X}_t^1|| \leq X_1$, and that we know bounds K and U such that for some $\mathbf{u}$ we have $L^\ell(\mathbf{u}, S) \leq K$ and $||\mathbf{w}_1 - \mathbf{u}|| \leq U$. Then if $TD^*(1)$ uses any start vector $\mathbf{w}_1$ and learning rates

$$\eta_t = \eta = \frac{U}{UX_1{}^2(1 + \gamma)^2 + X_1(1 + \gamma)\sqrt{K}},$$

then

$$L^\ell(TD^*(1), S) \leq L^\ell(\mathbf{u}, S) + 2\sqrt{K}(1 + \gamma)UX_1 + (1 + \gamma)^2||\mathbf{w}_1 - \mathbf{u}||^2X_1{}^2.$$

Proof: When $\lambda = 1$,

$$C_b = 2\eta - \eta^2X_1{}^2(1 + \gamma)^2 - \frac{\eta^2}{b}.$$

This is maximized, with respect to η, when

$$\eta = \frac{1}{X_1{}^2(1 + \gamma)^2 + 1/b}.$$

Proceeding as in Theorem 1, we see that the best choice for b is

$$b = \frac{U}{(1 + \gamma)X_1\sqrt{K}}.$$

Plugging into the bound in Lemma 1 completes the theorem.   ∎

The bound in Eq. (2) is obtained from Theorem 2 by setting $\mathbf{w}_1 = 0$, and noting that

$$||\mathbf{X}_t^1|| \leq \sum_{k=1}^{t}\gamma^{t-k}||\mathbf{x}_k|| \leq \frac{\max\{||\mathbf{x}_k|| : 1 \leq k \leq t\}}{1 - \gamma} \tag{14}$$

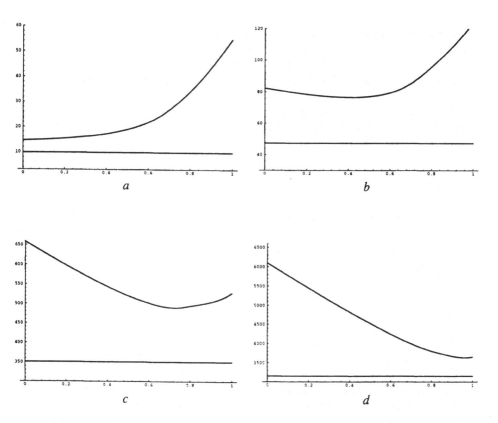

Figure 2. The loss bound given in Lemma 1 as a function of λ when η is chosen so as to minimize the bound.

by the triangle inequality; thus, X_1 can be replaced by $X_0/(1-\gamma)$. Note that the bounds in Eqs. (2) and (12) are incomparable in the sense that, depending on the values of the quantities involved, either bound can be better than the other. This suggests that $\mathrm{TD}^*(0)$ may or may not be better than $\mathrm{TD}^*(1)$ depending on the particular problem at hand; the bounds we have derived quantify those situations in which each will perform better than the other.

Ultimately, we hope to extend our analysis to facilitate the optimal choice of $\eta > 0$ and $\lambda \in [0, 1]$. In the meantime, we can numerically find the choices of η and λ that minimize the worst-case bound given in Lemma 1. Fig. 2 shows graphs of the worst-case bound given in Lemma 1 as a function of λ when η is chosen so as to minimize our worst-case bound and for fixed settings of the other parameters. More specifically, in all the graphs we have assumed $||\mathbf{w}_1 - \mathbf{u}|| = 1$, and $||\mathbf{x}_t|| \leq 1$ (which implies that $||\mathbf{X}_t^\lambda|| \leq 1/(1 - \gamma\lambda)$). We have also fixed $\gamma = 0.7$. Figs. 2a, b, c and d assume that $L^\ell(\mathbf{u}, S)$ equals 3, 30, 300 and 3000, respectively, and each curve shows the upper bound on $L^\ell(\mathrm{TD}^*(\lambda), S)$ given in Lemma 1. The straight solid line in each figure shows the lower bound obtained in Section 5. In each figure the x-axis crosses the y-axis at the value of $L^\ell(\mathbf{u}, S)$. Note that the gap between the lower bound and $L^\ell(\mathbf{u}, S)$ grows as $\theta(\sqrt{L^\ell(\mathbf{u}, S)})$ when all other variables are kept constant. (This is not visible from the figures because the scaling of the figures varies.) The figures were produced using Mathematica.

As the figures clearly indicate, the higher the loss $L^\ell(\mathbf{u}, S)$, the higher should be our choice for λ. It is interesting that in some intermediate cases, an intermediate value for λ in $(0, 1)$ is the best choice.

5. A lower bound

We next prove a lower bound on the performance of *any* learning algorithm in the model that we have been considering.

THEOREM 3 *Let* $\gamma \in [0, 1]$, $X_0 > 0$, $K \geq 0$, $U \geq 0$ *and* ℓ *a positive integer. For every algorithm A, there exists a sequence S such that the following hold:*

1. $\|\mathbf{x}_t\| \leq X_0$,

2. $K = \min\{L^\ell(\mathbf{u}, S) : \|\mathbf{u}\| \leq U\}$, *and*

3. $L^\ell(A, S) \geq (\sqrt{K} + U X_0 \sqrt{\sigma_\ell})^2$

where $\sigma_\ell := \sum_{k=0}^{\ell-1} \gamma^{2k}$.

Proof: The main idea of the proof is to construct a training sequence in which the learning algorithm A receives essentially no information until trial ℓ, at which time the adversary can force the learner to incur significant loss relative to the best linear predictor.

Without loss of generality, we prove the result in the one-dimensional case[4] (i.e., $N = 1$), so we write the instance $\mathbf{x}_t$ simply as x_t. The sequence S is defined as follows: We let $x_t = \gamma^{\ell-t} X_0$ for $t \leq \ell$, and $x_t = 0$ for $t > \ell$ (thus satisfying part 1). The reinforcement given is $r_t = 0$ if $t \neq \ell$, and $r_\ell = sz$ where $z = U X_0 + \sqrt{K/\sigma_\ell}$ and $s \in \{-1, +1\}$ is chosen adversarially after A has made predictions $\hat{y}_1, \ldots, \hat{y}_\ell$ on the first ℓ trials. Then

$$y_t = \sum_{k=0}^\infty \gamma^k r_{t+k} = \begin{cases} \gamma^{\ell-t} sz & \text{if } t \leq \ell \\ 0 & \text{otherwise.} \end{cases}$$

To see that part 2 holds, let $\mathbf{u} = u$ be any vector (scalar, really, since $N = 1$) with $|u| \leq U$. Then

$$L^\ell(u, S) = \sum_{t=1}^\ell (u x_t - y_t)^2$$

$$= \sum_{t=1}^\ell \gamma^{2(\ell-t)} (u X_0 - sz)^2$$

$$= (u X_0 - sz)^2 \sigma_\ell.$$

Since $|u| \le U$, it can be seen that this is minimized when $u = sU$, in which case $L^\ell(u, S) = K$ by z's definition.

Finally, consider the loss of A on this sequence:

$$L^\ell(A, S) = \sum_{t=1}^{\ell} (\hat{y}_t - y_t)^2 = \sum_{t=1}^{\ell} (\hat{y}_t - s\gamma^{\ell-t}z)^2.$$

For any real numbers p and q, we have $(p - q)^2 + (p + q)^2 = 2(p^2 + q^2) \ge 2q^2$. Thus, if $s \in \{-1, +1\}$ is chosen uniformly at random, then A's expected loss will be

$$\frac{1}{2} \left(\sum_{t=1}^{\ell} (\hat{y}_t - \gamma^{\ell-t}z)^2 + \sum_{t=1}^{\ell} (\hat{y}_t + \gamma^{\ell-t}z)^2 \right)$$

$$\ge \sum_{t=1}^{\ell} (\gamma^{\ell-t}z)^2 = z^2\sigma_\ell = (\sqrt{K} + UX_0\sqrt{\sigma_\ell})^2.$$

It follows that for the choice of $s \in \{-1, +1\}$ that maximizes A's loss, we will have that $L^\ell(A, S) \ge (\sqrt{K} + UX_0\sqrt{\sigma_\ell})^2$ as claimed. ∎

When $K = 0$, Theorem 3 gives a lower bound of $U^2X_0{}^2/\sigma_\ell$ which approaches $U^2X_0{}^2/(1 - \gamma^2)$ as ℓ becomes large. This lower bound matches the second bound of Theorem 1 in the corresponding case. Thus, in the "noise-free" case that there exists a vector $\mathbf{u}$ that perfectly matches the data (i.e., $\min\{L^\ell(\mathbf{u}, S) : \|\mathbf{u}\| \le U\} = 0$), this shows that TD$^*(0)$ is "optimal" in the sense that its worst-case performance is best possible.

6. An application to Markov processes

For the purposes of illustration, we show in this section how our results can be applied in the Markov-process setting more commonly used for studying temporal-difference algorithms. Specifically, we prove a kind of convergence theorem for TD$^*(1)$.

We consider Markov processes consisting of a finite set of states denoted $1, 2, \ldots, N$. An agent moves about the Markov process in the usual manner: An initial state i_1 is chosen stochastically. Then, at each time step t, the agent moves stochastically from state i_t to state i_{t+1} where i_{t+1} may depend only on the preceding state i_t. Upon exiting state i_t, the agent receives a probabilistic reward r_t which also may depend only on i_t.

Formally, the Markov process is defined by a transition matrix $\mathbf{Q} \in [0, 1]^{N \times N}$ and an initial state distribution matrix $\mathbf{p}_1 \in [0, 1]^N$. The entries of each column of $\mathbf{Q}$ sum to 1, as do the entries of $\mathbf{p}_1$. The interpretation here is that the initial state i_1 is distributed according to $\mathbf{p}_1$, and, if the agent is in state i_t at time t, then the next state i_{t+1} is distributed according to the $i_t{}^{\text{th}}$ column of $\mathbf{Q}$. Thus, state i_t has distribution $\mathbf{p}_t = \mathbf{Q}^{t-1}\mathbf{p}_1$.

The reward r_t received at time t depends only on the current state i_t so formally we can write $r_t = r(\omega_t, i_t)$ where $\omega_1, \omega_2, \ldots$ are independent identically distributed random variables from some event space Ω, and $r : \Omega \times \{1, \ldots, N\} \to \mathbb{R}$ is some fixed function.

Let V_i denote the expected discounted reward for a random walk produced by the Markov process that begins in state i. That is, we define the *value function*

$$V_i := \mathrm{E}\left[\sum_{k=0}^{\infty} \gamma^k r_{1+k} \mid i_1 = i\right]$$

where, as usual, $\gamma \in [0, 1)$ is a fixed parameter. Our goal is to estimate V_i, a problem often referred to as value-function approximation.

At each time step t, the learner computes an estimate $\hat{V}_i^t$ of the value function. Thus, learning proceeds as follows. At time $t = 1, 2, \ldots, \ell$:

1. The learner formulates an estimated value function $\hat{V}_i^t$.

2. The current state i_t is observed.

3. The current reward r_t is observed.

4. The learner moves to the next state i_{t+1}.

The states i_t and rewards r_t are random variables defined by the stochastic process described above. All expectations in this section are with respect to this random process.

Theorem 4, the main result of this section, gives a bound for $\mathrm{TD}^*(1)$ on the average expected squared distance of $\hat{V}_i^t$ to the correct values V_i. Specifically, we show that

$$\frac{1}{\ell}\sum_{t=1}^{\ell} \mathrm{E}\left[(\hat{V}_{i_t}^t - V_{i_t})^2\right] \leq O\left(\frac{1}{\sqrt{\ell}}\right)$$

for some setting of the learning rate, and given certain benign assumptions about the distribution of rewards. Note that $\mathrm{E}\left[(\hat{V}_{i_t}^t - V_{i_t})^2\right]$ is the expected squared distance between the tth estimate $\hat{V}_i^t$ and the true value function V_i where the expectation is with respect to the stochastic choice of the tth state i_t. Thus, the states more likely to be visited at step t receive the greatest weight under this expectation. Theorem 4 states that the average of these expectations (over the first ℓ time steps) rapidly drops to zero.

We apply $\mathrm{TD}^*(1)$ to this problem in the most natural manner. We define the observation vector $\mathbf{x}_t \in \mathbb{R}^N$ to have a 1 in component i_t, and 0 in all other components. (The generalization to other state representations is straightforward.) We then execute $\mathrm{TD}^*(1)$ using the sequence $\mathbf{x}_1, r_1, \mathbf{x}_2, r_2, \ldots, \mathbf{x}_\ell, r_\ell$ where these are random variables defined by the Markov process.

The estimated value function $\hat{V}_i^t$ is computed as follows: Recall that $\mathrm{TD}^*(1)$, at each time step t, generates an estimate $\hat{y}_t$ of the discounted sum $y_t = \sum_{k=0}^{\infty} \gamma^k r_{t+k}$. Note that if we are in state i at time t, then the expected value of y_t is exactly V_i, i.e.,

$$\mathrm{E}\left[y_t \mid i_t = i\right] = V_i.$$

So it makes sense to use the estimate $\hat{y}_t$ in computing the tth value-function approximation $\hat{V}_i^t$.

A minor difficulty arises from the fact that $\hat{y}_t$ is computed *after* x_t is observed (and therefore after i_t is observed), but $\hat{V}_i^t$ must be computed for all states i *before* i_t is observed. However, if we fix the history prior to trial t, then $\hat{y}_t$ is a function only of x_t, which in turn is determined by i_t. Therefore, for each state i, we can precompute what the value of $\hat{y}_t$ will be if i_t turns out to be i. We then define $\hat{V}_i^t$ to be this value. Note that, with this definition, the estimate $\hat{y}_t$ computed by TD*(1) once i_t is observed is equal to $\hat{V}_{i_t}^t$.

We now state and prove our convergence theorem. For this result, we assume a finite upper bound both on the value function and on the variance of the sum of discounted rewards.

THEOREM 4 *Suppose the Markov process is such that, for all i, $|V_i| \leq V$ and*

$$E\left[\left(\sum_{k=0}^{\infty} \gamma^k r_{1+k} - V_i\right)^2 \mid i_1 = i\right] \leq R \tag{15}$$

for finite and known V and R.

Suppose that TD*(1) *is executed as described above with $\mathbf{w}_1 = 0$ and learning rate*

$$\eta_t = \eta = \frac{V\sqrt{N}}{J^2 + J\sqrt{R\ell}}$$

where $J = V\sqrt{N}(1+\gamma)/(1-\gamma)$. Then

$$\frac{1}{\ell}\sum_{t=1}^{\ell} E\left[(\hat{V}_{i_t}^t - V_{i_t})^2\right] \leq 2J\sqrt{\frac{R}{\ell}} + \frac{J^2}{\ell}.$$

Proof: From Eq. (14), $\|\mathbf{X}_t^1\| \leq 1/(1-\gamma)$ so we choose $X_1 = 1/(1-\gamma)$. Let $\mathbf{u}$ be such that $u_i = V_i$. Then $\|\mathbf{u}\| \leq U$ where $U = V\sqrt{N}$. Finally, let $K = R\ell$. Our choice for η is identical to that in Theorem 2 where the appropriate substitutions have been made.

Note that

$$E\left[(y_t - V_{i_t})^2\right] = \sum_{i=1}^{N} \Pr\left[i_t = i\right] \cdot E\left[\left(\sum_{k=0}^{\infty} \gamma^k r_{t+k} - V_i\right)^2 \mid i_t = i\right] \leq R$$

by the assumption in Eq. (15). Thus, because $\mathbf{u} \cdot x_t = V_{i_t}$,

$$E\left[L^\ell(\mathbf{u}, S)\right] = \sum_{t=1}^{\ell} E\left[(\mathbf{u} \cdot x_t - y_t)^2\right] = \sum_{t=1}^{\ell} E\left[(V_{i_t} - y_t)^2\right] \leq R\ell = K.$$

Taking expectations of both sides of the bound in Lemma 1, we have that

$$E\left[L^\ell(\text{TD}^*(\lambda), S)\right] \leq \frac{bE\left[L^\ell(\mathbf{u}, S)\right] + \|\mathbf{u} - \mathbf{w}_1\|^2}{C_b}$$

for any $b > 0$ for which $C_b > 0$. Therefore, by a proof identical to the proof of Theorem 2 (except that we assume only that $L^\ell(\mathbf{u}, S)$ is bounded by K in *expectation*), we have

$$E\left[L^\ell(\text{TD}^*(1), S)\right] \leq E\left[L^\ell(\mathbf{u}, S)\right] + 2\sqrt{K}(1 + \gamma)U X_1 + (1 + \gamma)^2\|\mathbf{u}\|^2 X_1{}^2. \quad (16)$$

Since $\mathbf{u} \cdot \mathbf{x}_t = V_{i_t}$, $\hat{y}_t = \hat{V}_{i_t}^t$, and $E\left[y_t \mid i_t\right] = V_{i_t}$, it can be verified that

$$E\left[(y_t - \hat{y}_t)^2 - (y_t - \mathbf{u} \cdot \mathbf{x}_t)^2\right] = E\left[(V_{i_t} - \hat{V}_{i_t}^t)^2\right].$$

The theorem now follows by averaging over all time steps t and combining with Eq. (16).
∎

Unfortunately, we do not know how to prove a convergence result similar to Theorem 4 for $\text{TD}^*(\lambda)$ for general λ. This is because this proof technique requires a worst-case bound in which the term $L^\ell(\mathbf{u}, S)$ appears with a coefficient of 1.

Of course, Theorem 4 represents a considerable weakening of the worst-case results presented in Section 4. These worst-case bounds are stronger because (1) they are in terms of the *actual* discounted sum of rewards rather than its expectation, and (2) they do not depend on any statistical assumptions. Indeed, the generality of the results in Section 4 allows us to say something meaningful about the behavior of $\text{TD}^*(\lambda)$ for many similar but more difficult situations such as when

- there are a very large or even an infinite number of states (a state can be any vector in $\mathbb{R}^N$).

- some states are ambiguously represented so that two or more states are represented by the same vector.

- the underlying transition probabilities are allowed to change with time.

- each transition is chosen entirely or in part by an adversary (as might be the case in a game-playing scenario).

7. Algorithm for the batch model

In the usual supervised learning setting, the on-line learning proceeds as follows: In each trial $t \geq 1$ the learner receives an instance $\mathbf{x}_t \in \mathbb{R}^N$. Then, after producing a prediction $\hat{y}_t$ it gets a reinforcement y_t and incurs loss $(\hat{y}_t - y_t)^2$.

A classical algorithm for this problem is the Widrow-Hoff algorithm. It keeps a linear hypothesis represented by the vector $\mathbf{w}_t$ and predicts with $\hat{y}_t = \mathbf{w}_t \cdot \mathbf{x}_t$. The weight vector is updated using gradient descent:

$$\mathbf{w}_{t+1} := \mathbf{w}_t - 2\eta_t(\mathbf{w}_t \cdot \mathbf{x}_t - y_t)\mathbf{x}_t.$$

Note that $2(\mathbf{w}_t \cdot \mathbf{x}_t - y_t)\mathbf{x}_t$ is the gradient of the loss $(\mathbf{w}_t \cdot \mathbf{x}_t - y_t)^2$ with respect to $\mathbf{w}_t$.

There is a straightforward generalization of the above scenario when more than one instance is received in each trial t. In this generalization, the learner does the following in each trial:

1. receives a real-valued matrix $\mathbf{M}_t$ with N columns;

2. computes a prediction $\hat{\mathbf{y}}_t$;

3. gets reinforcement $\mathbf{y}_t$; both $\mathbf{y}_t$ and $\hat{\mathbf{y}}_t$ are real column vectors whose dimension is equal to the number of rows of $\mathbf{M}_t$;

4. incurs loss $||\hat{\mathbf{y}}_t - \mathbf{y}_t||^2$.

The rows of the matrix $\mathbf{M}_t$ can be viewed as a batch of instances received at trial t. The algorithm has to predict on all instances received in trial t before it gets the reinforcement vector $\mathbf{y}_t$ which contains one reinforcement per row. For each instance, the algorithm is charged for the usual square loss, and the loss in trial t is summed over all instances received in that trial.

The algorithm we study, called WHM, is a direct generalization of the Widrow-Hoff algorithm and was previously analyzed in the noise-free case by Cesa-Bianchi, Long and Warmuth (1993), The learner maintains a weight vector $\mathbf{w}_t \in \mathbb{R}^N$ and, on each trial t, computes its prediction as

$$\hat{\mathbf{y}}_t := \mathbf{M}_t\mathbf{w}_t.$$

After receiving reinforcement $\mathbf{y}_t$, the weight vector is updated using the rule

$$\mathbf{w}_{t+1} := \mathbf{w}_t - 2\eta_t\mathbf{M}_t^T(\mathbf{M}_t\mathbf{w}_t - \mathbf{y}_t).$$

Note that the Widrow-Hoff algorithm is a special case of WHM in which each matrix $\mathbf{M}_t$ contains exactly one row. Also, the update is standard gradient descent in that $2\mathbf{M}_t^T(\mathbf{M}_t\mathbf{w}_t - \mathbf{y}_t)$ is the gradient of the loss $||\mathbf{M}_t\mathbf{w}_t - \mathbf{y}_t||^2$ with respect to $\mathbf{w}_t$.

To model a particular reinforcement learning problem, we have the freedom to make up the matrices $\mathbf{M}_t$ and reinforcements $\mathbf{y}_t$ to suit our purpose. For example, Sutton (1988) and others have considered a model in which the learner takes a random walk on a Markov chain until it reaches a terminal state, whereupon it receives some feedback, and starts over with a new walk. The learner's goal is to predict the final outcome of each walk. This problem is really a special case of our model in which we let $\mathbf{M}_t$ contain the instances of a run and set $\mathbf{y}_t = (z_t, \cdots, z_t)^T$, where z_t is the reinforcement received for the tth run. (In this case, Sutton shows that the Widrow-Hoff algorithm is actually equivalent to a version of TD(1) in which updates are not made to the weight vector $\mathbf{w}_t$ until the final outcome is received.)

An *example* is a pair $(\mathbf{M}_t, \mathbf{y}_t)$, and, as before, we use S to denote a sequence of examples. We write $L^\ell(A, S)$ to denote the total loss of algorithm A on sequence S:

$$L^\ell(A, S) := \sum_{t=1}^{\ell} (\hat{\mathbf{y}}_t - \mathbf{y}_t)^2,$$

where $\hat{\mathbf{y}}_t$ is the prediction of A in the tth trial, and ℓ is the total length of the sequence. Similarly, the total loss of a weight vector $\mathbf{u} \in \mathbb{R}^N$ is defined as

$$L^\ell(\mathbf{u}, S) := \sum_{t=1}^{\ell} (\mathbf{M}_t \mathbf{u} - \mathbf{y}_t)^2.$$

The proof of the following lemma and theorem are a straightforward generalization of the worst-case analysis of the Widrow-Hoff algorithm given by Cesa-Bianchi, Long and Warmuth (1993). In the proof, we define, $||\mathbf{M}||$, the norm of any matrix $\mathbf{M}$, as

$$||\mathbf{M}|| = \max_{||\mathbf{x}||=1} ||\mathbf{Mx}||.$$

For comparison to the results in the first part of this paper, it is useful to note that $||\mathbf{M}|| \leq X\sqrt{m}$ where m is the number of rows of $\mathbf{M}$, and X is an upper bound on the norm of each row of $\mathbf{M}$.

For any vector $\mathbf{x}$, we write $\mathbf{x}^2$ to denote $\mathbf{x}^T \mathbf{x}$.

LEMMA 2 *Let* $(\mathbf{M}, \mathbf{y})$ *be an arbitrary example such that* $||\mathbf{M}|| \leq M$. *Let* $\mathbf{s}$ *and* $\mathbf{u}$ *be any weight vectors. Let* $b > 0$, *and let the learning rate be*

$$\eta = \frac{1}{2(||\mathbf{M}||^2 + 1/b)}.$$

Then

$$||\mathbf{Ms} - \mathbf{y}||^2 \leq (M^2 b + 1)||\mathbf{Mu} - \mathbf{y}||^2 + (M^2 + 1/b)(||\mathbf{u} - \mathbf{s}||^2 - ||\mathbf{u} - \mathbf{w}||^2), \tag{17}$$

where $\mathbf{w} = \mathbf{s} - 2\eta \mathbf{M}^T(\mathbf{Ms} - \mathbf{y})$ *denotes the weight vector of the algorithm* WHM *after updating its weight vector* $\mathbf{s}$.

Proof: Let $\mathbf{e} := \mathbf{y} - \mathbf{Ms}$ and $\mathbf{e}_\mathbf{u} := \mathbf{y} - \mathbf{Mu}$. Then inequality (17) holds if

$$f := ||\mathbf{u} - \mathbf{w}||^2 - ||\mathbf{u} - \mathbf{s}||^2 + 2\eta \mathbf{e}^2 - b\mathbf{e}_\mathbf{u}^2 \leq 0.$$

Since $\mathbf{w} = \mathbf{s} + 2\eta \mathbf{M}^T \mathbf{e}$, f can be rewritten as

$$\begin{aligned} f &= -4\eta(\mathbf{u} - \mathbf{s})^T \mathbf{M}^T \mathbf{e} + 4\eta^2 ||\mathbf{M}^T \mathbf{e}||^2 + 2\eta \mathbf{e}^2 - b\mathbf{e}_\mathbf{u}^2 \\ &= -4\eta(\mathbf{e} - \mathbf{e}_\mathbf{u})^T \mathbf{e} + 4\eta^2 ||\mathbf{M}^T \mathbf{e}||^2 + 2\eta \mathbf{e}^2 - b\mathbf{e}_\mathbf{u}^2 \\ &= -2\eta \mathbf{e}^2 + 4\eta \mathbf{e}_\mathbf{u}^T \mathbf{e} + 4\eta^2 ||\mathbf{M}^T \mathbf{e}||^2 - b\mathbf{e}_\mathbf{u}^2. \end{aligned}$$

Since $2\mathbf{e}_\mathbf{u}^T \mathbf{e} \leq \frac{b}{2\eta}\mathbf{e}_\mathbf{u}^2 + \frac{2\eta}{b}\mathbf{e}^2$ and since $||\mathbf{M}^T \mathbf{e}|| \leq ||\mathbf{M}||||\mathbf{e}||$, we can upper bound f by

$$\mathbf{e}^2(-2\eta + 4\eta^2(||\mathbf{M}||^2 + 1/b)) = 0.$$

∎

THEOREM 5 *Let S be any sequence of examples and let M be the largest norm $||\mathbf{M}_t||$. If the matrix algorithm WHM uses any start vector s and learning rates*

$$\eta_t = \eta = \frac{1}{2(||\mathbf{M}_t||^2 + M^2)},$$

then we have for any vector u the bound

$$L^\ell(\text{WHM}, S) \leq 2(L^\ell(\mathbf{u}, S) + M^2||\mathbf{s} - \mathbf{u}||^2). \tag{18}$$

Assume further that we know bounds K and U such that for some u we have $L^\ell(\mathbf{u}, S) \leq K$ and $||\mathbf{s} - \mathbf{u}|| \leq U$. Then for the learning rates

$$\eta_t = \eta = \frac{U}{2(||\mathbf{M}_t||^2 U + M\sqrt{K})}$$

we have

$$L^\ell(\text{WHM}, S) \leq L^\ell(\mathbf{u}, S) + 2MU\sqrt{K} + M^2||\mathbf{s} - \mathbf{u}||^2. \tag{19}$$

Proof: Assume that

$$\eta_t = \eta = \frac{1}{2(||\mathbf{M}||^2 + 1/b)}$$

for some $b > 0$ to be chosen later. By summing the inequality of Lemma 2 over all runs of S we get

$$L^\ell(\text{WHM}, S) \leq (bM^2 + 1)L^\ell(\mathbf{u}, S) + (M^2 + 1/b)(||\mathbf{u} - \mathbf{s}||^2 - ||\mathbf{u} - \mathbf{w}'||^2),$$

where $\mathbf{w}'$ is the weight vector after the last reinforcement of S is processed. Since $||\mathbf{u} - \mathbf{w}'||^2 \geq 0$, we have

$$L^\ell(\text{WHM}, S) \leq (bM^2 + 1)L^\ell(\mathbf{u}, S) + (M^2 + 1/b)||\mathbf{u} - \mathbf{s}||^2.$$

Now setting $b = 1/M^2$ gives the choice of η in the first part of the theorem and so yields the bound in Eq. (18).

Assuming further that $L^\ell(\mathbf{u}, S) \leq K$ and $||\mathbf{s} - \mathbf{u}|| \leq U$, we get

$$L^\ell(\text{WHM}, S) \leq L^\ell(\mathbf{u}, S) + M^2||\mathbf{s} - \mathbf{u}||^2 + bKM^2 + U^2/b. \tag{20}$$

The part of the right hand side that depends on b is $bKM^2 + U^2/b$ which is minimized when $b = U/(M\sqrt{K})$. Using this value of b in Eq. (20) gives the desired choice of η and the bound in Eq. (19).   ∎

In the special case that $K = 0$, setting $\eta_t = 1/(2||\mathbf{M}_t||^2)$ gives a bound of

$$L^\ell(\text{WHM}, S) \leq L^\ell(\mathbf{u}, S) + M^2||\mathbf{s} - \mathbf{u}||^2.$$

Note that to prove this, $b = \infty$ is used. The bound for $K = 0$ was previously proved by Cesa-Bianchi, Long and Warmuth (1993). An alternate proof of the above theorem via a reduction from the corresponding theorem for the original Widrow-Hoff algorithm was recently provided by Kivinen and Warmuth (1994).

The following lower bound shows that the bounds of the above theorem are best possible.

THEOREM 6 *Let $N, m \geq 1$, $K, U \geq 0$ and $M > 0$. For every prediction algorithm A there exists a sequence S consisting of a single example $(\mathbf{M}, \mathbf{y})$ such that the following hold:*

1. *$\mathbf{M}$ is an $m \times N$ matrix and $\|\mathbf{M}\| = M$;*

2. *$K = \min\{L^\ell(\mathbf{u}, S) : \|\mathbf{u}\| \leq U\}$; and*

3. *$L^\ell(A, S) \geq K + 2UM\sqrt{K} + U^2 M^2$.*

Proof: As in the proof of Theorem 3, we prove the result in the case that $N = 1$, without loss of generality. Thus, $\mathbf{M}$ is actually a column vector in $\mathbb{R}^m$.

Let each component of $\mathbf{M}$ be equal to $M/\sqrt{m}$ so that $\|\mathbf{M}\| = M$. Let each component of $\mathbf{y}$ be equal to sz where $z = (MU + \sqrt{K})/\sqrt{m}$ and $s \in \{-1, +1\}$ is chosen adversarially after A has made its prediction $\hat{\mathbf{y}} = (\hat{y}_1, \ldots, \hat{y}_m)^T$.

To see that part 2 holds, let $\mathbf{u} = u$ be a vector (scalar, really). Then

$$L^\ell(u, S) = \|\mathbf{M}u - \mathbf{y}\|^2 = m(Mu/\sqrt{m} - sz)^2$$

which is minimized when $u = sU$ for $|u| \leq U$. In this case, $L^\ell(u, S) = K$.

Finally, by choosing s adversarially to maximize algorithm A's loss, we have

$$
\begin{aligned}
L^\ell(A, S) &= \max_{s \in \{-1, +1\}} \sum_{i=1}^{m} (\hat{y}_i - sz)^2 \\
&\geq \frac{1}{2} \sum_{i=1}^{m} \left((\hat{y}_i - z)^2 + (\hat{y}_i + z)^2 \right) \\
&\geq \sum_{i=1}^{m} z^2 = K + 2MU\sqrt{K} + M^2 U^2.
\end{aligned}
$$

$\blacksquare$

8. Discussion

The primary contribution of this paper is the analysis of some simple temporal-difference algorithms using a worst-case approach. This method of analysis differs dramatically from the statistical approach that has been used in the past for such problems, and our approach has some important advantages.

First, the results that are obtained using the worst-case approach are quite robust. Obviously, any analysis of any learning algorithm is valid only when the assumed conditions actually hold in the real world. By making the most minimal of assumptions — and, in particular, by making no assumptions at all about the stochastic nature of the world — we hope to be able to provide analyses that are as robust and broadly applicable as possible.

Statistical methods for analyzing on-line learning algorithms are only necessary when worst-case bounds cannot be obtained. In this paper, we demonstrated that temporal-difference learning algorithms with simple linear models are highly amenable to worst-case analysis. Although one might expect such a pessimistic approach to give rather weak results, we have found, somewhat surprisingly, that very strong bounds can often be proved even in the worst case.

Worst-case bounds for on-line linear learning algorithms can be very tight even on artificial data (Kivinen & Warmuth, 1994). Good experimental performance of a particular algorithm might be seen as weak evidence for showing that the algorithm is good since every algorithm performs well on some data, particularly when the data is artificial. However, if we have a worst-case bound for a particular algorithm, then we can use experimental data to show how much worse the competitors can perform relative to the worst-case bound of the algorithm in question.

Another strength of the worst-case approach is its emphasis on the actual performance of the learning algorithm on the actually observed data. Breaking with more traditional approaches, we do not analyze how well the learning algorithm performs in expectation, or how well it performs asymptotically as the amount of training data becomes infinite, or how well the algorithm estimates the underlying parameters of some assumed stochastic model. Rather, we focus on the quality of the learner's predictions as measured against the finite sequence of data that it actually observes.

Finally, our method of analysis seems to be more fine-grained than previous approaches. As a result, the worst-case approach may help to resolve a number of open issues in temporal-difference learning, such as the following:

- *Which learning rules are best for which problems?* We use the total worst-case loss as our criterion. Minimizing this criterion led us to discover the modified learning rule $TD^*(\lambda)$. Unlike the original $TD(\lambda)$, this rule has a gradient descent interpretation for general λ. Our method can also be used to derive worst-case bounds for the original rule, but we were unable to obtain bounds for $TD(\lambda)$ stronger than those given for $TD^*(\lambda)$. It will be curious to see how the two rules compare experimentally.

  Also, the results in Section 4 provide explicit worst-case bounds on the performance of $TD^*(0)$ and $TD^*(1)$. These bounds show that one of the two algorithms may or may not be better than the other depending on the values of the parameters X_0, K, etc. Thus, using a priori knowledge we may have about a particular learning problem, we can use these bounds to guide us in deciding which algorithm to use.

- *How should a learning algorithm's parameters be tuned?* For instance, we have shown how the learning rate η should be chosen for $TD^*(0)$ and $TD^*(1)$ using knowledge which may be available about a particular problem. For the choice of λ,

Sutton showed experimentally that, in some cases, the learner's hypothesis got closest to the target when λ is chosen in $(0, 1)$ and that there is clearly one optimal choice. So far, our worst-case bounds for TD$^*(\lambda)$ are not in closed form when $\lambda \in (0, 1)$, but, numerically, we have found that our results are entirely consistent with Sutton's in this regard.

• *How does the performance of a learning algorithm depend on various parameters of the problem?* For instance, our bounds show explicitly how the performance of TD$^*(\lambda)$ degrades as γ approaches 1. Furthermore, the lower bounds that can sometimes be proved (such as in Section 5) help us to understand what performance is best possible as a function of these parameters.

Open problems.   There remain many open research problems in this area. The first of these is to reduce the bound given in Lemma 1 to closed form to facilitate the optimal choice of $\lambda \in [0, 1]$. However, as clearly indicated by Fig. 2, even when λ and η are chosen so as to minimize this bound, there remains a significant gap between the upper bounds proved in Section 4 and the lower bound proved in Section 5. This may be a weakness of our analysis, or this may be an indication that an algorithm better than either TD(λ) or TD$^*(\lambda)$ is waiting to be discovered.

So far, we have only been able to obtain results when the comparison class consists of linear predictors defined by a weight vector u which make predictions of the form $\mathbf{u} \cdot \mathbf{x}_t$. It is an open problem to prove worst-case loss bounds with respect to other comparison classes.

As described in Section 3, TD$^*(\lambda)$ can be motivated using gradient descent. Rules of this kind can alternatively be derived within a framework described by Kivinen and Warmuth (1994). Moreover, by modifying one of the parameters of their framework, they show that update rules having a qualitatively different flavor can be derived that use the approximation of the gradient $\nabla_{\mathbf{w}_t}(y_t - \hat{y}_t)^2$ in the exponent of a multiplicative update. (Note that the TD(λ) update is additive.) In particular, they analyze such an algorithm, which they call EG, for the same problem that we are considering in the special case that $\gamma = 0$. Although the bounds they obtain are generally incomparable with the bounds derived for gradient-descent algorithms, these new algorithms have great advantages in some very important cases. It is straightforward to generalize their update rule for $\gamma > 0$, but the analysis of the resulting update rule is an open problem (although we have made some preliminary progress in this direction).

Lastly, Sutton's TD(λ) algorithm can be viewed as a special case of Watkin's "Q-learning" algorithm (1989). This algorithm is meant to handle a setting in which the learner has a set of actions to choose from, and attempts to choose its actions so as to maximize its total payoff. A very interesting open problem is the extension of the worst-case approach to such a setting in which the learner has partial control over its environment and over the feedback that it receives.

Acknowledgments

We are very grateful to Rich Sutton for his continued feedback and guidance. Thanks also to Satinder Singh for thought-provoking discussions, and to the anonymous referees for their careful reading and feedback.

Manfred Warmuth acknowledges the support of NSF grant IRI-9123692 and AT&T Bell Laboratories. This research was primarily conducted while visiting AT&T Bell Laboratories.

Appendix

In this technical appendix, we complete the proof of Lemma 1 by bounding $\rho(\mathbf{M})$, the largest eigenvalue of the matrix $\mathbf{M}$.

Let $\mathbf{I}$ be the $\ell \times \ell$ identity matrix, and, for $i, j \geq 0$, define

$$
\begin{aligned}
\mathbf{S}_i &= \mathbf{Z}_i + \mathbf{Z}_i^T, \\
\mathbf{R}_i &= \mathbf{Z}_i^T \mathbf{Z}_i, \\
\mathbf{P}_{ij} &= \mathbf{Z}_i^T \mathbf{Z}_j + \mathbf{Z}_j^T \mathbf{Z}_i.
\end{aligned}
$$

Since $\mathbf{Z}_i$ is the zero matrix for $i \geq \ell$, we can rewrite $\mathbf{V}$ more conveniently as

$$
\mathbf{V} = \sum_{i \geq 0} (\gamma\lambda)^i \mathbf{Z}_i.
$$

By direct but tedious computations, we have that

$$
\mathbf{D}^T \mathbf{D} = \mathbf{I} - \gamma \mathbf{S}_1 + \gamma^2 \mathbf{R}_1,
$$

and

$$
\mathbf{V}\mathbf{D} = \mathbf{I} + \left(1 - \frac{1}{\lambda}\right) \sum_{i \geq 1} (\gamma\lambda)^i \mathbf{Z}_i
$$

since $\mathbf{Z}_i \mathbf{Z}_1 = \mathbf{Z}_{i+1}$ for $i \geq 0$. Also,

$$
\begin{aligned}
\mathbf{D}^T \mathbf{V}^T \mathbf{V}\mathbf{D} &= \mathbf{I} + \left(1 - \frac{1}{\lambda}\right)^2 \left[\sum_{i \geq 1} (\gamma\lambda)^{2i} \mathbf{R}_i + \sum_{j > i \geq 1} (\gamma\lambda)^{i+j} \mathbf{P}_{ij} \right] \\
&\quad + \left(1 - \frac{1}{\lambda}\right) \sum_{i \geq 1} (\gamma\lambda)^i \mathbf{S}_i.
\end{aligned}
$$

Thus, $\mathbf{M}$ can be written as:

$$
\left(\eta^2 {X_\lambda}^2 - 2\eta + \frac{\eta^2}{b}\right) \mathbf{I} + \left(-\eta^2 {X_\lambda}^2 \gamma + \left(\eta - \frac{\eta^2}{b}\right) \gamma(1 - \lambda)\right) \mathbf{S}_1 + \eta^2 {X_\lambda}^2 \gamma^2 \mathbf{R}_1
$$

$$+ \left(1 - \frac{1}{\lambda}\right) \left(\frac{\eta^2}{b} - \eta\right) \sum_{i \geq 2} (\gamma\lambda)^i \mathbf{S}_i$$

$$+ \frac{\eta^2}{b} \left(1 - \frac{1}{\lambda}\right)^2 \left[\sum_{i \geq 1} (\gamma\lambda)^{2i} \mathbf{R}_i + \sum_{j > i \geq 1} (\gamma\lambda)^{i+j} \mathbf{P}_{ij}\right].$$

It is known that $\rho(A+B) \leq \rho(A) + \rho(B)$ for real, symmetric matrices A and B. Further, it can be shown (for instance, using Eq. (10)) that

$$\rho(\mathbf{I}) = 1;$$
$$\rho(\mathbf{R}_i) \leq 1;$$
$$\rho(\pm\mathbf{S}_i) \leq 2;$$
$$\rho(\mathbf{P}_{ij}) \leq 2.$$

Applying these bounds gives that

$$\rho(\mathbf{M}) \leq \eta^2 X_\lambda^2 - 2\eta + \frac{\eta^2}{b} + 2\left|\left(\eta - \frac{\eta^2}{b}\right)\gamma(1 - \lambda) - \eta^2 X_\lambda^2 \gamma\right| + \eta^2 X_\lambda^2 \gamma^2$$

$$-2\left(1 - \frac{1}{\lambda}\right) \left|\frac{\eta^2}{b} - \eta\right| \sum_{i \geq 2} (\gamma\lambda)^i$$

$$+ \frac{\eta^2}{b} \left(1 - \frac{1}{\lambda}\right)^2 \left[\sum_{i \geq 1} (\gamma\lambda)^{2i} + 2 \sum_{j > i \geq 1} (\gamma\lambda)^{i+j}\right]$$

$$= -C_b.$$

Notes

1. In this paper we only use one vector norm, the L_2-norm: $\|\mathbf{u}\| = \sqrt{\sum_{i=1}^{N} u_i^2}$.
2. In some versions of TD(λ), this difficulty is overcome by replacing $\hat{y}_{t+1} = \mathbf{w}_{t+1} \cdot \mathbf{x}_{t+1}$ in the update rule (4) by the approximation $\mathbf{w}_t \cdot \mathbf{x}_{t+1}$.
3. The factor of two in front of η_t can be absorbed into η_t.
4. If $N > 1$, we can reduce to the one-dimensional case by zeroing all but one of the components of $\mathbf{x}_t$.

References

Nicolò Cesa-Bianchi, Philip M. Long, & Manfred K. Warmuth. (1993). Worst-case quadratic loss bounds for a generalization of the Widrow-Hoff rule. In *Proceedings of the Sixth Annual ACM Conference on Computational Learning Theory*, pages 429–438.

Peter Dayan. (1992). The convergence of $TD(\lambda)$ for general λ. *Machine Learning*, 8(3/4):341–362.

Peter Dayan & Terrence J. Sejnowski. (1994). $TD(\lambda)$ converges with probability 1. *Machine Learning*, 14(3):295–301.

Roger A. Horn & Charles R. Johnson. (1985). *Matrix Analysis*. Cambridge University Press.

Tommi Jaakkola, Michael I. Jordan, & Satinder P. Singh. (1993). On the convergence of stochastic iterative dynamic programming algorithms. Technical Report 9307, MIT Computational Cognitive Science.

Jyrki Kivinen & Manfred K. Warmuth. (1994). Additive versus exponentiated gradient updates for learning linear functions. Technical Report UCSC-CRL-94-16, University of California Santa Cruz, Computer Research Laboratory.

Richard S. Sutton. (1988). Learning to predict by the methods of temporal differences. *Machine Learning*, 3:9–44.

C. J. C. H. Watkins. (1989). *Learning from delayed rewards*. PhD thesis, University of Cambridge, England, 1989.

Received November 2, 1994
Accepted February 23, 1995
Final Manuscript September 29, 1995

Machine Learning, 22, 123–158 (1996)

Reinforcement Learning with Replacing Eligibility Traces

SATINDER P. SINGH   singh@psyche.mit.edu

Dept. of Brain and Cognitive Sciences
Massachusetts Institute of Technology, Cambridge, Mass. 02139

RICHARD S. SUTTON   rich@cs.umass.edu
Dept. of Computer Science
University of Massachusetts, Amherst, Mass. 01003

Editor: Leslie Pack Kaelbling

Abstract. The eligibility trace is one of the basic mechanisms used in reinforcement learning to handle delayed reward. In this paper we introduce a new kind of eligibility trace, the *replacing* trace, analyze it theoretically, and show that it results in faster, more reliable learning than the conventional trace. Both kinds of trace assign credit to prior events according to how recently they occurred, but only the conventional trace gives greater credit to repeated events. Our analysis is for conventional and replace-trace versions of the offline TD(1) algorithm applied to undiscounted absorbing Markov chains. First, we show that these methods converge under repeated presentations of the training set to the same predictions as two well known Monte Carlo methods. We then analyze the relative efficiency of the two Monte Carlo methods. We show that the method corresponding to conventional TD is biased, whereas the method corresponding to replace-trace TD is unbiased. In addition, we show that the method corresponding to replacing traces is closely related to the maximum likelihood solution for these tasks, and that its mean squared error is always lower in the long run. Computational results confirm these analyses and show that they are applicable more generally. In particular, we show that replacing traces significantly improve performance and reduce parameter sensitivity on the "Mountain-Car" task, a full reinforcement-learning problem with a continuous state space, when using a feature-based function approximator.

Keywords: reinforcement learning, temporal difference learning, eligibility trace, Monte Carlo method, Markov chain, CMAC

1. Eligibility Traces

Two fundamental mechanisms have been used in reinforcement learning to handle delayed reward. One is temporal-difference (TD) learning, as in the TD(λ) algorithm (Sutton, 1988) and in Q-learning (Watkins, 1989). TD learning in effect constructs an internal reward signal that is less delayed than the original, external one. However, TD methods can eliminate the delay completely only on fully Markov problems, which are rare in practice. In most problems some delay always remains between an action and its effective reward, and on all problems some delay is always present during the time before TD learning is complete. Thus, there is a general need for a second mechanism to handle whatever delay is not eliminated by TD learning.

The second mechanism that has been widely used for handling delay is the *eligibility trace*.[1] Introduced by Klopf (1972), eligibility traces have been used in a variety of reinforcement learning systems (e.g., Barto, Sutton & Anderson, 1983; Lin, 1992; Tesauro, 1992; Peng & Williams, 1994). Systematic empirical studies of eligibility traces in conjunction with TD methods were made by Sutton (1984), and theoretical results have been obtained by several authors (e.g., Dayan, 1992; Jaakkola, Jordan & Singh, 1994; Tsitsiklis, 1994; Dayan & Sejnowski, 1994; Sutton & Singh, 1994).

The idea behind all eligibility traces is very simple. Each time a state is visited it initiates a short-term memory process, a trace, which then decays gradually over time. This trace marks the state as *eligible* for learning. If an unexpectedly good or bad event occurs while the trace is non-zero, then the state is assigned credit accordingly. In a conventional *accumulating trace*, the trace builds up each time the state is entered. In a *replacing trace*, on the other hand, each time the state is visited the trace is reset to 1 regardless of the presence of a prior trace. The new trace replaces the old. See Figure 1.

Sutton (1984) describes the conventional trace as implementing the credit assignment heuristics of *recency*—more credit to more recent events—and *frequency*—more credit to events that have occurred more times. The new replacing trace can be seen simply as discarding the frequency heuristic while retaining the recency heuristic. As we show later, this simple change can have a significant effect on performance.

Typically, eligibility traces decay exponentially according to the product of a decay parameter, λ, $0 \leq \lambda \leq 1$, and a discount-rate parameter, γ, $0 \leq \gamma \leq 1$. The conventional accumulating trace is defined by:[2]

$$e_{t+1}(s) = \begin{cases} \gamma \lambda e_t(s) & \text{if } s \neq s_t; \\ \gamma \lambda e_t(s) + 1 & \text{if } s = s_t, \end{cases}$$

where $e_t(s)$ represents the trace for state s at time t, and s_t is the actual state at time t. The corresponding replacing trace is defined by:

$$e_{t+1}(s) = \begin{cases} \gamma \lambda e_t(s) & \text{if } s \neq s_t; \\ 1 & \text{if } s = s_t. \end{cases}$$

In a control problem, each state-action *pair* has a separate trace. When a state is visited and an action taken, the state's trace for that action is reset to 1 while the traces for the other actions are reset to zero (see Section 5).

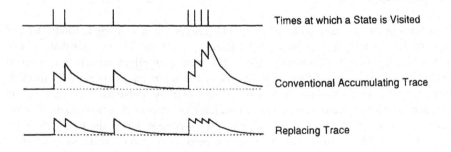

Figure 1. Accumulating and replacing eligibility traces.

For problems with a large state space it may be extremely unlikely for the exact same state ever to recur, and thus one might think replacing traces would be irrelevant. However, large problems require some sort of generalization between states, and thus some form of function approximator. Even if the same *states* never recur, states with the same *features* will. In Section 5 we show that replacing traces do indeed make a significant difference on problems with a large state space when the traces are done on a feature-by-feature basis rather than on a state-by-state basis.

The rest of this paper is structured as follows. In the next section we review the TD(λ) prediction algorithm and prove that its variations using accumulating and replacing traces are closely related to two Monte Carlo algorithms. In Section 3 we present our main results on the relative efficiency of the two Monte Carlo algorithms. Sections 4 and 5 are empirical and return to the general case.

2. TD(λ) and Monte Carlo Prediction Methods

The prediction problem we consider is a classical one in reinforcement learning and optimal control. A Markov chain emits on each of its transitions a *reward*, $r_{t+1} \in \Re$, according to a probability distribution dependent only on the pre-transition state, s_t, and the post-transition state, s_{t+1}. For each state, we seek to predict the expected total (cumulative) reward emitted starting from that state until the chain reaches a terminal state. This is called the *value* of the state, and the function mapping states s to their values $V(s)$ is called the *value function*. In this paper, we assume no discounting ($\gamma = 1$) and that the Markov chain always reaches a terminal state. Without loss of generality we assume that there is a single terminal state, T, with value $V(T) = 0$. A single trip from starting state to terminal state is called a *trial*.

2.1. TD(λ) Algorithms

The TD(λ) family of algorithms combine TD learning with eligibility traces to estimate the value function. The discrete-state form of the TD(λ) algorithm is defined by

$$\Delta V_t(s) = \alpha_t(s) \Big[r_{t+1} + V_t(s_{t+1}) - V_t(s_t) \Big] e_{t+1}(s) \qquad \forall s, \forall t \text{ s.t. } s_t \neq T, \qquad (1)$$

where $V_t(s)$ is the estimate at time t of $V(s)$, $\alpha_t(s)$ is a positive step-size parameter, $e_{t+1}(s)$ is the eligibility trace for state s, and $\Delta V_t(s)$ is the increment in the estimate of $V(s)$ determined at time t.[3] The value at the terminal state is of course defined as $V_t(T) = 0$, $\forall t$. In *online* TD(λ), the estimates are incremented on every time step: $V_{t+1}(s) = V_t(s) + \Delta V_t(s)$. In *offline* TD($\lambda$), on the other hand, the increments $\Delta V_t(s)$ are set aside until the terminal state is reached. In this case the estimates $V_t(s)$ are constant while the chain is undergoing state transitions, all changes being deferred until the end of the trial.

There is also a third case in which updates are deferred until after an entire set of trials have been presented. Usually this is done with a small fixed step size, $\alpha_t(s) = \alpha$, and with the training set (the set of trials) presented over and over again until convergence of the value estimates. Although this "repeated presentations" training paradigm is rarely used in practice, it can reveal telling theoretical properties of the algorithms. For example, Sutton (1988) showed that TD(0) (TD(λ) with $\lambda = 0$) converges under these conditions to a maximum likelihood estimate, arguably the best possible solution to this prediction problem (see Section 2.3). In this paper, for convenience, we refer to the repeated presentations training paradigm simply as *batch* updating. Later in this section we show that the batch versions of conventional and replace-trace TD(1) methods are equivalent to two Monte Carlo prediction methods.

2.2. Monte Carlo Algorithms

The total reward following a particular visit to a state is called the *return* for that visit. The value of a state is thus the expected return. This suggests that one might estimate a state's value simply by averaging all the returns that follow it. This is what is classically done in *Monte Carlo* (MC) prediction methods (Rubinstein, 1981; Curtiss, 1954; Wasow, 1952; Barto & Duff, 1994). We distinguish two specific algorithms:

Every-visit MC: Estimate the value of a state as the average of the returns that have followed all visits to the state.

First-visit MC: Estimate the value of a state as the average of the returns that have followed the *first visits* to the state, where a first visit is the first time during a trial that the state is visited.

Note that both algorithms form their estimates based entirely on actual, complete returns. This is in contrast to TD(λ), whose updates (1) are based in part on existing estimates. However, this is only in part, and, as $\lambda \to 1$, TD(λ) methods come to more and more closely approximate MC methods (Sections 2.4 and 2.5). In particular, the conventional, accumulate-trace version of TD(λ) comes to approximate every-visit MC, whereas replace-trace TD(λ) comes to approximate first-visit MC. One of the main points of this paper is that we can better understand the difference between replace and accumulate versions of TD(λ) by understanding the difference between these two MC methods. This naturally brings up the question that we focus on in Section 3: what are the relative merits of first-visit and every-visit MC methods?

2.3. A Simple Example

To help develop intuitions, first consider the very simple Markov chain shown in Figure 2a. On each step, the chain either stays in S with probability p, or goes on to terminate in T with probability $1 - p$. Suppose we wish to estimate the expected number of steps before termination when starting in S. To put this in the form of estimating a value function, we say that a reward of $+1$ is emitted on every step, in which case $V(S)$

is equal to the expected number of steps before termination. Suppose that the only data that has been observed is a single trial generated by the Markov chain, and that that trial lasted 4 steps, 3 passing from S to S, and one passing from S to T, as shown in Figure 2b. What do the two MC methods conclude from this one trial?

We assume that the methods do not know the structure of the chain. All they know is the one experience shown in Figure 2b. The first-visit MC method in effect sees a single traversal from the first time S was visited to T. That traversal lasted 4 steps, so its estimate of $V(S)$ is 4. Every-visit MC, on the other hand, in effect sees 4 separate traversals from S to T, one with 4 steps, one with 3 steps, one with 2 steps, and one with 1 step. Averaging over these four effective trials, every-visit MC estimates $V(S)$ as $\frac{1+2+3+4}{4} = 2.5$. The replace and accumulate versions of TD(1) may or may not form exactly these estimates, depending on their α sequence, but they will move their estimates in these directions. In particular, if the corresponding offline TD(1) method starts the trial with these estimates, then it will leave them unchanged after experiencing the trial. The batch version of the two TD(1) algorithms will compute exactly these estimates.

Which estimate is better, 4 or 2.5? Intuitively, the first answer appears better. The only trial observed took 4 steps, so 4 seems like the best estimate of its expected value. In any event, the answer 2.5 seems too low. In a sense, the whole point of this paper is to present theoretical and empirical analyses in support of this intuition. We show below that in fact the answer 4 is the only unbiased answer, and that 2.5, the answer of every-visit MC and of conventional TD(1), is biased in a statistical sense.

It is instructive to compare these two estimates of the value function with the estimate that is optimal in the maximum likelihood sense. Given some data, in this case a set of observed trials, we can construct the maximum-likelihood model of the underlying Markov process. In general, this is the model whose probability of generating the observed data is the highest. Consider our simple example. After the one trial has been observed, the maximum-likelihood estimate of the S-to-S transition probability is $\frac{3}{4}$, the fraction of the actual transitions that went that way, and the maximum-likelihood estimate of the S-to-T transition probability is $\frac{1}{4}$. No other transitions have been observed, so they are estimated as having probability 0. Thus, the maximum-likelihood model of the Markov chain is as shown in Figure 2c.

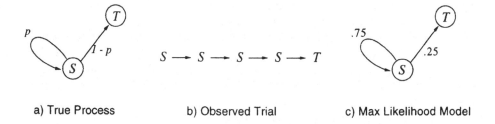

a) True Process   b) Observed Trial   c) Max Likelihood Model

Figure 2. A simple example of a Markov prediction problem. The objective is to predict the number of steps until termination.

We define the *ML estimate* of the value function as the value function that would be exactly correct if the maximum-likelihood model of the Markov process were exactly correct. That is, it is the estimate equal to the correct answer if the estimate of the Markov chain was not really an estimate, but was known with certainty.[4] Note that the ML estimate makes full use of all the observed data.

Let us compute the ML estimate for our simple example. If the maximum-likelihood model of the chain, as shown in Figure 2c, were exactly correct, what then would be the expected number of time steps before termination? For each possible number of steps, k, we can compute the probability of its occurring, and then the expected number, as

$$V^{ML}(S) = \sum_{k=1}^{\infty} \Pr(k)k$$
$$= \sum_{k=1}^{\infty} (0.75)^{k-1} \cdot 0.25 \cdot k$$
$$= 4.$$

Thus, in this simple example the ML estimate is the same as the first-visit MC estimate. In general, these two are not exactly the same, but they are closely related. We establish the relationship in the general case in Section 3.2.

Computing the ML estimate is in general very computationally complex. If the number of states is n, then the maximum-likelihood model of the Markov chain requires $O(n^2)$ memory, and computing the ML estimate from it requires roughly $O(n^3)$ computational operations.[5] The TD methods by contrast all use memory and computation per step that is only $O(n)$. It is in part because of these computational considerations that learning solutions are of interest while the ML estimate remains an ideal generally unreachable in practice. However, we can still ask how closely the various learning methods approximate this ideal.

2.4. *Equivalence of Batch TD(1) and MC Methods*

In this subsection we establish that the replace and accumulate forms of batch TD(1) are equivalent, respectively, to first-visit and every-visit MC. The next subsection proves a similar equivalence for the offline TD(1) algorithms.

The equivalence of the accumulate-trace version of batch TD(1) to every-visit MC follows immediately from prior results. Batch TD(1) is a gradient-descent procedure known to converge to the estimate with minimum mean squared error on the training set (Sutton, 1988; Dayan, 1992; Barnard, 1993). In the case of discrete states, the minimum MSE estimate for a state is the sample average of the returns from every visit to that state in the training set, and thus it is the same as the estimate computed by every-visit MC.

Showing the equivalence of replace-trace batch TD(1) and first-visit MC requires a little more work.

THEOREM 1: *For any training set of N trials and any fixed $\alpha_t(s) = \alpha < \frac{1}{N}$, batch replace TD(1) produces the same estimates as first-visit MC.*

Proof: In considering updates to the estimates for any state s we need only consider trials in which s occurs. On trials in which s does not occur, the estimates of both methods are obviously unchanged. We index the trials in which state s occurs from 1 to $N(s)$. Let $t_s(n)$ be the time at which state s is first visited in trial n, and let $t_T(n)$ be the time at which the terminal state is reached. Let $V_i^R(s)$ represent the replace TD(1) estimate of the value of state s after i passes through the training set, for $i \geq 1$:

$$
\begin{aligned}
V_{i+1}^R(s) &= V_i^R(s) + \sum_{n=1}^{N(s)} \sum_{t=t_s(n)}^{t_T(n)-1} \Delta V_t(s) \\
&= V_i^R(s) + \alpha \sum_{n=1}^{N(s)} \sum_{t=t_s(n)}^{t_T(n)-1} \left[r_{t+1} + V_i^R(s_{t+1}) - V_i^R(s_t) \right] \\
&= V_i^R(s) + \alpha \sum_{n=1}^{N(s)} \left[-V_i^R(s_{t_s(n)}) + \sum_{t=t_s(n)}^{t_T(n)-1} r_{t+1} \right] \\
&= V_i^R(s) + \alpha \sum_{n=1}^{N(s)} \left[R(t_s(n)) - V_i^R(s) \right] \\
&= (1 - N(s)\alpha) V_i^R(s) + \alpha \sum_{n=1}^{N(s)} R(t_s(n)),
\end{aligned}
$$

where $R(t)$ is the return following time t to the end of the trial. This in turn implies that

$$
V_i^R(s) = (1 - N(s)\alpha)^i V_0^R(s) + \alpha \sum_{n=1}^{N(s)} R(t_s(n)) \left[1 + (1 - N(s)\alpha) + \ldots (1 - N(s)\alpha)^{i-1} \right].
$$

Therefore,

$$
\begin{aligned}
V_\infty^R(s) &= (1 - N(s)\alpha)^\infty V_0^R(s) + \alpha \sum_{n=1}^{N(s)} R(t_s(n)) \sum_{j=0}^{\infty} (1 - N(s)\alpha)^j \\
&= \alpha \sum_{n=1}^{N(s)} R(t_s(n)) \frac{1}{1 - (1 - N(s)\alpha)} \qquad \text{(because } N(s)\alpha < 1) \\
&= \frac{\sum_{n=1}^{N(s)} R(t_s(n))}{N(s)},
\end{aligned}
$$

which is the first-visit MC estimate. ∎

2.5. *Equivalence of Offline TD(1) and MC Methods by Choice of α*

In this subsection we establish that the replace and accumulate forms of *offline* TD(1) can also be made equivalent to the corresponding MC methods by suitable choice of the step-size sequence $\alpha_t(s)$.

THEOREM 2: *Offline replace TD(1) is equivalent to first-visit MC under the step-size schedule*

$$\alpha_t(s) = \frac{1}{number\ of\ first\ visits\ to\ s\ up\ through\ time\ t}.$$

Proof: As before, in considering updates to the estimates of $V(s)$, we need only consider trials in which s occurs. The cumulative increment in the estimate of $V(s)$ as a result of the ith trial in which s occurs is

$$\sum_{t=t_s(i)}^{t_T(i)} \Delta V_t(s) = \sum_{t=t_s(i)}^{t_T(i)} \alpha_t(s) \left[r_{t+1} + V_{i-1}^R(s_{t+1}) - V_{i-1}^R(s_t)\right]$$

$$= \frac{1}{i}(R(t_s(i)) - V_{i-1}^R(s)).$$

Therefore, the update for offline replace TD(1), after a complete trial, is

$$V_i^R(s) = V_{i-1}^R(s) + \frac{1}{i}(R(t_s(i)) - V_{i-1}^R(s)),$$

which is just the iterative recursive equation for incrementally computing the average of the first-visit returns, $\{R(t_s(1)), R(t_s(2)), R(t_s(3)), \ldots\}$. ∎

THEOREM 3: *Offline accumulate TD(1) is equivalent to every-visit MC under the step-size schedule*

$$\alpha_t(s) = \frac{1}{number\ of\ visits\ to\ s\ up\ through\ the\ entire\ trial\ containing\ time\ t}$$

Proof: Once again we consider only trials in which state s occurs. For this proof we need to use the time index of every visit to state s, complicating notation somewhat. Let $t_s(i; k)$ be the time index of the kth visit to state s in trial i. Also, let $K_s(i)$ be the total number of visits to state s in trial i. The essential idea behind the proof is to again show that the offline TD(1) equation is an iterative recursive averaging equation, only this time of the returns from every visit to state s.

Let $\alpha_i(s)$ be the step-size parameter used in processing trial i. The cumulative increment in the estimate of $V(s)$ as a result of trial i is

$$\sum_{t=t_s(i;1)}^{t_T(i)} \Delta V_t(s) = \alpha_i(s) \left[\sum_{t=t_s(i;1)}^{t_s(i;2)-1} \Delta_{i-1}(s_t) + 2 \sum_{t=t_s(i;2)}^{t_s(i;3)-1} \Delta_{i-1}(s_t) \right.$$

$$+ \cdots + K_s(i) \sum_{t=t_s(i;K_s(i))}^{t_T(i)} \Delta_{i-1}(s_t) \Bigg]$$

$$= \alpha_i(s) \left[\sum_{j=1}^{K_s(i)} R(t_s(i;j)) - K_s(i)V_{i-1}^A(s) \right],$$

where $\Delta_i(s_t) = r_{t+1} + V_i^A(s_{t+1}) - V_i^A(s_t)$, and $V_i^A(s)$ is the accumulate-trace estimate at trial i. Therefore,

$$V_i^A(s) = V_{i-1}^A(s) + \alpha_i(s) \left[\sum_{j=1}^{K_s(i)} R(t_s(i;j)) - K_s(i)V_{i-1}^A(s) \right].$$

Because $\alpha_i(s) = \frac{1}{\sum_{j=1}^i K_s(j)}$, this will compute the sample average of all the actual returns from every visit to state s up to and including trial i. ∎

3. Analytic Comparison of Monte Carlo Methods

In the previous section we established close relationships of replace and accumulate TD(1) to first-visit and every-visit MC methods respectively. By better understanding the difference between the MC methods, then, we might hope to better understand the difference between the TD methods. Accordingly, in this section we evaluate analytically the quality of the solutions found by the two MC methods. In brief, we explore the asymptotic correctness of all methods, the bias of the MC methods, the variance and mean-squared error of the MC methods, and the relationship of the MC methods to the maximum-likelihood estimate. The results of this section are summarized in Table 1.

3.1. Asymptotic Convergence

In this subsection we briefly establish the asymptotic correctness of the TD methods. The asymptotic convergence of accumulate TD(λ) for general λ is well known (Dayan, 1992; Jaakkola, Jordan & Singh, 1994; Tsitsiklis, 1994; Peng, 1993). The main results appear to carry over to the replace-trace case with minimal modifications. In particular:

THEOREM 4: *Offline (online) replace TD(λ) converges to the desired value function w.p.1 under the conditions for w.p.1 convergence of offline (online) conventional TD(λ) stated by Jaakkola, Jordan and Singh (1994).*

Proof: Jaakkola, Jordan and Singh (1994) proved that online and offline TD(λ) converges $w.p.1$ to the correct predictions, under natural conditions, as the number of trials goes to infinity (or as the number of time steps goes to infinity in the online, non-absorbing case, with $\gamma < 1$). Their proof is based on showing that the offline TD(λ) estimator is a contraction mapping in expected value. They show that it is a weighted sum of n-step corrected truncated returns,

$$V_t^{(n)}(s_t) = r_{t+1} + \gamma r_{t+2} + \ldots + \gamma^{n-1} r_{t+n} + \gamma^n V_t(s_{t+n}),$$

that, for all $n \geq 1$, are better estimates (in expected value) of $V(s_t)$ than is $V_t(s_t)$. The eligibility trace collects successive n-step estimators, and its magnitude determines their weighting. The TD(λ) estimator is

$$\sum_{k=0}^{\tau-1} \Big[r_{t+k+1} + \gamma V_t(s_{t+k+1}) - V_t(s_{t+k}) \Big] e_{t+k+1}(s_t) + V_t(s_t) =$$

$$(1 - \lambda) \left[\sum_{n=1}^{\tau} \lambda^{n-1} V_t^{(n)}(s_t) + V_t^{(\tau)}(s_t) \sum_{n=\tau+1}^{\infty} \lambda^{n-1} \right],$$

where, for the accumulating trace, τ is the number of time steps until termination, whereas, for the replacing trace, τ is the number of time steps until the next revisit to state s_t. Although the weighted sum is slightly different in the replace case, it is still a contraction mapping in expected value and meets all the conditions of Jaakkola *et al.*'s proofs of convergence for online and offline updating.   ∎

3.2. Relationship to the ML Estimate

In the simple example in Figure 2, the first-visit MC estimate is the same as the ML estimate. However, this is true in general only for the *starting state*, assuming all trials start in the same state. One way of thinking about this is to consider for any state s just the subset of the training trials that include s. For each of these trials, discard the early part of the trial before s was visited for the first time. Consider the remaining "tails" of the trials as a new training set. This reduced training set is really all the MC methods ever see in forming their estimates for s. We refer to the ML estimate of $V(s)$ based on this reduced training set as the *reduced-ML* estimate. In this subsection we show that the reduced ML estimate is equivalent in general to the first-visit MC estimate.

THEOREM 5: *For any undiscounted absorbing Markov chain, the estimates computed by first-visit MC are the reduced-ML estimates, for all states and after all trials.*

Proof: The first-visit MC estimate is the average of the returns from first visits to state s. Because the maximum-likelihood model is built from the partial experience rooted in state s, the sum over all t of the probability of making a particular transition at time step t according to the maximum-likelihood model is equal to the ratio of the number of times that transition was actually made to the number of trials. Therefore, the reduced-ML estimate for state s is equal to the first-visit MC estimate. See Appendix A.1 for a complete proof.   ∎

Theorem 5 shows the equivalence of the first-visit MC and reduced-ML estimates. *Every*-visit MC in general produces an estimate different from the reduced-ML estimate.

3.3. Reduction to a Two-State Abstracted Markov Chain

In this subsection we introduce a conceptual reduction of arbitrary undiscounted absorbing Markov chains to a two-state *abstracted Markov chain* that we then use in the rest of this paper's analyses of MC methods. The reduction is based on focusing on each state individually. Assume for the moment that we are interested in the value only of one state, s. We assume that all training trials start in state s. We can do this without loss of generality because the change in the value of a state after a trial is unaffected by anything that happens before the first visit to the state on that trial.

For any Markov chain, a trial produces two sequences, a random sequence of states, $\{s\}$, beginning with s and ending in T, and an associated random sequence of rewards, $\{r\}$. Partition sequence $\{s\}$ into contiguous subsequences that begin in s and end just before the next revisit to s. The subsequence starting with the ith revisit to s is denoted $\{s\}_i$. The last such subsequence is special in that it ends in the terminal state and is denoted $\{s\}_T$. The corresponding reward sequences are similarly denoted $\{r\}_i$ and $\{r\}_T$. Because of the Markov property, $\{s\}_i$ is independent of $\{s\}_j$, for all $i \neq j$, and similarly $\{r\}_i$ is independent of $\{r\}_j$. This is useful because it means that the precise sequence of states that actually occurs between visits to s does not play a role in the first-visit MC or the every-visit MC estimates for $V(s)$. Similarly, the precise sequence of rewards, $\{r\}_i$, does not matter, as only the sum of the rewards in between visits to s are used in the MC methods.

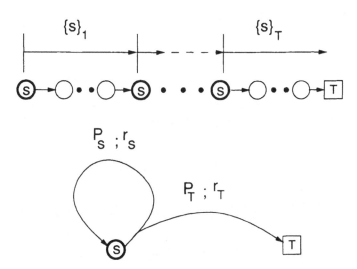

Figure 3. Abstracted Markov chain. At the top is a typical sequence of states comprising a training trial. The sequence can be divided into contiguous subsequences at the visits to start state s. For our analyses, the precise sequence of states and rewards in between revisits to s does not matter. Therefore, in considering the value of s, arbitrary undiscounted Markov chains can be abstracted to the two-state chain shown in the lower part of the figure.

Therefore, for the purpose of analysis, arbitrary undiscounted Markov chains can be reduced to the two-state abstract chain shown in the lower part of Figure 3. The associated probabilities and rewards require careful elaboration. Let P_T and P_s denote the probabilities of terminating and looping respectively in the abstracted chain. Let r_s and r_T represent the random rewards associated with a $s \rightsquigarrow s$ transition and a $s \rightsquigarrow T$ transition in Figure 3. We use the quantities, $R_s = E\{r_s\}$, $Var(r_s) = E\{(r_s - R_s)^2\}$, $R_T = E\{r_T\}$, and $Var(r_T) = E\{(r_T - R_T)^2\}$ in the following analysis. Precise definitions of these quantities are given in Appendix A.2.

first-visit MC:
Let $\{x\}$ stand for the paired random sequence $(\{s\}, \{r\})$. The first-visit MC estimate for $V(s)$ after one trial, $\{x\}$, is

$$V_1^F(s) = f(\{x\}) = r_{s_1} + r_{s_2} + r_{s_3} + \ldots + r_{s_k} + r_T,$$

where k is the *random* number of *revisits* to state s, r_{s_i} is the sum of the individual rewards in the sequence $\{r\}_i$, and r_T is the random total reward received after the last visit to state s. For all i, $E\{r_{s_i}\} = R_s$. The first-visit MC estimate of $V(s)$ after n trials, $\{x\}^1, \{x\}^2, \ldots, \{x\}^n$, is

$$V_n^F(s) = f(\{x\}^1, \{x\}^2, \ldots, \{x\}^n) = \frac{\sum_{i=1}^n f(\{x\}^i)}{n}. \tag{2}$$

In words, $V_n^F(s)$ is simply the average of the estimates from the n sample trajectories, $\{x\}^1, \{x\}^2, \ldots, \{x\}^n$, all of which are independent of each other because of the Markov property.

every-visit MC:
The every-visit MC estimate for one trial, $\{x\}$, is

$$V_1^E(s) = t(\{x\}) = \frac{t_{num}(\{x\})}{k+1} = \frac{r_{s_1} + 2r_{s_2} + \ldots + kr_{s_k} + (k+1)r_T}{k+1},$$

where k is the random number of revisits to state s in the sequence $\{x\}$. Every visit to state s effectively starts another trial. Therefore, the rewards that occur in between the ith and $(i+1)^{st}$ visits to state s are included i times in the estimate.

The every-visit MC estimate after n trials, $\{x\}^1, \{x\}^2, \ldots, \{x\}^n$, is

$$V_n^E(s) = t(\{x\}^1, \{x\}^2, \ldots, \{x\}^n) = \frac{\sum_{i=1}^n t_{num}(\{x\}^i)}{\sum_{i=1}^n (k_i + 1)}, \tag{3}$$

where k_i is the number of revisits to s in the ith trial $\{x\}^i$. Unlike the first-visit MC estimator, the every-visit MC estimator for n trials is not simply the average of the estimates for individual trials, making its analysis more complex.

We derive the bias ($Bias$) and variance (Var) of first-visit MC and every-visit MC as a function of the number of trials, n. The mean squared error (MSE) is $Bias^2 + Var$.

3.4. Bias Results

First consider the true value of state s in Figure 3. From Bellman's equation (Bellman, 1957):

$$V(s) = P_s(R_s + V(s)) + P_T(R_T + V_T)$$

or

$$(1 - P_s)V(s) = P_s R_s + P_T R_T,$$

and therefore

$$V(s) = \frac{P_s}{P_T} R_s + R_T.$$

THEOREM 6: *First-visit MC is unbiased, i.e.,* $Bias_n^F(s) = V(s) - E\{V_n^F(s)\} = 0$ *for all $n > 0$.*

Proof: The first-visit MC estimate is unbiased because the total reward on a sample path from the start state s to the terminal state T is by definition an unbiased estimate of the expected total reward across all such paths. Therefore, the average of the estimates from n independent sample paths is also unbiased. See Appendix A.3 for a detailed proof.

■

THEOREM 7: *Every-visit MC is biased and, after n trials, its bias is*

$$Bias_n^E(s) = V(s) - E\{V_n^E(s)\} = \frac{2}{n+1} Bias_1^E(s) = \frac{2}{(n+1)} \left[\frac{P_s}{2P_T} R_s \right].$$

Proof: See Appendix A.4.

■

One way of understanding the bias in the every-visit MC estimate is to note that this method averages many returns for each trial. Returns from the same trial share many of the same rewards and are thus not independent. The bias becomes smaller as more trials are observed because the returns from different trials *are* independent. Another way of understanding the bias is to note that the every-visit MC estimate (3) is the ratio of two random variables. In general, the expected value of such a ratio is not the ratio of the expected values of the numerator and denominator.

COROLLARY 7a: *Every-visit MC is unbiased in the limit as $n \to \infty$.*

3.5. *Variance and MSE Results*

THEOREM 8: *The variance of first-visit MC is*

$$Var_n^F(s) = \frac{Var_1^F(s)}{n} = \frac{1}{n} \left[Var(r_T) + \frac{P_s}{P_T} Var(r_s) + \frac{P_s}{P_T^2} R_s^2 \right].$$

Proof: See Appendix A.5.

■

Because the first-visit MC estimate is the sample average of estimates derived from independent trials, the variance goes down as $\frac{1}{n}$. The first two terms in the variance are due to the variance of the rewards, and the third term is the variance due to the random number of revisits to state s in each trial.

COROLLARY 8a: *The MSE of first-visit MC is*

$$MSE_n^F(s) = (Bias_n^F(s))^2 + Var_n^F(s) = \frac{1}{n} \left[Var(r_T) + \frac{P_s}{P_T} Var(r_s) + \frac{P_s}{P_T^2} R_s^2 \right].$$

THEOREM 9: *The variance of every-visit MC after one trial is bounded by*

$$Var(r_s)\left[\frac{P_s}{3P_T} - \frac{1}{6}\right] + Var(r_T) + \frac{1}{4}\frac{P_s}{P_T^2}R_s^2 \leq Var_1^E(s)$$

and

$$Var_1^E(s) \leq Var(r_s)\left[\frac{P_s}{3P_T}\right] + Var(r_T) + \frac{1}{4}\frac{P_s}{P_T^2}R_s^2.$$

Proof: See Appendix A.6. ∎

We were able to obtain only these upper and lower bounds on the variance of every-visit MC. For a single trial, every-visit MC produces an estimate that is closer to zero than the estimate produced by first-visit MC; therefore $Var_1^E \leq Var_1^F$. This effect was seen in the simple example of Figure 2, in which the every-visit MC estimator underestimated the expected number of revisits.

Of course, a low variance is not of itself a virtue. For example, an estimator that returns a constant independent of the data has zero variance, but is not a good estimator. Of greater importance is to be low in mean squared error (MSE):

COROLLARY 9a: *After one trial, $MSE_1^E(s) \leq MSE_1^F(s)$ because $(Bias_1^E(s))^2 + Var_1^E(s) \leq MSE_1^F(s)$.*

Thus, after one trial, every-visit MC is always as good or better than first-visit MC in terms of both variance and MSE. Eventually, however, this relative advantage always reverses itself:

THEOREM 10: *There exists an $N < \infty$, such that for all $n \geq N$, $Var_n^E(s) \geq Var_n^F(s)$.*

Proof: The basic idea of the proof is that the $O(\frac{1}{n})$ component of Var_n^E is larger than that of Var_n^F. The other $O\left(\frac{1}{n^{\frac{3}{2}}}\right)$ components of Var_n^E fall off much more rapidly than the $O\left(\frac{1}{n}\right)$ component, and can be ignored for large enough n. See Appendix A.7 for a complete proof. ∎

COROLLARY 10a: *There exists an $N < \infty$, such that, for all $n \geq N$,*

$$MSE_n^E(s) = (Bias_n^E(s))^2 + Var_n^E(s) \geq MSE_n^F(s) = Var_n^F(s).$$

Figure 4 shows an empirical example of this *crossover* of MSE. These data are for the two MC methods applied to an instance of the example task of Figure 2a. In this case crossover occurred at trial $N = 5$. In general, crossover can occur as early as the first trial. For example, if the only non-zero reward in a problem is on termination, then $R_s = 0$, and $Var(r_s) = 0$, which in turn implies that $Bias_n^E = 0$, for all n, and that $Var_1^E(s) = Var_1^F(s)$, so that $MSE_1^E(s) = MSE_1^F(s)$.

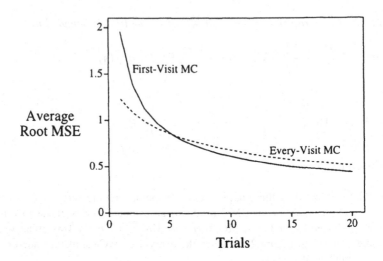

Figure 4. Empirical demonstration of crossover of MSE on the example task shown in Figure 2a. The S-to-S transition probability was $p = 0.6$. These data are averages over 10, 000 runs.

3.6. *Summary*

Table 1 summarizes the results of this section comparing first-visit and every-visit MC methods. Some of the results are unambiguously in favor of the first-visit method over the every-visit method: only the first-visit estimate is unbiased and related to the ML estimate. On the other hand, the MSE results can be viewed as mixed. Initially, every-visit MC is of better MSE, but later it is always overtaken by first-visit MC. The implications of this are unclear. To some it might suggest that we should seek a combination of the two estimators that is always of lowest MSE. However, that might be a mistake. We suspect that the first-visit estimate is always the more useful one, even when it is worse in terms of MSE. Our other theoretical results are consistent with this view, but it remains a speculation and a topic for future research.

Table 1. Summary of Statistical Results

Algorithm	Convergent	Unbiased	Short MSE	Long MSE	Reduced-ML
First-Visit MC	Yes	Yes	Higher	Lower	Yes
Every-Visit MC	Yes	No	Lower	Higher	No

4. Random-Walk Experiment

In this section we present an empirical comparison of replacing and accumulating eligibility traces. Whereas our theoretical results are limited to the case of $\lambda = 1$ and either offline or batch updating, in this experiment we used online updating and general λ. We used the random-walk process shown in Figure 5. The rewards were zero everywhere except upon entering the terminal states. The reward upon transition into State 21 was $+1$ and upon transition into State 1 was -1. The discount factor was $\gamma = 1$. The initial value estimates were 0 for all states. We implemented online TD(λ) with both kinds of traces for ten different values of λ: 0.0, 0.2, 0.4, 0.6, 0.8, 0.9, 0.95, 0.975, 0.99, and 1.0.

The step-size parameter was held constant, $\alpha_t(s) = \alpha, \forall t, \forall s$. For each value of λ, we used α values between 0 and 1.0 in increments of 0.01. Each (λ, α) pair was treated as a separate algorithm, each of which we ran for 10 trials. The performance measure for a trial was the root mean squared error (RMSE) between the correct predictions and the predictions made at the end of the trial from states that had been visited at least once in that or a previous trial. These errors were then averaged over the 10 trials, and then over 1000 separate runs to obtain the performance measure for each algorithm plotted in Figures 6 and 7. The random number generator was seeded such that all algorithms experienced exactly the same trials.

Figure 6 shows the performance of each method as a function of α and λ. For each value of λ, both kinds of TD method performed best at an intermediate value of α, as is typically the case for such learning algorithms. The larger the λ value, the smaller the α value that yielded best performance, presumably because the eligibility trace multiplies the step-size parameter in the update equation.

The critical results are the differences between replace and accumulate TD methods. Replace TD was much more robust to the choice of the step-size parameter than accumulate TD. Indeed, for $\lambda \geq 0.9$, accumulate TD(λ) became unstable for $\alpha \geq 0.6$. At large λ, accumulate TD built up very large eligibility traces for states that were revisited many times before termination. This caused very large changes in the value estimates and led to instability. Figure 7 summarizes the data by plotting, for each λ, only the performance at the best α for that λ. For every λ, the best performance of replace TD was better than or equal to the best performance of accumulate TD. We conclude that, at least for the problem studied here, replace TD(λ) is faster and more robust than accumulate TD(λ).

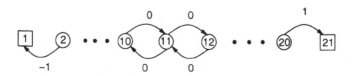

Figure 5. The random-walk process. Starting in State 11, steps are taken left or right with equal probability until either State 1 or State 21 is entered, terminating the trial and generating a final non-zero reward.

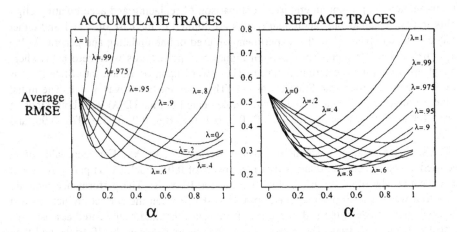

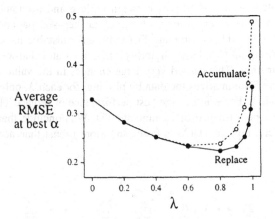

Figure 6. Performance of replace and accumulate TD(λ) on the random-walk task, for various values of λ and α. The performance measure was the RMSE per state per trial over the first 10 trials. These data are averages over 1000 runs.

Figure 7. Best performances of accumulate and replace TD(λ) on the random-walk task.

5. Mountain-Car Experiment

In this section we describe an experimental comparison of replacing and accumulating traces when used as part of a reinforcement learning system to solve a control problem. In this case, the methods learned to predict the value not of a state, but of a state-action pair, and the approximate value function was implemented as a set of CMAC neural networks, one for each action.

The control problem we used was Moore's (1991) *mountain car* task. A car drives along a mountain track as shown in Figure 8. The objective is to drive past the top of the mountain on the righthand side. However, gravity is stronger than the engine, and even at full thrust the car cannot accelerate up the steep slope. The only way to solve the problem is to first accelerate *backwards*, away from the goal, and then apply full thrust forwards, building up enough speed to carry over the steep slope even while slowing down the whole way. Thus, one must initially move away from the goal in order to reach it in the long run. This is a simple example of a task where things must get worse before they can get better. Many control methodologies have great difficulties with tasks of this kind unless explicitly aided by a human designer.

The reward in this problem is -1 for all time steps until the car has passed to the right of the mountain top. Passing the top ends the trial and ends this punishment. The reinforcement learning agent seeks to maximize its total reward prior to the termination of the trial. To do so, it must drive to the goal in minimum time. At each time step the learning agent chooses one of three actions: full thrust forward, full thrust reverse, or no thrust. This action, together with the effect of gravity (dependent on the steepness of the slope), determines the next velocity and position of the car. The complete physics of the mountain-car task are given in Appendix B.

The reinforcement learning algorithm we applied to this task was the *Sarsa* algorithm studied by Rummery and Niranjan (1994) and others. The objective in this algorithm is to learn to estimate the action-value function $Q^\pi(s, a)$ for the current policy π. The action-

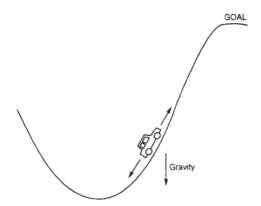

Figure 8. The Mountain-Car task. The force of gravity is stronger than the motor.

---

1. Initially: $w_a(f) := -20$, $e_a(f) := 0$, $\forall a \in Actions$, $\forall f \in CMAC\text{-}tiles$.

2. Start of Trial: $s := random\text{-}state()$;
   $\qquad\qquad\quad F := features(s)$;
   $\qquad\qquad\quad a := greedy\text{-}policy(F)$.

3. Eligibility Traces: $e_b(f) := \lambda e_b(f)$, $\forall b$, $\forall f$;
   $\qquad$ 3a. Accumulate algorithm: $e_a(f) := e_a(f) + 1$, $\forall f \in F$.
   $\qquad$ 3b. Replace algorithm: $\qquad e_a(f) := 1$, $e_b(f) := 0$, $\forall f \in F$, $\forall b \neq a$.

4. Environment Step:
   $\qquad$ Take action a; observe resultant reward, r, and next state s'.

5. Choose Next Action:
   $\qquad F' := features(s')$, unless s' is the terminal state, then $F' := \emptyset$;
   $\qquad a' := greedy\text{-}policy(F')$.

6. Learn: $w_b(f) := w_b(f) + \frac{\alpha}{m}[r + \sum_{f \in F'} w_{a'} - \sum_{f \in F} w_a]e_b(f)$, $\forall b, \forall f$.

7. Loop: $a := a'$; $s := s'$; $F := F'$; if s' is the terminal state, go to 2; else go to 3.

Figure 9. The Sarsa Algorithm used on the Mountain-Car task. The function *greedy-policy(F)* computes $\sum_{f \in F} w_a$ for each action a and returns the action for which the sum is largest, resolving any ties randomly. The function *features(s)* returns the set of CMAC tiles corresponding to the state s. Programming optimizations can reduce the expense per iteration to a small multiple (dependent on λ) of the number of features, m, present on a typical time step. Here m is 5.

value $Q^\pi(s, a)$ gives, for any state, s, and action, a, the expected return for starting from state s, taking action a, and thereafter following policy π. In the case of the mountain-car task the return is simply the sum of the future reward, i.e., the negative of the number of time steps until the goal is reached. Most of the details of the Sarsa algorithm we used are given in Figure 9. The name "Sarsa" comes from the quintuple of actual events involved in the update: $(s_t, a_t, r_{t+1}, s_{t+1}, a_{t+1})$. This algorithm is closely related to Q-learning (Watkins, 1989) and to various simplified forms of the bucket brigade (Holland, 1986; Wilson, to appear). It is also identical to the TD(λ) algorithm applied to state-action pairs rather than to states.[6]

The mountain-car task has a continuous two-dimensional state space with an infinite number of states. To apply reinforcement learning requires some form of function approximator. We used a set of three CMACs (Albus, 1981; Miller, Glanz, & Kraft, 1990), one for each action. These are simple functions approximators using repeated overlapping tilings of the state space to produce a feature representation for a final linear mapping. In this case we divided the two state variables, the position and velocity of the car, each into eight evenly spaced intervals, thereby partitioning the state space into 64 regions, or boxes. A ninth row and column were added so that the tiling could be offset by a random fraction of an interval without leaving any states uncovered. We repeated this five times, each with a different, randomly selected offset. For example, Figure 10 shows two tilings superimposed on the 2D state space. The result was a total of $9 \times 9 \times 5 = 405$ boxes. The state at any particular time was represented by the five boxes, one per tiling, within which the state resided. We think of the state representation as a feature vector with 405 features, exactly 5 of which are present (non-zero) at any point in time. The approximate action-value function is linear in this feature representation. Note that this representation of the state causes the problem to no longer be Markov: many different nearby states produce exactly the same feature representation.

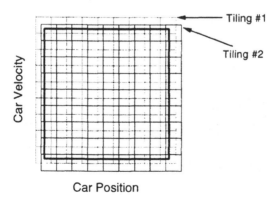

Figure 10. Two 9×9 CMAC tilings offset and overlaid over the continuous, two-dimensional state space of the Mountain-Car task. Any state is in exactly one tile/box/feature of each tiling. The experiments used 5 tilings, each offset by a random fraction of a tile width.

The eligibility traces were implemented on a feature-by-feature basis. Corresponding to each feature were three traces, one per action. The features are treated in essence like states. For replace algorithms, whenever a feature occurs, its traces are reset to 1 (for the action selected) or 0 (for all the other actions). This is not the only possibility, of course. Another would be to allow the traces for each state-action pair to continue until that *pair* occurred again. This would be more in keeping with the idea of replacing traces as a mechanism, but the approach we chose seems like the appropriate way to generalize the idea of first-visit MC to the control case: after a state has been revisited, it no longer matters what action was taken on the previous visit. A comparison of these two possibilities (and perhaps others) would make a good extension to this work.

The greedy policy was used to select actions. The initial weights were set to produce a uniform, optimistic initial estimate of value (-100) across the state space.[7] See Figure 9 for further details.

We applied replace and accumulate Sarsa algorithms to this task, each with a range of values for λ and α. Each algorithm was run for 20 trials, where a trial was one passage from a randomly selected starting state to the goal. All algorithms used the same sets of random starting states. The performance measure for the run was the average trial length over the 20 trials. This measure was then averaged over 30 runs to produce the results shown in Figures 11 and 12. Figure 11 shows the detailed results for each value of λ and α, whereas Figure 12 is a summary showing only the best performance of each algorithm at each λ value.

Several interesting results are evident from this data. First, the replace-trace method performed better than the accumulate-trace method at all λ values. The accumulate method performed particularly poorly relative to the replace method at high values of λ. For both methods, performance appeared to be best at an intermediate λ value. These

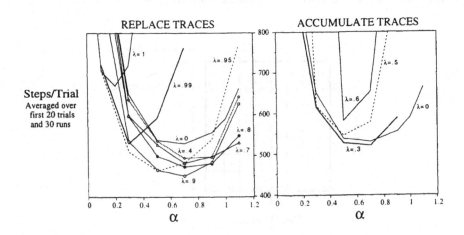

Figure 11. Results on the Mountain-Car task for each value of λ and α. Each data point is the average duration of the first 20 trials of a run, averaged over 30 runs. The standard errors are omitted to simplify the graph; they ranged from about 10 to about 50.

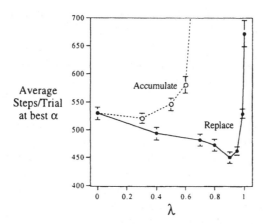

Figure 12. Summary of results on the Mountain-Car task. For each value of λ we show its performance at its best value of α. The error bars indicate one standard error.

results are all consistent with those presented for the random-walk task in the previous section. On the mountain-car task, accumulating traces at best improved only slightly over no traces ($\lambda = 0$) and at worst dramatically degraded performance. Replacing traces, on the other hand, significantly improved performance at all except the very longest trace lengths ($\lambda > .99$). Traces that do not decay ($\lambda = 1$) resulted in significantly worse performance than all other λ values tried, including no traces at all ($\lambda = 0$).

Much more empirical experience is needed with trace mechanisms before a definitive conclusion can be drawn about their relative effectiveness, particularly when function approximators are used. However, these experiments do provide significant evidence for two key points: 1) that replace-trace methods can perform much better than conventional, accumulate-trace methods, particularly at long trace lengths, and 2) that although long traces may help substantially, best performance is obtained when the traces are not infinite, that is, when intermediate predictions are used as targets rather than actual sample returns.

6. Conclusions

We have presented a variety of analytical and empirical evidence supporting the idea that replacing eligibility traces permit more efficient use of experience in reinforcement learning and long-term prediction.

Our analytical results concerned a special case closely related to that used in classical studies of Monte Carlo methods. We showed that methods using conventional traces are biased, whereas replace-trace methods are unbiased. While the conclusions of our mean-squared-error analysis are mixed, the maximum likelihood analysis is clearly in favor of replacing traces. As a whole, these analytic results strongly support the conclusion that replace-trace methods make better inferences from limited data than conventional accumulate-trace methods.

On the other hand, these analytic results concern only a special case quite different from those encountered in practice. It would be desirable to extend our analyses to the case of $\lambda < 1$ and to permit other step-size schedules. Analysis of cases involving function approximators and violations of the Markov assumption would also be useful further steps.

Our empirical results treated a much more realistic case, including in some cases all of the extensions listed above. These results showed consistent, significant, and sometimes large advantages of replace-trace methods over accumulate-trace methods, and of trace methods generally over trace-less methods. The mountain-car experiment showed that the replace-trace idea can be successfully used in conjunction with a feature-based function approximator. Although it is not yet clear how to extend the replace-trace idea to other kinds of function approximators, such as back-propagation networks or nearest-neighbor methods, Sutton and Whitehead (1993) and others have argued that feature-based function approximators are actually preferable for online reinforcement learning.

Our empirical results showed a sharp drop in performance as the trace parameter λ approached 1, corresponding to very long traces. This drop was much less severe with replacing traces but was still clearly present. This bears on the long-standing question of the relative merits of TD(1) methods versus true temporal-difference ($\lambda < 1$) methods. It might appear that replacing traces make TD(1) methods more capable competitors; the replace TD(1) method is unbiased in the special case, and more efficient than conventional TD(1) in both theory and practice. However, this is at the cost of losing some of the theoretical advantages of conventional TD(1). In particular, conventional TD(1) converges in many cases to a minimal mean-squared-error solution when function approximators are used (Dayan, 1992) and has been shown to be useful in non-Markov problems (Jaakkola, Singh & Jordan, 1995). The replace version of TD(1) does not share these theoretical guarantees. Like $\lambda < 1$ methods, it appears to achieve greater efficiency in part by relying on the Markov property. In practice, however, the relative merits of different $\lambda = 1$ methods may not be of great significance. All of our empirical results suggest far better performance is obtained with $\lambda < 1$, even when function approximators are used that create an apparently non-Markov task.

Replacing traces are a simple modification of existing discrete-state or feature-based reinforcement learning algorithms. In cases in which a good state representation can be obtained they appear to offer significant improvements in learning speed and reliability.

Acknowledgments

We thank Peter Dayan for pointing out and helping to correct errors in the proofs. His patient and thorough reading of the paper and his participation in our attempts to complete the proofs were invaluable. Of course, any remaining errors are our own. We thank Tommi Jaakkola for providing the central idea behind the proof of Theorem 10, an anonymous reviewer for pointing out an error in our original proof of Theorem 10, and Lisa White for pointing out another error. Satinder P. Singh was supported by grants to

Michael I. Jordan (Brain and Cognitive Sciences, MIT) from ATR Human Information Processing Research and from Siemens Corporation.

Appendix A
Proofs of Analytical Results

A.1. Proof of Theorem 5: First-Visit MC is Reduced-ML

In considering the estimate $V_t(s)$, we can assume that all trials start in s, because both first-visit MC and reduced-ML methods ignore transitions prior to the first-visit to s. Let n_i be the number of times state i has been visited, and let n_{ij} be the number of times transition $i \rightsquigarrow j$ has been encountered. Let R_{jk} be the average of the rewards seen on the $j \rightsquigarrow k$ transitions.

Then $V_N^F(s)$, the first-visit MC estimate after N trials with start state s, is

$$V_N^F(s) = \frac{1}{N} \sum_{k \in S, j \in S} n_{jk} R_{jk}.$$

This is identical to (2) because $\sum_{k \in S, j \in S} n_{jk} R_{jk}$ is the total summed reward seen during the N trials. Because $N = n_s - \sum_{i \in S} n_{is}$, we can rewrite this as

$$V_N^F(s) = \sum_{k \in S, j \in S} \frac{n_{jk}}{n_s - \sum_{i \in S} n_{is}} R_{jk} = \sum_{j,k} U'_{jk} R_{jk}. \tag{A.1}$$

The maximum-likelihood model of the Markov process after N trials has transition probabilities $P(ij) = \frac{n_{ij}}{n_i}$ and expected rewards $R(ij) = R_{ij}$. Let $V_N^{ML}(s)$ denote the reduced-ML estimate after N trials. By definition $V_N^{ML}(s) = E_N\{r_1 + r_2 + r_3 + r_4 + \ldots\}$, where E_N is the expectation operator for the maximum-likelihood model after N trials, and r_i is the payoff at step i. Therefore

$$V_N^{ML}(s) = \sum_{j,k} R_{jk} \left[Prob_1(j \rightsquigarrow k) + Prob_2(j \rightsquigarrow k) + Prob_3(j \rightsquigarrow k) + \ldots \right]$$

$$= \sum_{j,k} R_{jk} U_{jk}, \tag{A.2}$$

where $Prob_i(j \rightsquigarrow k)$ is the probability of a j-to-k transition at the ith step according to the maximum-likelihood model. We now show that for all j, k, U_{jk} of (A.2) is equal to U'_{jk} of (A.1).

Consider two special cases of j in (A.2):

Case 1, $j = s$:

$$U_{sk} = P(sk) + P(ss)P(sk) + \sum_m P(sm)P(ms)P(sk) + \ldots$$

$$= \frac{n_{sk}}{n_s} \left[1 + \frac{n_{ss}}{n_s} + \sum_m \frac{n_{sm}}{n_s} \frac{n_{ms}}{n_m} + \ldots \right]$$

$$= \frac{n_{sk}}{n_s}\left[1 + P^1(ss) + P^2(ss) + P^3(ss) + \ldots\right]$$

$$= \frac{n_{sk}}{n_s}(1 + N_{ss}), \tag{A.3}$$

where $P^n(ij)$ is the probability of going from state i to state j in exactly n steps, and N_{ss} is the expected number of revisits to state s as per our current maximum-likelihood model.

Case 2, $j \neq s$:

$$U_{jk} = P(sj)P(jk) + \sum_m P(sm)P(mj)P(jk) + \ldots$$

$$= \frac{n_{jk}}{n_j}\left[\frac{n_{sj}}{n_s} + \sum_m \frac{n_{sm}}{n_s}\frac{n_{mj}}{n_m} + \ldots\right]$$

$$= \frac{n_{jk}}{n_j}\left[P^1(sj) + P^2(sj) + P^3(sj) + \ldots\right]$$

$$= \frac{n_{jk}}{n_j}N_{sj}, \tag{A.4}$$

where N_{sj} is the expected number of visits to state j.

For all j, the N_{sj} satisfy the recursions

$$N_{sj} = P(sj) + \sum_m N_{sm}P(mj) = \frac{n_{sj}}{n_s} + \sum_m N_{sm}\frac{n_{mj}}{n_m}. \tag{A.5}$$

We now show that $N_{sj} = \frac{n_j}{n_s - \sum_i n_{is}}$ for $j \neq s$, and $N_{ss} = \frac{n_s}{n_s - \sum_i n_{is}} - 1$, by showing that these quantities satisfy the recursions (A.5).

$$N_{sj} = \frac{n_{sj}}{n_s} + \sum_{m \neq s}\left(\frac{n_{mj}}{n_m}\frac{n_m}{n_s - \sum_i n_{is}}\right) + \frac{n_{sj}}{n_s}\left(\frac{n_s}{n_s - \sum_i n_{is}} - 1\right)$$

$$= \frac{n_{sj}}{n_s} + \sum_m \frac{n_{mj}}{n_s - \sum_i n_{is}} - \frac{n_{sj}}{n_s}$$

$$= \frac{\sum_m n_{mj}}{n_s - \sum_i n_{is}} = \frac{n_j}{n_s - \sum_i n_{is}}.$$

and, again from (A.5),

$$N_{ss} = \frac{n_{ss}}{n_s} + \sum_{m \neq s}\left(\frac{n_{ms}}{n_m}\frac{n_m}{n_s - \sum_i n_{is}}\right) + \frac{n_{ss}}{n_s}\left(\frac{n_s}{n_s - \sum_i n_{is}} - 1\right)$$

$$= \frac{\sum_m n_{ms}}{n_s - \sum_i n_{is}} = \frac{n_s}{n_s - \sum_i n_{is}} - 1.$$

Plugging the above values of N_{ss} and N_{sj} into (A.3) and (A.4), we obtain $U_{jk} = \frac{n_{jk}}{n_s - \sum_i n_{is}} = U'_{jk}$. ∎

A.2. Facts Used in Proofs of Theorems 6-10

The proofs for Theorems 6-10 assume the abstract chain of Figure 3 with just two states, s and T. The quantities $R_s = E\{r_s\}$, $Var(r_s) = E\{(r_s - R_s)^2\}$, $R_T = E\{r_T\}$, and $Var(r_T) = E\{(r_T - R_T)^2\}$ are of interest for the analysis and require careful elaboration. Let S_s be the set of all state sequences that can occur between visits to state s (including state s at the head), and let S_T be the set of all state sequences that can occur on the final run from s to T (including state s). The termination probability is $P_T = \sum_{\{s\}_T \in S_T} P(\{s\}_T)$, where the probability of a sequence of states is the product of the probabilities of the individual state transitions. By definition $P_s = 1 - P_T$. The reward probabilities are defined as follows: $Prob\{r_s = q\} = \Pi_{\{s\} \in S_s} P(\{s\}) P(r_{\{s\}} = q|\{s\})$, and $Prob\{r_T = q\} = \Pi_{\{s\}_T \in S_T} P(\{s\}_T) P(r_T = q|\{s\}_T)$. Therefore, $R_s = \sum_q q Prob\{r_s = q\}$, $R_T = \sum_q q Prob\{r_T = q\}$. Similarly, $Var(r_s) = \sum_q Prob\{r_s = q\}(q - R_s)^2$, and $Var(r_T) = \sum_q Prob\{r_T = q\}(q - R_T)^2$.

If the rewards in the original Markov chain are deterministic functions of the state transitions, then there will be a single r_i associated with each $\{s\}_i$. If the rewards in the original problem are stochastic, however, then there is a set of possible random r_i's associated with each $\{s\}_i$. Also note that even if all the individual rewards in the original Markov chain are deterministic, $Var(r_s)$ and $Var(r_T)$ can still be greater than zero because r_s and r_T will be stochastic because of the many different paths from s to s and from s to T.

The following fact is used throughout:

$$E[f(\{x\})] = \sum_{\{x\}} P(\{x\}) f(\{x\})$$
$$= \sum_k P(k) E_{\{r\}}\{f(\{x\})|k\}, \tag{A.6}$$

where k is the number of revisits to state s. We also use the facts that, if $r < 1$, then,

$$\sum_{i=0}^{\infty} i r^i = \frac{r}{(1-r)^2} \quad \text{and} \quad \sum_{i=0}^{\infty} i^2 r^i = \frac{r(1+r)}{(1-r)^3}.$$

A.3. Proof of Theorem 6: First-Visit MC is Unbiased

First we show that first-visit MC is unbiased for one trial. From (A.6)

$$E\{V_1^F(s)\} = E_{\{x\}}[f(\{x\})] = \sum_k P(k) E_{\{r\}}\{f(\{x\})|k\}$$
$$= \sum_k P_T P_s^k (k R_s + R_T)$$
$$= P_T \left[\frac{P_s}{(1-P_s)^2} R_s + \frac{1}{1-P_s} R_T \right]$$

$$= \frac{P_s}{P_T} R_s + R_T$$

$$= V(s).$$

Because the estimate after n trials, $V_n^F(s)$, is the sample average of n independent estimates each of which is unbiased, the n-trial estimate itself is unbiased. ∎

A.4.   Proof of Theorem 7: Every-Visit MC is Biased

For a single trial, the bias of the every-visit MC algorithm is

$$E\{V_1^E(s)\} = E_{\{x\}}[t(\{x\})] = \sum_k P(k) E_{\{r\}}\{t(\{x\})|k\}$$

$$= \sum_k P_T P_s^k \left(\frac{R_s + 2R_s + \ldots + kR_s + (k+1)R_T}{k+1} \right)$$

$$= \sum_k P_T P_s^k \left(\frac{k}{2} R_s + R_T \right)$$

$$= \frac{P_s}{2P_T} R_s + R_T.$$

Therefore, $Bias_1^E = V(s) - E\{V_1^E(s)\} = \frac{P_s}{2P_T} R_s$.

Computing the bias after n trials is a bit more complex, because of the combinatorics of getting k revisits to state s in n trials, denoted $B(n; k)$. Equivalently, one can think of $B(n; k)$ as the number of different ways of factoring k into n non-negative integer additive factors *with the order considered important*. Therefore, $B(n; k) = \binom{k+n-1}{n-1}$. Further let $B(n; k_1, k_2, \ldots, k_n | k)$ be the number of different ways one can get k_1 to k_n as the factors *with the order ignored*. Note that $\sum_{\text{factors of } k} B(n; k_1, k_2, \ldots, k_n | k) = B(n; k)$. We use superscripts to distinguish the rewards from different trials, e.g., r_j^2 refers to the random total reward received between the jth and $(j+1)^{st}$ visits to start state s in the second trial.

$$E\{V_n^E(s)\} = E_{\{x\}} \left\{ \frac{\sum_{i=1}^n t_{num}(\{x\}^i)}{\sum_{i=1}^n (k_i + 1)} \right\}$$

$$= \sum_{k_1, k_2, \ldots, k_n} P(k_1, k_2, \ldots, k_n) E_{\{r\}} \left\{ \frac{\sum_{i=1}^n \sum_{j=1}^{k_i} j r_{s_j}^i}{\sum_{i=1}^n (k_i + 1)} \,\middle|\, k_1, k_2, \ldots, k_n \right\}$$

$$= \sum_k P(k) \sum_{\text{factors of } k} B(n; k_1, k_2, \ldots, k_n | k) \left[\frac{\sum_{i=1}^n k_i (k_i + 1)}{2(k+n)} R_s + R_T \right]$$

$$= R_T + R_s \left(\sum_k P_T^n P_s^k \sum_{\text{factors of } k} B(n; k_1, k_2, \ldots, k_n | k) \left[\frac{\sum_{i=1}^n (k_i)^2 + k}{2(k+n)} \right] \right).$$

This, together with

$$\sum_{\text{factors of } k} B(n; k_1, k_2, \ldots, k_n | k) \left[\frac{\sum_{i=1}^{n}(k_i)^2 + k}{2(k+n)} \right] = \frac{k}{n+1} B(n; k), \qquad \text{(A.7)}$$

which we show below, and the fact that

$$\sum_{k=0}^{\infty} P_s^k P_T^n B(n; k) \frac{k}{n} = \frac{P_s}{P_T},$$

leads to the conclusion that

$$E\{V_n^E(s)\} = R_T + \frac{n}{n+1} \frac{P_s}{P_T} R_s.$$

Therefore $Bias_n^E(s) = V(s) - E\{V_n^E(s)\} = \frac{1}{n+1} \frac{P_s}{P_T} R_s$. This also proves that the every-visit MC algorithm is unbiased in the limit as $n \to \infty$. $\blacksquare$

Proof of (A.7):
Define $T(n; k)$ as the sum over all factorizations of the squared factors of k:

$$T(n; k) = \sum_{\text{factors of } k} B(n; k_1, k_2, \ldots, k_n | k) \sum_{i=1}^{n}(k_i)^2.$$

We know the following facts from first principles:

Fact 1: $\sum_{j=0}^{k} B(n; k - j) = B(n + 1; k)$;

Fact 2: $\sum_{j=0}^{k} j B(n; k - j) = \frac{k B(n + 1; k)}{n + 1}$;

Fact 3: $\sum_{j=0}^{k} j^2 B(n; k - j) = \frac{T(n + 1; k)}{n + 1}$;

and also that, by definition,

Fact 4: $T(n + 1; k) = \sum_{j=0}^{k} T(n; k - j) + j^2 B(n; k - j).$

Facts 3 and 4 imply that

$$T(n + 1; k) = \frac{n + 1}{n} \sum_{j=0}^{k} T(n; k - j). \qquad \text{(A.8)}$$

Using Facts 1-3 we can show by substitution that $T(n;k) = \frac{k(2k+n-1)}{n+1}B(n;k)$ is a solution of the recursion (A.8), hence proving that

$$\sum_{\text{factors of } k} B(n;k_1,k_2,\ldots,k_n|k)\left[\sum_{i=1}^{n}(k_i)^2\right] = \frac{k(2k+n-1)}{n+1}B(n;k) \Rightarrow$$

$$\sum_{\text{factors of } k} B(n;k_1,k_2,\ldots,k_n|k)\left[\frac{\sum_{i=1}^{n}(k_i)^2}{2(k+n)}\right] + \frac{k}{2(k+n)}B(n;k) = \frac{k}{n+1}B(n;k) \Rightarrow$$

$$\sum_{\text{factors of } k} B(n;k_1,k_2,\ldots,k_n|k)\left[\frac{\sum_{i=1}^{n}(k_i)^2 + k}{2(k+n)}\right] = \frac{k}{n+1}B(n;k). \qquad\blacksquare$$

A.5. Proof of Theorem 8: Variance of First-Visit MC

We first compute the variance of first-visit MC after one trial $\{x\}$:

$$Var_1^F(s) = E_{\{x\}}\{(f(\{x\}) - V(s))^2\}$$

$$= \sum_k P(k)E_{\{r\}}\{(f(\{x\}) - V(s))^2|k\}$$

$$= \sum_k P_T P_s^k E_{\{r\}}\left\{\left((\sum_{i=1}^{k} r_{s_i} + r_T) - (\frac{P_s}{P_T}R_s + R_T)\right)^2 \middle| k\right\}$$

$$= \sum_k P_T P_s^k E_{\{r\}}\left\{\sum_{i=1}^{k}(r_{s_i}^2 + 2r_T r_{s_i}) + 2\sum_{i\neq j} r_{s_i} r_{s_j} + r_T^2 + \frac{P_s^2}{P_T^2}R_s^2 \right.$$

$$+ 2\frac{P_s}{P_T}R_s R_T + R_T^2 - \sum_{i=1}^{k} 2r_{s_i}\left(\frac{P_s}{P_T}R_s + R_T\right) - 2r_T\left(\frac{P_s}{P_T}R_s + R_T\right)\middle| k\Bigg\}$$

$$= \sum_k P_T P_s^k \left[E_{\{r\}}\left\{\sum_{i=1}^{k}(r_{s_i}^2 - R_s^2) + (r_T^2 - R_T^2)\middle| k\right\} + k^2 R_s^2 + 2kR_s R_T\right.$$

$$+ \frac{P_s^2}{P_T^2}R_s^2 + R_T^2 + 2\frac{P_s}{P_T}R_s R_T - 2k\frac{P_s}{P_T}R_s^2 - 2\frac{P_s}{P_T}R_s R_T - R_T^2 - 2kR_s R_T\Bigg]$$

$$= \sum_k P_T P_s^k \left[kVar(r_s) + Var(r_T) + R_s^2\left(k - \frac{P_s}{P_T}\right)^2\right]$$

$$= \frac{P_s}{P_T^2}R_s^2 + Var(r_T) + \frac{P_s}{P_T}Var(r_s).$$

The first term is the variance due to the variance in the number of revisits to state s, and the second term is the variance due to the random rewards. The first-visit MC estimate after n trials is the sample average of the n independent trials; therefore, $Var_n^F(s) = \frac{Var_1^F(s)}{n}$. $\qquad\blacksquare$

A.6. Proof of Theorem 9: Variance of Every-Visit MC After 1 Trial

$$
\begin{aligned}
Var_1^E(s) &= E_{\{x\}}\{(t(\{x\}) - E\{t(\{x\})\})^2\} \\
&= \sum_{\{x\}} P(\{x\}) \left[\frac{r_{s_1} + 2r_{s_2} + \ldots + kr_{s_k} + (k+1)r_T}{k+1} - \left(\frac{P_s}{2P_T} R_s + R_T \right) \right]^2 \\
&= \sum_k P(k) E_{\{r\}} \left\{ \sum_{i=1}^k \left(\frac{i^2}{(k+1)^2} r_{s_i}^2 + 2\frac{i}{k+1} r_{s_i} r_T \right) + r_T^2 \right. \\
&\quad + \sum_{i\neq j} \frac{2ij}{(k+1)^2} r_{s_i} r_{s_j} + \frac{P_s^2}{4P_T^2} R_s^2 + \frac{P_s}{P_T} R_s R_T + R_T^2 \\
&\quad \left. -2\sum_{i=1}^k \frac{i}{k+1} r_{s_i} \left(\frac{P_s}{2P_T} R_s + R_T \right) - 2\frac{P_s}{2P_T} R_s r_T - 2R_T r_T \,\Big|\, k \right\} \\
&= \sum_k P_T P_s^k \left[Var(r_T) + \frac{k(2k+1)}{6(k+1)} Var(r_s) + \frac{R_s^2}{4}\left(k - \frac{P_s}{P_T}\right)^2 \right] \\
&= Var(r_T) + \frac{R_s^2}{4}\frac{P_s}{P_T^2} + Var(r_s) \sum_k P_T P_s^k \frac{k(2k+1)}{6(k+1)}.
\end{aligned}
$$

Note that $\frac{k}{3} - \frac{1}{6} \leq \frac{k(2k+1)}{6(k+1)} = \frac{k}{3} - \frac{k}{6(k+1)} \leq \frac{k}{3}$. Therefore,

$$
Var(r_T) + \frac{1}{4}\frac{P_s}{P_T^2} R_s^2 + Var(r_s)\left[\frac{P_s}{3P_T} - \frac{1}{6}\right] \leq Var_1^E(s),
$$

and

$$
Var_1^E(s) \leq Var(r_T) + \frac{1}{4}\frac{P_s}{P_T^2} R_s^2 + Var(r_s)\left[\frac{P_s}{3P_T}\right]. \qquad \blacksquare
$$

A.7. Proof of Theorem 10: First-Visit MC is Eventually Lower in Variance and MSE

The central idea for this proof was provided to us by Jaakkola (personal communication). The every-visit MC estimate,

$$
V_n^E(s) = \frac{\sum_{i=1}^n t_{num}(\{x\}^i)}{\sum_{i=1}^n (k_i + 1)},
$$

can be rewritten as

$$
V_n^E(s) = \frac{\frac{1}{n}\sum_{i=1}^n \frac{t_{num}(\{x\}^i)}{E\{(k_i+1)\}}}{\frac{1}{n}\sum_{i=1}^n \frac{(k_i+1)}{E\{k_i+1\}}} = \frac{\frac{1}{n}\sum_{i=1}^n P_T t_{num}(\{x\}^i)}{\frac{1}{n}\sum_{i=1}^n P_T (k_i + 1)}
$$

because, for all i, $E\{k_i+1\} = \frac{1}{P_T}$. It is also easy to show that for all i, $E\{P_T t_{num}(\{x\}^i)\} = V(s)$, and $E\{P_T(k_i + 1)\} = 1$.

Consider the sequence of functions

$$f_n(\delta) = \frac{V(s) + \delta \bar{T}_n}{1 + \delta \bar{K}_n},$$

where $\bar{T}_n = \frac{1}{\sqrt{n}} \sum_{i=1}^{n} (P_T t_{num}(\{x\}^i) - V(s))$, and $\bar{K}_n = \frac{1}{\sqrt{n}} \sum_{i=1}^{n} (P_T(k_i + 1) - 1)$.

Note that $\forall n$, $E\{\bar{T}_n\} = 0$, $E\{\bar{K}_n\} = 0$, and $f_n\left(\frac{1}{\sqrt{n}}\right) = V_n^E(s)$. Therefore, $Var_n^E(s) = E\{(f_n(\frac{1}{\sqrt{n}}))^2\} - (E\{V_n^E(s)\})^2$. Using Taylor's expansion,

$$f_n^2\left(\frac{1}{\sqrt{n}}\right) = f_n^2(0) + \frac{1}{\sqrt{n}} \frac{\partial}{\partial \delta} f_n^2(\delta) \Big|_{\delta=0} + \frac{1}{n2!} \frac{\partial^2}{\partial \delta^2} f_n^2(\delta) \Big|_{\delta=0} + \cdots.$$

Therefore,

$$Var_n^E(s) = E\left\{ f_n^2(0) + \frac{1}{\sqrt{n}} \frac{\partial}{\partial \delta} f_n^2(\delta) \Big|_{\delta=0} + \frac{1}{2n} \frac{\partial^2}{\partial \delta^2} f_n^2(\delta) \Big|_{\delta=0} \right\}$$

$$+ E\left\{ \frac{1}{6n^{\frac{3}{2}}} \frac{\partial^3}{\partial \delta^3} f_n^2(\delta) \Big|_{\delta=0} + \cdots \right\} - \left(E\{V_n^E(s)\} \right)^2. \qquad (A.9)$$

We prove below that $E\left\{ \frac{1}{6n^{\frac{3}{2}}} \frac{\partial^3}{\partial \delta^3} f_n^2(\delta) \Big|_{\delta=0} + \cdots \right\}$ is $O\left(\frac{1}{n^{\frac{3}{2}}}\right)$ by showing that for all $i > 2$, $E\{\frac{\partial^i}{\partial \delta^i} f_n^2(\delta)|_{\delta=0}\}$ is O(1). The $O\left(\frac{1}{n^{\frac{3}{2}}}\right)$ term in (A.9) can be ignored, because $Var_n^F(s)$ decreases as $\frac{1}{n}$, and the goal here is to show that for large N, $Var_n^E(s) \geq Var_n^F(s)$ for all $n > N$.

We use the following facts without proof (except 12 which is proved subsequently):

Fact 5: $E\{f_n^2(0)\} = V^2(s)$;

Fact 6: $E\left\{ \frac{\partial}{\partial \delta} f_n^2(\delta) \Big|_{\delta=0} \right\} = 0$ because $\frac{\partial}{\partial \delta} f_n^2(\delta) \Big|_{\delta=0} = 2V(s)(\bar{T}_n - V(s)\bar{K}_n)$;

Fact 7: $\frac{\partial^2}{\partial \delta^2} f_n^2(\delta) \Big|_{\delta=0} = 2\bar{T}_n^2 - 8V(s)\bar{T}_n\bar{K}_n + 6V^2(s)\bar{K}_n^2$;

Fact 8: $E\{\bar{K}_n^2\} = P_s$;

Fact 9: $E\{\bar{K}_n\bar{T}_n\} = \frac{2P_s + P_s^2}{P_T} R_s + R_T(P_s + 1) - V(s)$:

Fact 10: $E\{\bar{T}_n^2\} = \frac{P_s + P_s^2}{P_T} Var(r_s) + (P_s + 1)Var(r_T) - V^2(s)$

$$+ (P_s + 1)R_T^2 + \frac{4P_s + 2P_s^2}{P_T}R_sR_T + \frac{P_s + 4P_s^2 + P_s^3}{P_T^2}R_s^2;$$

Fact 11: $V^2(s) - (E\{V_n^E(s)\})^2 = \dfrac{(2n+1)P_s^2}{(n+1)^2 P_T^2}R_s^2 + \dfrac{2P_s}{(n+1)P_T}R_sR_T;$

Fact 12: $E\left\{\dfrac{1}{2n}\dfrac{\partial^2}{\partial\delta^2}f_n^2(\delta)\Big|_{\delta=0}\right\} = \dfrac{P_s + P_s^2}{nP_T}Var(r_s) + \dfrac{(P_s+1)}{n}Var(r_T)$

$$+ \frac{P_s - P_s^2}{nP_T^2}R_s^2 - \frac{2P_s}{nP_T}R_sR_T.$$

Therefore, the $O(\frac{1}{n})$ behavior of the variance of the every-visit MC estimate is

$$Var_n^E(s) \approx \frac{P_s + P_s^2}{nP_T}Var(r_s) + \frac{(P_s+1)}{n}Var(r_T) + \frac{P_s}{nP_T^2}R_s^2 + \frac{n^2 - n - 1}{n(n+1)^2}\frac{P_s^2}{P_T^2}R_s^2.$$

Finally, comparing with

$$Var_n^F(s) = \frac{P_s}{nP_T}Var(r_s) + \frac{1}{n}Var(r_T) + \frac{P_s}{nP_T^2}R_s^2$$

proves Theorem 10.   ■

Ignoring higher order terms:

Note that $f_n^2(\delta)$ is of the form $\left(\frac{V(s)+\delta\bar{T}_n}{1+\delta\bar{K}_n}\right)^2$. Therefore, for all $i > 0$, the denominator of $\frac{\partial^i f_n^2(\delta)}{\partial\delta^i}$ is of the form $(1 + \delta\bar{K}_n)^j$ for some $j > 0$. On evaluating at $\delta = 0$, the denominator will always be 1, leaving in the numerator terms of the form $c\bar{T}_n^m\bar{K}_n^z$, where c is some constant and $m, z \geq 0$. For example, Fact 6 shows that $\frac{\partial}{\partial\delta}f_n^2(\delta)\big|_{\delta=0} = 2V(s)(\bar{T}_n - V(s)\bar{K}_n)$, and Fact 7 shows that $\frac{\partial^2}{\partial\delta^2}f_n^2(\delta)\big|_{\delta=0} = 2\bar{T}_n^2 - 8V(s)\bar{T}_n\bar{K}_n + 6V^2(s)\bar{K}_n^2$. We show that $E\{\bar{T}_n^m\bar{K}_n^z\}$ is $O(1)$.

Both $\bar{T}_n$ and $\bar{K}_n$ are sums of n independent mean-zero random variables. $\bar{T}_n^m\bar{K}_n^z$ contains terms that are products of random variables for the n trials. On taking expectation, any term that contains a random variable from some trial exactly once drops out. Therefore all terms that remain should contain variables from no more than $\lfloor\frac{m+z}{2}\rfloor$ trials. This implies that the number of terms that survive are $O(n^{\lfloor\frac{m+z}{2}\rfloor})$ (the constant is a function of i, but is independent of n). Therefore,

$$E\{\bar{T}_n^m\bar{K}_n^z\} = \frac{1}{n^{\frac{m+z}{2}}}n^{\lfloor\frac{m+z}{2}\rfloor}c_1 \text{ which is } O(1),$$

where c_1 is some constant. This implies that the expected value of the terms in the Taylor expansion corresponding to $i > 2$ are $O\left(\frac{1}{n^{\frac{3}{2}}}\right)$.   ■

Proof of Fact 12:

$$E\left\{\frac{1}{2n}\frac{\partial^2}{\partial\delta^2}f_n^2(\delta)\Big|_{\delta=0}\right\} = \frac{E\{2\bar{T}_n^2 - 8V(s)\bar{T}_n\bar{K}_n + 6V^2(s)\bar{K}_n^2\}}{2n}. \tag{A.10}$$

From Fact 10:

$$
\begin{aligned}
E\{2\bar{T}_n^2\} &= 2\frac{P_s + P_s^2}{P_T}Var(r_s) + 2(P_s + 1)Var(r_T) - 2V^2(s) \\
&\quad + 2(1 + P_s)\left(\frac{P_s^2}{P_T^2}R_s^2 + 2\frac{P_s}{P_T}R_sR_T + R_T^2\right) \\
&\quad + \frac{4P_s}{P_T}R_sR_T + \frac{2P_s}{P_T^2}R_s^2 + \frac{6P_s^2}{P_T^2}R_s^2 \\
&= 2\frac{P_s + P_s^2}{P_T}Var(r_s) + 2(P_s + 1)Var(r_T) \\
&\quad + 2P_sV^2(s) + \frac{4P_s}{P_T}R_sR_T + \frac{2P_s}{P_T^2}R_s^2 + \frac{6P_s^2}{P_T^2}R_s^2.
\end{aligned}
$$

Similarly, from Fact 9:

$$
\begin{aligned}
E\{-8V(s)\bar{T}_n\bar{K}_n\} &= -8V(s)\left(\frac{P_s}{P_T}R_s + (1 + P_s)(\frac{P_s}{P_T}R_s + R_T) - V(s)\right) \\
&= -8V^2(s)P_s - \frac{8P_s^2}{P_T^2}R_s^2 - \frac{8P_s}{P_T}R_sR_T,
\end{aligned}
$$

and from Fact 8:

$$E\{6V^2(s)\bar{K}_n^2\} = 6P_sV_s^2.$$

Therefore, from (A.10), we see that

$$E\{\tfrac{1}{2n}\tfrac{\partial^2}{\partial\delta^2}f^2(\delta)\big|_{\delta=0}\} = \tfrac{P_s+P_s^2}{nP_T}Var(r_s) + \tfrac{P_s+1}{n}Var(r_T) + \tfrac{P_s-P_s^2}{nP_T^2}R_s^2 - \tfrac{2P_s}{nP_T}R_sR_T. \quad\blacksquare$$

Appendix B

Details of the Mountain-Car Task

The mountain-car task (Figure 8) has two continuous state variables, the position of the car, p_t, and the velocity of the car, v_t. At the start of each trial, the initial state is chosen randomly, uniformly from the allowed ranges: $-1.2 \le p \le 0.5$, $-0.07 \le v \le 0.07$. The mountain geography is described by $altitude = \sin(3p)$. The action, a_t, takes on values in $\{+1, 0, -1\}$ corresponding to forward thrust, no thrust, and reverse thrust. The state evolution was according to the following simplified physics:

$$v_{t+1} = bound\left[v_t + 0.001a_t - g\cos(3p_t)\right]$$

and

$$p_{t+1} = bound\,[p_t + v_{t+1}]\,,$$

where $g = -0.0025$ is the force of gravity and the *bound* operation clips each variable within its allowed range. If p_{t+1} is clipped in this way, then v_{t+1} is also reset to zero. Reward is -1 on all time steps. The trial terminates with the first position value that exceeds $p_{t+1} > 0.5$.

Notes

1. Arguably, yet a third mechanism for managing delayed reward is to change representations or world models (e.g., Dayan, 1993; Sutton, 1995).
2. In some previous work (e.g., Sutton & Barto, 1987, 1990) the traces were normalized by a factor of $1 - \gamma\lambda$, which is equivalent to replacing the "1" in these equations by $1 - \gamma\lambda$. In this paper, as in most previous work, we absorb this linear normalization into the step-size parameter, α, in equation (1).
3. The time index here is assumed to continue increasing across trials. For example, if one trial reaches a terminal state at time τ, then the next trial begins at time $\tau + 1$.
4. For this reason, this estimate is sometimes also referred to as the *certainty equivalent* estimate (e.g., Kumar and Varaiya, 1986).
5. In theory it is possible to get this down to $O(n^{2.376})$ operations (Baase, 1988), but, even if practical, this is still far too complex for many applications.
6. Although this algorithm is indeed identical to TD(λ), the theoretical results for TD(λ) on stationary prediction problems (e.g., Sutton, 1988; Dayan, 1992) do not apply here because the policy is continually changing, creating a *nonstationary* prediction problem.
7. This is a very simple way of assuring initial exploration of the state space. Because most values are better than they should be, the learning system is initially disappointed no matter what it does, which causes it to try a variety of things even though its policy at any one time is deterministic. This approach was sufficient for this task, but of course we do not advocate it in general as a solution to the problem of assuring exploration.

References

Albus, J. S., (1981). *Brain, Behavior, and Robotics*, chapter 6, pages 139–179. Byte Books.

Baase, S., (1988). *Computer Algorithms: Introduction to design and analysis*. Reading, MA: Addison-Wesley.

Barnard, E., (1993). Temporal-difference methods and Markov models. *IEEE Transactions on Systems, Man, and Cybernetics, 23*(2), 357–365.

Barto, A. G. & Duff, M., (1994). Monte Carlo matrix inversion and reinforcement learning. In *Advances in Neural Information Processing Systems 6*, pages 687–694, San Mateo, CA. Morgan Kaufmann.

Barto, A. G., Sutton, R. S., & Anderson, C. W., (1983). Neuronlike elements that can solve difficult learning control problems. *IEEE Trans. on Systems, Man, and Cybernetics, 13*, 835–846.

Bellman, R. E., (1957). *Dynamic Programming*. Princeton, NJ: Princeton University Press.

Curtiss, J. H., (1954). A theoretical comparison of the efficiencies of two classical methods and a Monte Carlo method for computing one component of the solution of a set of linear algebraic equations. In Meyer, H. A. (Ed.), *Symposium on Monte Carlo Methods*, pages 191–233, New York: Wiley.

Dayan, P., (1992). The convergence of TD(λ) for general λ. *Machine Learning, 8*(3/4), 341–362.

Dayan, P., (1993). Improving generalization for temporal difference learning: The successor representation. *Neural Computation, 5*(4), 613–624.

Dayan, P. & Sejnowski, T., (1994). TD(λ) converges with probability 1. *Machine Learning, 14*, 295–301.

Holland, J. H., (1986). *Escaping brittleness: The possibilities of general-purpose learning algorithms applied to parallel rule-based systems*, Volume 2 of *Machine Learning: An Artificial Intelligence Approach*, chapter 20. Morgan Kaufmann.

Jaakkola, T., Jordan, M. I., & Singh, S. P., (1994). On the convergence of stochastic iterative dynamic programming algorithms. *Neural Computation*, 6(6), 1185–1201.

Jaakkola, T., Singh, S. P., & Jordan, M. I., (1995). Reinforcement learning algorithm for partially observable Markov decision problems. In *Advances in Neural Information Processing Systems 7*. Morgan Kaufmann.

Klopf, A. H., (1972). Brain function and adaptive systems—A heterostatic theory. Technical Report AFCRL-72-0164, Air Force Cambridge Research Laboratories, Bedford, MA.

Kumar, P. R. & Varaiya, P. P., (1986). *Stochastic Systems: Estimation, Identification, and Adaptive Control*. Englewood Cliffs, N.J.: Prentice Hall.

Lin, L. J., (1992). Self-improving reactive agents based on reinforcement learning, planning and teaching. *Machine Learning*, 8(3/4), 293–321.

Miller, W. T., Glanz, F. H., & Kraft, L. G., (1990). CMAC: An associative neural network alternative to backpropagation. *Proc. of the IEEE*, 78, 1561–1567.

Moore, A. W., (1991). Variable resolution dynamic programming: Efficiently learning action maps in multivariate real-valued state-spaces. In *Machine Learning: Proceedings of the Eighth International Workshop*, pages 333–337, San Mateo, CA. Morgan Kaufmann.

Peng, J., (1993). *Dynamic Programming-based Learning for Control*. PhD thesis, Northeastern University.

Peng, J. & Williams, R. J., (1994). Incremental multi-step Q-learning. In *Machine Learning: Proceedings of the Eleventh International Conference*, pages 226–232. Morgan Kaufmann.

Rubinstein, R., (1981). *Simulation and the Monte Carlo method*. New York: John Wiley Sons.

Rummery, G. A. & Niranjan, M., (1994). On-line Q-learning using connectionist systems. Technical Report CUED/F-INFENG/TR 166, Cambridge University Engineering Dept.

Sutton, R. S., (1984). *Temporal Credit Assignment in Reinforcement Learning*. PhD thesis, University of Massachusetts, Amherst, MA.

Sutton, R. S., (1988). Learning to predict by the methods of temporal differences. *Machine Learning*, 3, 9–44.

Sutton, R. S., (1995). TD models: Modeling the world at a mixture of time scales. In *Machine Learning: Proceedings of the Twelfth International Conference*, pages 531–539. Morgan Kaufmann.

Sutton, R. S. & Barto, A. G., (1987). A temporal-difference model of classical conditioning. In *Proceedings of the Ninth Annual Conference of the Cognitive Science Society*, pages 355–378, Hillsdale, NJ. Erlbaum.

Sutton, R. S. & Barto, A. G., (1990): Time-derivative models of Pavlovian conditioning. In Gabriel, M. & Moore, J. W. (Eds.), *Learning and Computational Neuroscience*, pages 497–537. Cambridge, MA: MIT Press.

Sutton, R. S. & Singh, S. P., (1994). On step-size and bias in temporal-difference learning. In *Eighth Yale Workshop on Adaptive and Learning Systems*, pages 91–96, New Haven, CT.

Sutton, R. S. & Whitehead, S. D., (1993). Online learning with random representations. In *Machine Learning: Proceedings of the Tenth Int. Conference*, pages 314–321. Morgan Kaufmann.

Tesauro, G. J., (1992). Practical issues in temporal difference learning. *Machine Learning*, 8(3/4), 257–277.

Tsitsiklis, J., (1994). Asynchronous stochastic approximation and Q-learning. *Machine Learning*, 16(3), 185–202.

Wasow, W. R., (1952). A note on the inversion of matrices by random walks. *Math. Tables Other Aids Comput.*, 6, 78–81.

Watkins, C. J. C. H., (1989). *Learning from Delayed Rewards*. PhD thesis, Cambridge Univ., Cambridge, England.

Wilson, S. W., (to appear). Classifier fitness based on accuracy. *Evolutionary Computation*.

Received November 7, 1994
Accepted March 10, 1995
Final Manuscript October 4, 1995

Machine Learning, 22, 159–195 (1996)

Average Reward Reinforcement Learning: Foundations, Algorithms, and Empirical Results

SRIDHAR MAHADEVAN   mahadeva@samuel.csee.usf.edu
Department of Computer Science and Engineering, University of South Florida, Tampa, Florida 33620

Editor: Leslie Pack Kaelbling

Abstract. This paper presents a detailed study of average reward reinforcement learning, an undiscounted optimality framework that is more appropriate for cyclical tasks than the much better studied discounted framework. A wide spectrum of average reward algorithms are described, ranging from synchronous dynamic programming methods to several (provably convergent) asynchronous algorithms from optimal control and learning automata. A general sensitive discount optimality metric called *n-discount-optimality* is introduced, and used to compare the various algorithms. The overview identifies a key similarity across several asynchronous algorithms that is crucial to their convergence, namely independent estimation of the average reward and the relative values. The overview also uncovers a surprising limitation shared by the different algorithms: while several algorithms can provably generate *gain-optimal* policies that maximize average reward, none of them can reliably filter these to produce *bias-optimal* (or *T-optimal*) policies that also maximize the finite reward to absorbing goal states. This paper also presents a detailed empirical study of R-learning, an average reward reinforcement learning method, using two empirical testbeds: a stochastic grid world domain and a simulated robot environment. A detailed sensitivity analysis of R-learning is carried out to test its dependence on learning rates and exploration levels. The results suggest that R-learning is quite sensitive to exploration strategies, and can fall into sub-optimal limit cycles. The performance of R-learning is also compared with that of Q-learning, the best studied discounted RL method. Here, the results suggest that R-learning can be fine-tuned to give better performance than Q-learning in both domains.

Keywords: Reinforcement learning, Markov decision processes

1. Introduction

Machine learning techniques that enable autonomous agents to adapt to dynamic environments would find use in many applications, ranging from flexible robots for automating chores (Engelberger, 1989) to customized software apprentices for managing information (Dent, et al., 1992). Recently, an adaptive control paradigm called reinforcement learning (RL) has received much attention (Barto, et al., 1995, Kaelbling, 1993b, Sutton, 1992). Here, an agent is placed in an initially unknown task environment, and learns by trial and error to choose actions that maximize, over the long run, rewards that it receives. RL has been shown to scale much better than related dynamic programming (DP) methods for solving Markov decision processes, such as policy iteration (Howard, 1960). For example, Tesauro (Tesauro, 1992) recently produced a grandmaster backgammon program using Sutton's TD(λ) temporal difference learning algorithm (Sutton, 1988).

Most of the research in RL has studied a problem formulation where agents maximize the cumulative sum of rewards. However, this approach cannot handle infinite horizon tasks, where there are no absorbing goal states, without discounting future rewards. Traditionally, discounting has served two purposes. In some domains, such as economics,

discounting can be used to represent "interest" earned on rewards, so that an action that
generates an immediate reward will be preferred over one that generates the same reward
some steps into the future. However, the typical domains studied in RL, such as robotics
or games, do not fall in this category. In fact, many RL tasks have absorbing goal states,
where the aim of the agent is to get to a given goal state as quickly as possible. As
Kaelbling showed (Kaelbling, 1993a), such tasks can be solved using undiscounted RL
methods.

Clearly, discounting is only really necessary in cyclical tasks, where the cumulative
reward sum can be unbounded. Examples of such tasks include a robot learning to
avoid obstacles (Mahadevan & Connell, 1992) or an automated guided vehicle (AGV)
serving multiple queues (Tadepalli & Ok, 1994). Discounted RL methods can and have
been applied to such tasks, but can lead to sub-optimal behavior if there is a short term
mediocre payoff solution that looks more attractive than a more long-term high reward
one (see Figure 1).

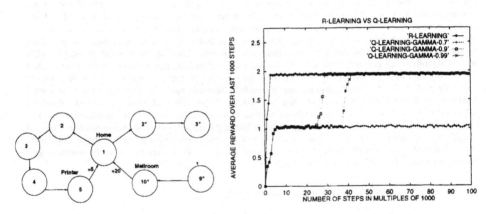

Figure 1. An example to illustrate why an average-reward RL method (R-learning) is preferable to a discounted
RL method (Q-learning) for cyclical tasks. Consider a robot that is rewarded by +5 if it makes a roundtrip
service from "Home" to the "Printer", but rewarded +20 for servicing the more distant "Mailroom". The only
action choice is in the "Home" state, where the robot must decide the service location. The accompanying
graph shows that if the discount factor γ is set too low (0.7), Q-learning converges to a suboptimal solution of
servicing the "Printer". As γ is increased, Q-learning infers the optimal policy of servicing the "Mailroom",
but converges much more slowly than R-learning.

A more natural long-term measure of optimality exists for such cyclical tasks, based on
maximizing the *average* reward per action. There are well known classical DP algorithms
for finding optimal average reward policies, such as policy iteration (Howard, 1960) and
value iteration (White, 1963). However, these algorithms require knowledge of the state
transition probability matrices, and are also computationally intractable. RL methods for
producing optimal average reward policies should scale much better than classical DP
methods. Unfortunately, the study of average reward RL is currently at an early stage.
The first average-reward RL method (R-learning) was proposed only relatively recently
by Schwartz (Schwartz, 1993).

Schwartz hypothesized several reasons why R-learning should outperform discounted methods, such as Q-learning (Watkins, 1989), but did not provide any detailed experimental results to support them. Instead, he used simplified examples to illustrate his arguments, such as the one in Figure 1. Here, Q-learning converges much more slowly than R-learning, because it initially chooses the closer but more mediocre reward over the more distant but better reward.[1] An important question, however, is whether the superiority of R-learning will similarly manifest itself on larger and more realistic problems. A related question is whether average-reward RL algorithms can be theoretically shown to converge to optimal policies, analogous to Q-learning.

This paper undertakes a detailed examination of average reward reinforcement learning. First, a detailed overview of average reward Markov decision processes is presented, covering a wide range of algorithms from dynamic programming, adaptive control, learning automata, and reinforcement learning. A general optimality metric called *n-discount-optimality* (Veinott, 1969) is introduced to relate the discounted and average reward frameworks. This metric is used to compare the strengths and limitations of the various average reward algorithms. The key finding is that while several of the algorithms described can provably yield optimal average reward (or *gain-optimal* (Howard, 1960)) policies, none of them can also reliably discriminate among these to obtain a *bias-optimal* (Blackwell, 1962) (or what Schwartz (Schwartz, 1993) referred to as *T-optimal*) policy. Bias-optimal policies maximize the finite reward incurred in getting to absorbing goal states.

While we do not provide a convergence proof for R-learning, we lay the groundwork for such a proof. We discuss a well known counter-example by Tsitsiklis (described in (Bertsekas, 1982)) and show why it does not rule out provably convergent gain-optimal RL methods. In fact, we describe several provably convergent asynchronous algorithms, and identify a key difference between these algorithms and the counter-example, namely independent estimation of the average reward and the relative values. Since R-learning shares this similarity, it suggests that a convergence proof for R-learning is possible.

The second contribution of this paper is a detailed empirical study of R-learning using two testbeds: a stochastic grid world domain with membranes, and a simulated robot environment. The sensitivity of R-learning to learning rates and exploration levels is also studied. The performance of R-learning is compared with that of Q-learning, the best studied discounted RL method. The results can be summarized into two findings: R-learning is more sensitive than Q-learning to exploration strategies, and can get trapped in limit cycles; however, R-learning can be fine-tuned to outperform Q-learning in both domains.

The rest of this paper is organized as follows. Section 2 provides a detailed overview of average reward Markov decision processes, including algorithms from DP, adaptive control, and learning automata, and shows how R-learning is related to them. Section 3 presents a detailed experimental test of R-learning using two domains, a stochastic grid world domain with membranes, and a simulated robot domain. Section 4 draws some conclusions regarding average reward methods, in general, and R-learning, in particular. Finally, Section 5 outlines some directions for future work.

2. Average Reward Markov Decision Processes

This section introduces the reader to the study of average reward Markov decision processes (MDP). It describes a wide spectrum of algorithms ranging from synchronous DP algorithms to asynchronous methods from optimal control and learning automata. A general sensitive discount optimality metric is introduced, which nicely generalizes both average reward and discounted frameworks, and which is useful in comparing the various algorithms. The overview clarifies the relationship of R-learning to the (very extensive) previous literature on average reward MDP. It also highlights a key similarity between R-learning and several provably convergent asynchronous average reward methods, which may help in its further analysis.

The material in this section is drawn from a variety of sources. The literature on average reward methods in DP is quite voluminous. Howard (Howard, 1960) pioneered the study of average reward MDP's, and introduced the policy iteration algorithm. Bias optimality was introduced by Blackwell (Blackwell, 1962), who pioneered the study of average reward MDP as a limiting case of the discounted MDP framework. The concept of n-discount-optimality was introduced by Veinott (Veinott, 1969). We have drawn much of our discussion, including the key definitions and illustrative examples, from Puterman's recent book (Puterman, 1994), which contains an excellent treatment of average reward MDP.

By comparison, the work on average reward methods in RL is in its infancy. Schwartz's original paper on R-learning (Schwartz, 1993) sparked interest in this area in the RL community. Other more recent work include that of Singh (Singh, 1994b), Tadepalli and Ok (Tadepalli & Ok, 1994), and Mahadevan (Mahadevan, 1994). There has been much work in the area of adaptive control on algorithms for computing optimal average reward policies; we discuss only one algorithm from (Jalali & Ferguson, 1989). Finally, we have also included here a provably convergent asynchronous algorithm from learning automata (Narendra & Thathachar, 1989), which has only recently come to our attention, and which to our knowledge has not been discussed in the previous RL literature. Average reward MDP has also drawn attention in recent work on decision-theoretic planning (e.g. see Boutilier and Puterman (Boutilier & Puterman, 1995)).

2.1. Markov Decision Processes

A Markov Decision Process (MDP) consists of a (finite or infinite) set of states S, and a (finite or infinite) set of actions A for moving between states. In this paper we will assume that S and A are finite. Associated with each action a is a state transition matrix $P(a)$, where $P_{xy}(a)$ represents the probability of moving from state x to y under action a. There is also a reward or payoff function $r : S \times A \rightarrow \mathcal{R}$, where $r(x, a)$ is the *expected* reward for doing action a in state x.

A *stationary deterministic policy* is a mapping $\pi : S \rightarrow A$ from states to actions. In this paper we restrict our discussion to algorithms for such policies, with the exception of the learning automata method, where we allow stationary randomized policies. In any case, we do not allow nonstationary policies. [2] Any policy induces a state transition matrix $P(\pi)$, where $P_{xy}(\pi) = P_{xy}(\pi(x))$. Thus, any policy yields a Markov chain $(S, P(\pi))$. A key difference between discounted and average reward frameworks is that the policy chain structure plays a critical role in average reward methods. All the algorithms described here incorporate some important assumptions about the underlying MDP, which need to be defined before presenting the algorithms.

Two states x and y *communicate* under a policy π if there is a positive probability of reaching (through zero or more transitions) each state from the other. It is easy to see that the communication relation is an equivalence relation on states, and thus partitions the state space into communicating classes. A state is *recurrent* if starting from the state, the probability of eventually reentering it is 1. Note that this implies that recurrent states will be visited forever. A non-recurrent state is called *transient*, since at some finite point in time the state will never be visited again. Any finite MDP must have recurrent states, since not all states can be transient. If a recurrent state x communicates with another state y, then y has to be recurrent also.

An *ergodic* or recurrent class of states is a set of recurrent states that all communicate with each other, and do not communicate with any state outside this class. If the set of all states forms an ergodic class, the Markov chain is termed *irreducible*. Let $p_{s,s}^n(\pi)$ denote the probability of reaching state s from itself in n steps using policy π. The period of a state s under policy π is the greatest common divisor of all n for which $p_{s,s}^n(\pi) > 0$. A state is termed periodic if its period exceeds 1, else it is aperiodic. States in a given recurrence class all have the same period.

An *ergodic or recurrent* MDP is one where the transition matrix corresponding to every (deterministic stationary) policy has a single recurrent class. An MDP is termed *unichain* if the transition matrix corresponding to every policy contains a single recurrent class, and a (possibly empty) set of transient states. An MDP is *communicating* if every pair of states communicate under some stationary policy. Finally, an MDP is *multichain* if there is at least one policy whose transition matrix has two or more recurrent classes. We will not discuss algorithms for multichain MDP's in this paper.

Figure 2 contains two running examples that we will use to illustrate these concepts, as well as to explain the following algorithms. These examples originally appeared in (Puterman, 1994). For simplicity, in both MDP's, there is a choice of action only in state **A**. Each transition is labeled with the action taken, and a pair of numbers. The first number is the immediate reward, and the second number represents the transition probability. Both MDP's are unichain. In the two-state MDP, state **A** is transient under either policy (that is, doing action **a1** or **a2** in state **A**), and state **B** is recurrent. Such recurrent single-state classes are often called *absorbing* states. Both states are aperiodic under the policy that selects **a1**.

In the three-state MDP, there are two recurrent classes, one for each policy. If action **a1** is taken in state **A**, the recurrent class is formed by **A** and **B**, and is periodic with period 2. If action **a2** is taken, the recurrent class is formed by **A** and **C**, which is also periodic with period 2. As we will show below, these differences between the two MDP's will highlight themselves in the operation of the algorithms.

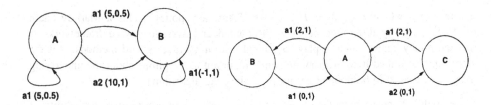

Figure 2. Two simple MDP's that illustrate the underlying concepts and algorithms in this section. Both policies in the two MDP's maximize average reward and are gain-optimal. However, only the policy that selects action a1 in state A is bias-optimal (T-optimal) in both MDP's. While most of the algorithms described here can compute the bias-optimal policy for the 2 state MDP, none of them can do so for the 3 state MDP.

2.2.   Gain and Bias Optimality

The aim of average reward MDP is to compute policies that yield the highest expected payoff per step. The average reward $\rho^\pi(x)$ associated with a particular policy π at a state x is defined as [3]

$$\rho^\pi(x) = \lim_{N \to \infty} \frac{E(\sum_{t=0}^{N-1} R_t^\pi(x))}{N},$$

where $R_t^\pi(x)$ is the reward received at time t starting from state x, and actions are chosen using policy π. $E(.)$ denotes the expected value. We define a *gain-optimal* policy π^* as one that maximizes the average reward over all states, that is $\rho^{\pi^*}(x) \geq \rho^\pi(x)$ over all policies π and states x.

A key observation that greatly simplifies the design of average reward algorithms is that for unichain MDP's, the average reward of any policy is state independent. That is,

$$\rho^\pi(x) = \rho^\pi(y) = \rho^\pi.$$

The reason is that unichain policies generate a single recurrent class of states, and possibly a set of transient states. Since states in the recurrent class will be visited forever, the expected average reward cannot differ across these states. Since the transient states will eventually never be reentered, they can at most accumulate a finite total expected reward before entering a recurrent state, which vanishes under the limit.

For the example MDP's in Figure 2, note that the average rewards of the two policies in the two-state MDP are both -1. In the three-state MDP, once again, the average rewards of the two policies are identical and equal to 1.

2.2.1.   Bias Optimality

Since the aim of solving average reward MDP's is to compute policies that maximize the expected payoff per step, this suggests defining an optimal policy π^* as one that is gain-optimal. Figure 3 illustrates a problem with this definition. MDP's of this type naturally occur in goal-oriented tasks. For example, consider a two-dimensional grid-world problem, where the agent is rewarded by +1 if an absorbing goal state is reached, and is rewarded −1 otherwise. Clearly, in this case, all policies that reach the goal will have the same average reward, but we are mainly interested in policies that reach the goal in the shortest time, i.e. those that maximize the finite reward incurred in reaching the goal.

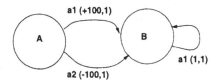

Figure 3. In this MDP, both policies yield the same average reward, but doing action **a1** is clearly preferrable to doing **a2** in state **A**.

The notion of *bias optimality* is needed to address such problems. We need to define a few terms before introducing this optimality metric. A *value* function $V : S \to \mathcal{R}$ is any mapping from states to real numbers. In the traditional discounted framework, any policy π induces a value function that represents the cumulative discounted sum of rewards earned following that policy starting from any state s

$$V_\gamma^\pi(s) = \lim_{N \to \infty} E(\sum_{t=0}^{N-1} \gamma^t R_t^\pi(s)),$$

where $\gamma < 1$ is the discount factor, and $R_t^\pi(s)$ is the reward received at time t starting from state s under policy π. An *optimal discounted policy* π^* maximizes the above value function over all states x and policies π, i.e. $V_\gamma^{\pi^*}(x) \geq V_\gamma^\pi(x)$. In average reward MDP, since the undiscounted sum of rewards can be unbounded, a policy π is measured using a different value function, namely the *average adjusted* sum of rewards earned following that policy, [4]

$$V^\pi(s) = \lim_{N \to \infty} E(\sum_{t=0}^{N-1} (R_t^\pi(s) - \rho^\pi)).$$

where ρ^π is the average reward associated with policy π. The term $V^\pi(x)$ is usually referred to as the *bias* value, or the *relative* value, since it represents the relative difference in total reward gained from starting in state s as opposed to some other state x. In particular, it can be easily shown that

$$V^\pi(s) - V^\pi(x) = \lim_{N \to \infty} \left(E(\sum_{t=0}^{N-1} R_t^\pi(s)) - E(\sum_{t=0}^{N-1} R_t^\pi(x)) \right).$$

A policy π^* is termed *bias-optimal* if it is gain-optimal, and it also maximizes bias values, that is $V^{\pi^*}(x) \geq V^\pi(x)$ over all $x \in S$ and policies π. Bias optimality was

referred to as *T-optimality* by Schwartz (Schwartz, 1993). The notion of bias-optimality
was introduced by Blackwell (Blackwell, 1962), who also pioneered the study of average
reward MDP as the limiting case of the discounted framework. In particular, he showed
how the gain and bias terms relate to the total expected discounted reward using a Laurent
series expansion. A key result is the following truncated Laurent series expansion of the
discounted value function V_γ^π (Puterman, 1994):

LEMMA 1 *Given any policy π and state s,*

$$V_\gamma^\pi(s) = \frac{\rho^\pi(s)}{1 - \gamma} + V^\pi(s) + e^\pi(s, \gamma)$$

where $\lim_{\gamma \to 1} e^\pi(s, \gamma) = 0$.

We will use this corollary shortly. First, we define a general optimality metric that
relates discounted and average reward RL, which is due to Veinott (Veinott, 1969):

DEFINITION 1 *A policy π^* is n-discount-optimal, for $n = -1, 0, 1, ...$, if for each state
s, and over all policies π,*

$$\lim_{\gamma \to 1} (1 - \gamma)^{-n} \left(V_\gamma^{\pi^*}(s) - V_\gamma^\pi(s) \right) \geq 0.$$

We now consider some special cases of this general optimality metric to illustrate how it
nicely generalizes the optimality metrics used previously in both standard discounted RL
and average reward RL. In particular, we show below (see Lemma 2) that gain optimality
is equivalent to *-1-discount-optimality*. Furthermore, we also show (see Lemma 3) that
bias optimality is equivalent to *0-discount-optimality*. Finally, the traditional discounted
MDP framework can be viewed as studying 0-discounted-optimality for a *fixed* value of
γ.

In the two-state MDP in Figure 2, the two policies have the same average reward of
-1, and are both gain-optimal. However, the policy π that selects action **a1** in state **A**
results in bias values $V^\pi(A) = 12$ and $V^\pi(B) = 0$. The policy π' that selects action
a2 in state **A** results in bias values $V^{\pi'}(A) = 11$ and $V^{\pi'}(B) = 0$. Thus, π is better
because it is also bias-optimal. In the 3-state MDP, both policies are again gain-optimal
since they yield an average reward of 1. However, the policy π that selects action **a1**
in state **A** generates bias values $V^\pi(A) = 0.5$, $V^\pi(B) = -0.5$, and $V^\pi(C) = 1.5$. The
policy π is bias-optimal because the only other policy is π' that selects action **a2** in state
A, and generates bias values $V^{\pi'}(A) = -0.5$, $V^\pi(B) = -1.5$, and $V^\pi(C) = 0.5$.

It is easy to see why we might prefer bias-optimal policies, but when do we need to
consider a more selective criterion than bias optimality? Figure 4 shows a simple MDP
that illustrates why *n-discount-optimality* is useful for $n > 0$.

For brevity, we only include proofs showing one side of the equivalences for bias
optimality and gain optimality. More detailed proofs can be found in (Puterman, 1994).

LEMMA 2 *If π^* is a -1-discount-optimal policy, then π^* is a gain-optimal policy.*

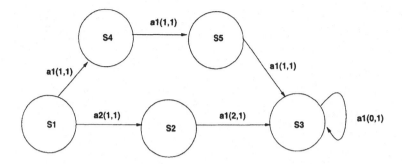

Figure 4. A simple MDP that clarifies the meaning of *n-discount-optimality*. There are only 2 possible policies, corresponding to the two actions in state **S1**. Both policies yield an average reward of 0, and are *-1-discount-optimal*. Both policies yield a cumulative reward of 3 in reaching the absorbing goal state **S3**, and are also *0-discount-optimal*. However, the bottom policy reaches the goal quicker than the top policy, and is *1-discount-optimal*. Finally, the bottom policy is also Blackwell optimal, since it is *n-discount-optimal*, for all $n \geq -1$.

Proof: Let π^* be -1-discount-optimal, and let π be any other policy. It follows from Definition 1 that over all states s

$$\lim_{\gamma \to 1} (1 - \gamma)(V_\gamma^{\pi^*}(s) - V_\gamma^{\pi}(s)) \geq 0.$$

Using Lemma 1, we can transform the above equation to

$$\lim_{\gamma \to 1} (1 - \gamma) \left(\frac{\rho^{\pi^*}(s) - \rho^{\pi}(s)}{1 - \gamma} + V^{\pi^*}(s) + e^{\pi^*}(s, \gamma) - V^{\pi}(s) - e^{\pi}(s, \gamma) \right) \geq 0.$$

Noting that $e^{\pi^*}(s, \gamma)$ and $e^{\pi}(s, \gamma)$ both approach 0 as $\gamma \to 1$, the above equation implies that

$$(\rho^{\pi^*}(s) - \rho^{\pi}(s)) \geq 0.$$

■

LEMMA 3 *If π^* is a 0-discount-optimal policy, then π^* is a bias-optimal (or T-optimal) policy.*

Proof: Note that we need only consider gain-optimal policies, since as $\gamma \to 1$, the first term on the right hand side in Lemma 1 dominates, and hence we can ignore all policies where $\rho^{\pi} < \rho^{\pi^*}$. From the definition, it follows that if π^* is a 0-discount-optimal policy, then for all gain-optimal policies π, over all states s

$$\lim_{\gamma \to 1} (V_\gamma^{\pi^*}(s) - V_\gamma^{\pi}(s)) \geq 0.$$

As before, we can expand this to yield

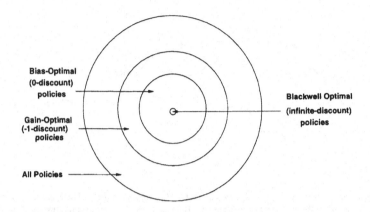

Figure 5. This diagram illustrates the concept of *n-discount-optimal* policies. Gain-optimal policies are -*1-discount-optimal* policies. Bias-optimal (or T-optimal) policies are *0-discount-optimal* policies. Note that *n-discount-optimal* policies get increasingly selective as n increases. Blackwell optimal policies are *n-discount-optimal* for all $n \geq -1$.

$$\lim_{\gamma \to 1} \left(\left(\frac{\rho^{\pi^*}(s)}{1-\gamma} + V^{\pi^*}(s) + e^{\pi^*}(s, \gamma) \right) - \left(\frac{\rho^{\pi}(s)}{1-\gamma} + V^{\pi}(s) + e^{\pi}(s, \gamma) \right) \right) \geq 0.$$

Since π^* and π are both gain-optimal, $\rho^{\pi^*} = \rho^{\pi}$, and hence

$$\lim_{\gamma \to 1} \left(\left((V^{\pi^*}(s) + e^{\pi^*}(s, \gamma)) - (V^{\pi}(s) + e^{\pi}(s, \gamma)) \right) \right) \geq 0.$$

Since both $e^{\pi^*}(s, \gamma)$ and $e^{\pi}(s, \gamma)$ approach 0 as $\gamma \to 1$, the above equation implies that

$$(V^{\pi^*}(s) - V^{\pi}(s)) \geq 0.$$

■

Figure 5 illustrates the structure of *n-discount-optimal* policies. The most selective optimality criterion is *Blackwell optimality*, and corresponds to a policy π^* that is *n-discount-optimal* for all $n \geq -1$. It is interesting to note that a policy that is *m-discount-optimal* will also be *n-discount-optimal* for all $n < m$. Thus, bias optimality is a more selective optimality metric than gain optimality. This suggests a computational procedure for producing Blackwell-optimal policies, by starting with *-1-discount-optimal* policies, and successively pruning these to produce *i-discount-optimal* policies, for $i > -1$. Exactly such a policy iteration procedure for multichain MDP's is described in (Puterman, 1994). In this paper, we restrict our attention to -1-discount-optimal and 0-discount-optimal policies.

2.3. Average Reward Bellman Equation

Clearly, it is useful to know the class of finite state MDP's for which (gain or bias) optimal policies exist. A key result, which essentially provides an average reward Bellman optimality equation, is the following theorem (see (Bertsekas, 1987, Puterman, 1994) for a proof):

THEOREM 1 *For any MDP that is either unichain or communicating, there exist a value function V^* and a scalar ρ^* satisfying the equation*

$$V^*(x) + \rho^* = \max_a \left(r(x, a) + \sum_y P_{xy}(a) V^*(y) \right) \tag{1}$$

such that the greedy policy π^* resulting from V^* achieves the optimal average reward $\rho^* = \rho^{\pi^*}$ where $\rho^{\pi^*} \geq \rho^\pi$ over all policies π.

Here, "greedy" policy means selecting actions that maximize the right hand side of the above Bellman equation. If the MDP is multichain, there is an additional optimality equation that needs to be specified to compute optimal policies, but this is outside the scope of this paper (see (Howard, 1960, Puterman, 1994)). Although this result assures us that stationary deterministic optimal policies exist that achieve the maximum average reward, it does not tell us how to find them. We turn to discuss a variety of algorithms for computing optimal average reward policies.

2.4. Average Reward DP Algorithms

2.4.1. Unichain Policy Iteration

The first algorithm, introduced by Howard (Howard, 1960), is called policy iteration. Policy iteration iterates over two phases: policy evaluation and policy improvement.

1. Initialize $k = 0$ and set π^0 to some arbitrary policy.

2. *Policy Evaluation:* Given a policy π^k, solve the following set of $|S|$ linear equations for the average reward ρ^{π^k} and relative values $V^{\pi^k}(x)$, by setting the value of a reference state $V(s) = 0$.

$$V^{\pi^k}(x) + \rho^{\pi^k} = r(s, \pi^k(x)) + \sum_y P_{xy}(\pi^k(x)) V^{\pi^k}(y).$$

3. *Policy Improvement:* Given a value function V^{π^k}, compute an improved policy π^{k+1} by selecting an action maximizing the following quantity at each state,

$$\max_a \left(r(x, a) + \sum_y P_{xy}(a) V^{\pi^k}(y) \right).$$

setting, if possible, $\pi^{k+1}(x) = \pi^k(x)$.

4. If $\pi^k(x) \neq \pi^{k+1}(x)$ for some state x, increment k and return to step 2.

$V(s)$ is set to 0 because there are $|S| + 1$ unknown variables, but only $|S|$ equations. Howard (Howard, 1960) also provided a policy iteration algorithm for multichain MDP's and proved that both algorithms would converge in finitely many steps to yield a gain-optimal policy. More recently, Haviv and Puterman (Haviv & Puterman, 1991) have developed a more efficient variant of Howard's policy iteration algorithm for communicating MDP's.

The interesting question now is: Does policy iteration produce bias-optimal policies, or only gain-optimal policies? It turns out that policy iteration will find bias-optimal policies if the *first* gain-optimal policy it finds has the same set of recurrent states as a bias-optimal policy (Puterman, 1994). Essentially, policy iteration first searches for a gain-optimal policy that achieves the highest average reward. Subsequent improvements in the policy can be shown to improve only the relative values, or bias, over the transient states.

An example will clarify this point. Consider the two-state MDP in Figure 2. Let us start with the initial policy π^0, which selects action **a2** in state **A**. The policy evaluation phase results in the relative values $V^{\pi^0}(A) = 11$ and $V^{\pi^0}(B) = 0$ (since **B** is chosen as the reference state), and the average reward $\rho^{\pi^0} = -1$. Clearly, policy iteration has already found a gain-optimal policy. However, in the next step of policy improvement, a better policy π^1 is found, which selects action **a1** in state **A**. Evaluating this policy subsequently reveals that the relative values in state **A** has been improved to $V^{\pi^1}(A) = 12$, with the average reward staying unchanged at $\rho^{\pi^1} = -1$. Policy iteration will converge at this point because no improvement in gain or bias is possible.

However, policy iteration will not always produce bias-optimal policies, as the three-state MDP in Figure 2 shows. Here, if we set the value of state **A** to 0, then both policies evaluate to identical bias values $V(A) = 0$, $V(B) = -1$, and $V(C) = 1$. Thus, if we start with the policy π^0 which selects action **a2** in state **A**, and carry out the policy evaluation and subsequent policy improvement step, no change in policy occurs. But, as we explained above, policy π^0 is only gain-optimal and not bias-optimal. The policy that selects action **a1** in state **A** is better since it is bias-optimal.

2.4.2. *Value Iteration*

The difficulty with policy iteration is that it requires solving $|S|$ equations at every iteration, which is computationally intractable when $|S|$ is large. Although some shortcuts have been proposed that reduce the computation (Puterman, 1994), a more attractive approach is to iteratively solve for the relative values and the average reward. Such algorithms are typically referred to in DP as *value iteration* methods.

Let us denote by $T(V)(x)$ the mapping obtained by applying the right hand side of Bellman's equation:

$$T(V)(x) = \max_a \left(r(x, a) + \sum_y P_{xy}(a)V(y) \right).$$

It is well known (e.g. see (Bertsekas, 1987)) that T is a *monotone* mapping, i.e. given two value functions $V(x)$ and $V'(x)$, where $V(x) \leq V'(x)$ over all states x, it follows that $T(V)(x) \leq T(V')(x)$. Monotonicity is a crucial property in DP, and it is the basis for showing the convergence of many algorithms (Bertsekas, 1987). The value iteration algorithm is as follows:

1. Initialize $V^0(t) = 0$ for all states t, and select an $\epsilon > 0$. Set $k = 0$.

2. Set $V^{k+1}(x) = T(V^k)(x)$ over all $x \in S$.

3. If $sp(V^{k+1} - V^k) > \epsilon$, increment k and go to step 2.

4. For each $x \in S$, choose $\pi(x) = a$ to maximize $\left(r(x, a) + \sum_y P_{xy}(a)V^k(y) \right)$.

The stopping criterion in step 3 uses the *span* semi-norm function $sp(f(x)) = max_x(f(x)) - min_x(f(x))$. Note that the value iteration algorithm does not explicitly compute the average reward, but this can be estimated as $V^{n+1}(x) - V^n(x)$ for large n.

The above value iteration algorithm can also be applied to communicating MDP's, where only optimal policies are guaranteed to have state-independent average reward. However, value iteration has the disadvantage that the values $V(x)$ can grow very large, causing numerical instability. A more stable *relative value iteration* algorithm proposed by White (White, 1963) subtracts out the value of a reference state at every step from the values of other states. That is, in White's algorithm, step 2 in value iteration is replaced by

$$V^{k+1}(x) = T(V^k)(x) - T(V^k)(s) \quad \forall x \in S$$

where s is some reference state. Note that $V^k(s) = 0$ holds for all time steps k.

Another disadvantage of (relative) value iteration is that it cannot be directly applied to MDP's where states are periodic under some policies (such as the 3-state MDP in Figure 2). Such MDP's can be handled by a modified value iteration procedure proposed by Hordijk and Tijms (Hordijk & Tijms, 1975), namely

$$V^{k+1}(x) = \max_a \left(r(x, a) + \alpha_k \sum_y P_{xy}(a)V^k(y) \right),$$

where $\alpha_k \to 1$ as $k \to \infty$. Alternatively, periodic MDP's can be transformed using an aperiodicity transformation (Bertsekas, 1987, Puterman, 1994) and then solved using value iteration.

Value iteration can be shown to converge to produce an ϵ-optimal policy in finitely many iterations, but the conditions are stricter than for policy iteration. In particular, it will converge if the MDP is aperiodic under all policies, or if there exists a state $s \in S$ that is reachable from every other state under all stationary policies. Like policy iteration, value iteration finds the bias-optimal policy for the 2 state MDP in Figure 2. Unfortunately, like policy iteration, value iteration cannot discriminate between the bias-optimal and gain-optimal policies in the 3-state example MDP.

2.4.3. Asynchronous Value Iteration

Policy iteration and value iteration are both *synchronous*, meaning that they operate by conducting an exhaustive sweep over the whole state space at each step. When updating the relative value of a state, only the old values of the other states are used. In contrast, RL methods are *asynchronous* because states are updated at random intervals, depending on where the agent happens to be. A general model of asynchronous DP has been studied by Bertsekas (Bertsekas, 1982). Essentially, we can imagine one processor for every state, which keeps track of the relative value of the state. Each processor can be in one of three states: idle, compute the next iterate, or broadcast its value to other processors. It turns out that the asynchronous version of value iteration for discounted MDP's converges under very general protocols for communication and computation.

A natural question, therefore, is whether the natural asynchronous version of value iteration and relative value iteration will similarly converge for the average reward framework. Tsitsiklis provided a counter-example (described in (Bertsekas, 1982)) to show that, quite surprisingly, asynchronous relative value iteration diverges. We discuss this example in some detail because it calls into question whether provably convergent asynchronous algorithms can exist at all for the average reward framework. As we show below, such provably convergent algorithms do exist, and they differ from asynchronous relative value iteration in a crucial detail.

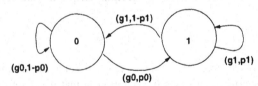

Figure 6. A simple MDP to illustrate that the asynchronous version of White's relative value iteration algorithm diverges, because the underlying mapping is non-monotonic. Both p0 and p1 can lie anywhere in the interval (0,1). g0 and g1 are arbitrary expected rewards. Asynchronous value iteration can be made to converge, however, if it is implemented by a monotonic mapping.

Figure 6 shows a simple two-state MDP where there are no action choices, and hence the problem is to evaluate the single policy to determine the relative values. For this particular MDP, the mapping T can be written as

$$T(V)(0) = g0 + p0V(1) + (1 - p0)V(0)$$
$$T(V)(1) = g1 + p1V(1) + (1 - p1)V(0).$$

If we let state 0 be the reference state, White's algorithm can be written as

$$T(V^{k+1})(0) = g0 + p0(T(V^k)(1) - T(V^k)(0))$$

$$T(V^{k+1})(1) = g1 + p1(T(V^k)(1) - T(V^k)(0)).$$

Now, it can be shown that this mapping has a fixpoint V^* such that $V^*(0)$ yields the optimal average reward. However, it is easy to show that this mapping is not monotonic! Thus, although the above iteration will converge when it is solved synchronously, divergence occurs even if we update the values in a Gauss-Seidel fashion, that is use the just computed value of $T(V^k(0))$ when computing for $T(V^k(1))$! Note that the asynchronous algorithm diverges, not only because the mapping is not monotonic, but also because it is not a *contraction* mapping with respect to the maximum norm.[5] For a more detailed analysis, see (Bertsekas, 1982). There are many cases when divergence can occur, but we have experimentally confirmed that divergence occurs if $g0 = g1 = 1$ and $p0 = p1 > \frac{1}{\sqrt{2}}$.

However, this counter-example should not be viewed as ruling out provably convergent asynchronous average reward algorithms. The key is to ensure that the underlying mapping remains monotonic or a contraction with respect to the maximum norm. Jalali and Ferguson (Jalali & Ferguson, 1990) describe an asynchronous value iteration algorithm that provably converges to produce a gain-optimal policy if there is a state $s \in S$ that is reachable from every other state under all stationary policies. The mapping underlying their algorithm is

$$V^{t+1}(x) = T(V^t)(x) - \rho^t \quad \forall x \neq s$$
$$V^t(s) = 0 \quad \forall t.$$

where ρ^t, the estimate of the average reward at time t, is *independently estimated* without using the relative values. Note that the mapping is monotonic since T is monotonic and ρ^t is a scalar value.

There are several ways of independently estimating average reward. We discuss below two provably convergent adaptive algorithms, where the average reward is estimated by online averaging over sample rewards. This similarity is critical to the convergence of the two asynchronous algorithms below, and since R-learning shares this similarity, it suggests that a convergence proof for R-learning is possible.

2.5. An Asynchronous Adaptive Control Method

One fundamental limitation of the above algorithms is that they require complete knowledge of the state transition matrices $P(a)$, for each action a, and the expected payoffs $r(x, a)$ for each action a and state x. This limitation can be partially overcome by inferring the matrices from sample transitions. This approach has been studied very thoroughly in the adaptive control literature. We focus on one particular algorithm by Jalali and Ferguson (Jalali & Ferguson, 1989). They describe two algorithms (named A and B), which both assume that the underlying MDP can be identified by a maximum likelihood estimator (MLE), and differ only in how the average reward is estimated. We will focus mainly on Algorithm B, since that is most relevant to this paper. Algorithm B, which is provably convergent under the assumption that the MDP is ergodic under all stationary policies, is as follows:

1. At time step $t = 0$, initialize the current state to some state x, cumulative reward to $K^0 = 0$, relative values $V^0(x) = 0$ for all states x, and average reward $\rho^0 = 0$. Fix $V^t(s) = 0$ for some reference state s for all t. The expected rewards $r(x, a)$ are assumed to be known.

2. Choose an action a that maximizes $\left(r(x, a) + \sum_y P^t_{xy}(a) V^t(y) \right)$.

3. Update the relative value $V^{t+1}(x) = T(V^t)(x) - \rho^t$, if $x \neq s$.

4. Update the average reward $\rho^{t+1} = \frac{K^{t+1}}{t+1}$, where $K^{t+1} = K^t + r(x, a)$.

5. Carry out action a, and let the resulting state be z. Update the probability transition matrix entry $P^t_{xz}(a)$ using a maximum-likelihood estimator. Increment t, set the current state x to z, and go to step 2.

This algorithm is asynchronous because relative values of states are updated only when they are visited. As discussed above, a key reason for the convergence of this algorithm is that the mapping underlying the algorithm is monotonic. However, some additional assumptions are also necessary to guarantee convergence, such as that the MDP is *identifiable* by an MLE-estimator (for details of the proof, see (Jalali & Ferguson, 1989)). Tadepalli and Ok (Tadepalli & Ok, 1994) have undertaken a detailed study of a variant of this algorithm, which includes two modifications. The first is to estimate the expected rewards $r(x, a)$ from sample rewards. The second is to occasionally take random actions in step 2 (this strategy is called *semi-uniform* exploration – see Section 3.3). This modification allows the algorithm to be applied to unichain MDP's (assuming of course that we also break up a learning run into a sequence of trials, and start in a random state in each trial). In our experiments, which have been confirmed by Tadepalli (Tadepall), we found that the modified algorithm cannot discriminate between the gain-optimal and bias-optimal policy for the 3-state MDP in Figure 2.

2.6. An Asynchronous Learning Automata Method

Thus far, all the algorithms we have discussed are *model-based*; that is, they involve transition probability matrices (which are either known or estimated). *Model-free* methods eliminate this requirement, and can learn optimal policies directly without the need for a model. We now discuss a provably convergent model-free algorithm by Wheeler and Narendra (Wheeler & Narendra, 1986) that learns an optimal average reward policy for any ergodic MDP with unknown state transition probabilities and payoff functions. Thus, this algorithm tackles the same learning problem addressed by R-learning, but as we will see below, is quite different from most RL algorithms.

We need to first briefly describe the framework underlying learning automata (Narendra & Thathachar, 1989). Unlike the previous algorithms, most learning automata algorithms work with randomized policies. In particular, actions are chosen using a vector of probabilities. The probability vector is updated using a learning algorithm. Although many different algorithms have been studied, we will focus on one particular algorithm called *linear reward-inaction* or L_{R-I} for short. Let us assume that there are n actions $a_1, \ldots, a_n$. Let the current probability vector be $p = (p_1, \ldots, p_n)$, where we of course require that $\sum_{i=1}^{n} p_i = 1$. The automata carries out an action a according to the probability distribution p, and receives a response from the environment $0 \leq \beta \leq 1$. The probability vector is updated as follows:

1. If action a is carried out, then $p_a^{t+1} = p_a^t + \alpha \beta^t (1 - p_a^t)$.

2. For all actions $b \neq a$, $p_b^{t+1} = p_b^t - \alpha \beta^t p_b^t$.

Here α is a learning rate in the interval $(0, 1)$. Thus, the probability of doing the chosen action is updated in proportion to the probability of not doing the action, weighted by the environmental response β^t at time step t. So far, we have not considered the influence of state. Let us now assume that for each state an independent copy of the L_{R-I} algorithm exists. So for each state, a different probability vector is maintained, which is used to select the best action. Now the key (and surprising) step in the Wheeler-Narendra algorithm is in computing the environmental response β. In particular, the procedure ignores the immediate reward received, and instead computes the average reward over repeated visits to a state under a particular action. The following global and local statistics need to be maintained.

1. Let c^t be the cumulative reward at time step t, and let the current state be s.

2. The local learning automaton in state s uses c^t as well as the global time t to update the following local statistics: $\delta_r^s(a)$, the incremental reward received since state s was last exited under action a, and $\delta_t^s(a)$ the elapsed time since state s was last exited under action a. These incremental values are used to compute the cumulative statistics, that is $r^s(a) \leftarrow r^s(a) + \delta_r^s(a)$, and $t^s(a) \leftarrow t^s(a) + \delta_t^s(a)$.

3. Finally, the learning automaton updates the action probabilities with the L_{R-I} algorithm using as environmental response the estimated average reward $r^s(a)/t^s(a)$.

Wheeler and Narendra prove that this algorithm converges to an optimal average reward policy with probability arbitrarily close to 1 (i.e. w.p. $1 - \epsilon$, where ϵ can be made as small as desired). The details of the proof are in (Wheeler & Narendra, 1986), which uses some interesting ideas from game theory, such as Nash equilibria. Unfortunately, this restriction to ergodic MDP's means it cannot handle either of the example MDP's in Figure 2. Also, for convergence to be guaranteed, the algorithm requires using a very small learning rate α which makes it converge very slowly.

2.7. R-learning: A Model-Free Average Reward RL Method

Schwartz (Schwartz, 1993) proposed an average-reward RL technique called R-learning. Like its counterpart, Q-learning (Watkins, 1989) (see page 193 for a description of Q-learning), R-learning uses the *action value* representation. The action value $R^\pi(x, a)$

represents the average adjusted value of doing an action a in state x once, and then following policy π subsequently. That is,

$$R^\pi(x,a) = r(x,a) - \rho^\pi + \sum_y P_{xy}(a)V^\pi(y),$$

where $V^\pi(y) = \max_{a \in A} R^\pi(y,a)$, and ρ^π is the average reward of policy π. R-learning consists of the following steps:

1. Let time step $t = 0$. Initialize all the $R_t(x,a)$ values (say to 0). Let the current state be x.

2. Choose the action a that has the highest $R_t(x,a)$ value with some probability, else let a be a random exploratory action.

3. Carry out action a. Let the next state be y, and the reward be $r_{imm}(x,y)$. Update the R values and the average reward ρ using the following rules:

$$R_{t+1}(x,a) \leftarrow R_t(x.a)(1-\beta) + \beta(r_{imm}(x,y) - \rho_t + \max_{a \in A} R_t(y,a))$$

$$\rho_{t+1} \leftarrow \rho_t(1-\alpha) + \alpha[r_{imm}(x,a) + \max_{a \in A} R_t(y,a) - \max_{a \in A} R_t(x,a)].$$

4. Set the current state to y and go to step 2.

Here $0 \le \beta \le 1$ is the learning rate controlling how quickly errors in the estimated action values are corrected, and $0 \le \alpha \le 1$ is the learning rate for updating ρ. One key point is that ρ is updated only when a non-exploratory action is performed.

Singh (Singh, 1994b) proposed some variations on the basic R-learning method, such as estimating average reward as the sample mean of the actual rewards, updating the average reward on every step, and finally, grounding a reference $R(x,a)$ value to 0. We have not as yet conducted any systematic experiments to test the effectiveness of these modifications.

2.7.1. R-learning on Sample MDP Problems

We found that in the 2-state problem, R-learning reliably learns the bias-optimal policy of selecting action **a1** in state **A**. However, in the 3-state problem, like all the preceding algorithms, it is unable to differentiate the bias-optimal policy from the gain-optimal policy.

Baird (Baird, personal communication) has shown that there exists a fixpoint of the $R(x,a)$ values for the 3-state MDP. Consider the following assignment of $R(x,a)$ values for the 3-state MDP in Figure 2. Assume $R(A.a1) = 100$, $R(A.a2) = 100$, $R(B,a1) = 99$, and $R(C,a1) = 101$. Also, assume that $\rho = 1$. These values satisfy the definition of $R(x,a)$ for all states and actions in the problem, but the resulting greedy policy does not discriminate between the two actions in state **A**. Thus, this is a clear counter-example that demonstrates that R-learning can converge to a policy with *sub-optimal* bias for unichain MDP's.[6]

Table 1. Summary of the average reward algorithms described in this paper. Only the convergence conditions for gain optimality are shown.

ALGORITHM	GAIN OPTIMALITY
Unichain Policy Iteration (Section 2.4.1) (Howard, 1960)	MDP is unichain
(Relative) Value Iteration (Section 2.4.2) (Puterman, 1994, White, 1963)	MDP is communicating
Asynchronous Relative Value Iteration (Section 2.4.3) (Bertsekas, 1982)	Does not converge
Asynchronous Value Iteration (Section 2.4.3) (Jalali & Ferguson, 1990) with Online Gain Estimation	A state s is reachable under every policy
Asynchronous Adaptive Control (Section 2.5) (Jalali & Ferguson, 1989) with Online Gain Estimation	MDP is ergodic and MLE-Identifiable
Asynchronous Learning Automata (Section 2.6) (Wheeler & Narendra, 1986)	MDP is ergodic
R-Learning (Section 2.7) (Schwartz, 1993)	Convergence unknown

2.7.2. *Convergence Proof for R-learning*

Of course, the question of whether R-learning can be guaranteed to produce gain-optimal policies remains unresolved. From the above discussion, it is clear that to show convergence, we need to determine when the mapping underlying R-learning satisfies some key properties, such as monotonicity and contraction. We are currently developing such a proof, which will also exploit the existing convergence results underlying Jalali and Ferguson's B algorithm (see Section 2.5) and Q-learning (for the undiscounted case).

2.8. **Summary of Average Reward Methods**

Table 1 summarizes the average reward algorithms described in this section, and lists the known convergence properties. While several convergent synchronous and asynchronous algorithms for producing gain-optimal policies exist, none of them is guaranteed to find bias-optimal policies. In fact, they all fail on the 3-state MDP in Figure 2 for the same reason, namely that bias optimality really requires solving an additional optimality equation (Puterman, 1994). An important problem for future research is to design RL algorithms that can yield bias-optimal policies. We discuss this issue in more detail in Section 5.

3. **Experimental Results**

In this section we present an experimental study of R-learning. We use two empirical testbeds, a stochastic grid-world domain with one-way membranes, and a simulated robot environment. The grid-world task involves reaching a specific goal state, while the robot task does not have any specific goal states. We use an idealized example to show how R-learning can get into limit cycles given insufficient exploration, and illustrate where such limit-cycle situations occur in the grid-world domain and the robot domain. We show, however, that provided sufficient exploration is carried out, R-learning can perform better than Q-learning in both these testbeds. Finally, we present a detailed sensitivity analysis of R-learning using the grid world domain.

We first presented the limit cycle behavior of R-learning in an earlier paper (Mahadevan, 1994). However, that paper mainly reported negative results for R-learning using the robot domain. This paper contains several new experimental results, including positive results where R-learning outperforms Q-learning, and a detailed sensitivity analysis of R-learning with respect to exploration and learning rate decay. Since we will not repeat our earlier experimental results, the interested reader is referred to our earlier paper for additional results on R-learning for the robot domain.

3.1. A Grid World Domain with One-Way Membranes

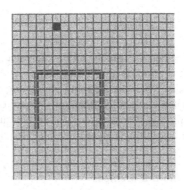

Figure 7. A grid world environment with "one-way" membranes.

Figure 7 illustrates a grid-world environment, similar to that used in many previous RL systems (Kaelbling, 1993a, Singh, 1994a, Sutton, 1990). Although such grid-world tasks are somewhat simplistic "toy" domains, they are quite useful for conducting controlled experimental tests of RL algorithms. The domain parameters can be easily varied, allowing a detailed sensitivity analysis of any RL algorithm.

The agent has to learn a policy that will move it from any initial location to the goal location (marked by the black square in the figure). The starting state can thus be any location. At each step, the agent can move to any square adjacent (row-wise or column-wise) to its current location. We also include "one way membranes", which allow the agent to move in one direction but not in the other (the agent can move from the lighter side of the wall to the darker side). The membrane wall is shown in the figure as an "inverted cup" shape. If the agent "enters" the cup, it cannot reach the goal by going up, but has to go down to leave the membrane. Thus, the membrane serves as a sort of local maxima. The environment is made stochastic by adding a controlled degree of randomness to every transition. In particular, the agent moves to the correct square with probability p, and either stays at its current position or moves to an incorrect adjacent square with probability $(1 - p)/N_a$, where N_a is the number of adjacent squares. Thus, if $p = 0.75$ and the robot is in a square with 4 adjacent squares, and it decides to move "up", then it may stay where it currently is, or move "left" or "down" or "right" with equal probability 0.25/4.

The agent receives a reward of +100 for reaching the goal, and a reward of +1 for traversing a membrane. The default reward is -1. Upon reaching the goal, the agent is "transported" to a random starting location to commence another trial. Thus, although this task involves reaching a particular goal state, the average reward obtained during a learning run does not really reflect the true average reward that would result if the goal were made absorbing. In the latter case, since every policy would eventually reach the goal, all policies result in a single recurrent class (namely, the goal state). Since all non-goal states are transient, this task is more closely related to the 2 state example MDP in Figure 2 than the 3 state MDP. This suggests that R-learning should be able to perform quite well on this task, as the experimental results demonstrate.

3.2. A Simulated Robot Environment

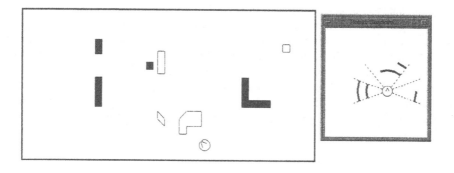

Figure 8. A simulated robot environment modeled after a real robot environment.

We also compare R-learning and Q-learning on a more realistic task, namely avoiding obstacles in a simulated robot environment illustrated in Figure 8. This task provides a nice contrast with the above grid world task because it does not involve reaching any goal states, but instead requires learning a policy that maximizes the average reward.

This testbed provides a good benchmark because we previously studied Q-learning using this environment (Mahadevan & Connell, 1992). The robot is shown as a circular figure; the "nose" indicates the orientation of the robot. Dark objects represent immovable obstacles. Outline figures represent movable boxes. The robot uses eight simulated "sonar" sensors arranged in a ring. For the experiment described below, the sonar data was compressed as follows. The eight sonar values were thresholded to 16 sonar bits, one "near" bit and one "far" bit in each of 8 radial directions. The figure illustrates what the robot actually "sees" at its present location. Dark bars represent the near and far bits that are on. The 16 sonar bits were further reduced to 10 bits by disjoining adjacent near and far bits.[7]

There are also 2 bits of non-sonar information, BUMP and STUCK, which indicate whether the robot is touching something and whether it is wedged. Thus, there are a total of 12 state bits, and 4096 resulting states. The robot moves about the simulator environment by taking one of five discrete actions: forward or turn left and right by either 45 or 90 degrees. The reinforcement function for teaching the robot to avoid obstacles is to reward it by +10 for moving forward to "clear" states (i.e., where the front near sonar bits are off), and punish it by -3 for becoming wedged (i.e., where STUCK is on). This task is very representative of many recurrent robot tasks which have no absorbing (i.e. terminal) goal states.

A learning run is broken down into a sequence of trials. For the obstacle avoidance task, a trial consists of 100 steps. After 100 steps, the environment and the robot are reset, except that at every trial the robot is placed at a random unoccupied location in the environment facing a random orientation. The location of boxes and obstacles does not vary across trials. A learning run lasts for 300 trials. We typically carried out 30 runs to get average performance estimates.

The simulator is a simplification of the real world situation (which we studied previously (Mahadevan & Connell, 1992)) in several important respects. First, boxes in the simulator can only move translationally. Thus a box will move without rotation even if a robot pushes the box with a force which is not aligned with the axis of symmetry. Second, the sonars on the real robot are prone to various types of noise, whereas the sensors on a simulator robot are "clean". Finally, the underlying dynamics of robot motion and box movement are deterministic, although given its impoverished sensors, the world does appear stochastic to the robot. Even though this simulator is a fairly inaccurate model of reality, it is sufficiently complex to illustrate the issues raised in this paper.

3.3. Exploration Strategies

RL methods usually require that all actions be tried in all states infinitely often for asymptotic convergence. In practice, this is usually implemented by using an "exploration" method to occasionally take sub-optimal actions. We can divide exploration methods into *undirected* and *directed* methods. Undirected exploration methods do not use the results of learning to guide exploration; they merely select a random action some of the time. Directed exploration methods use the results of learning to decide where to concentrate the exploration efforts. We will be using four exploration methods in our experimental tests of R-learning: two undirected exploration methods, Boltzmann exploration and semi-uniform exploration, and two directed exploration methods, recency-based (Sutton, 1990) and uncertainty exploration (UE). A detailed comparison of undirected and directed exploration methods is given in (Thrun).

3.3.1. Semi-Uniform Exploration

Let $p(x, a)$ denote the probability of choosing action a in state x. Let $U(x, a)$ denote a generic state action value, which could be a $Q(x, a)$ value or a $R(x, a)$ value. Denote by a_{best} the action that maximizes $U(x, a)$. Semi-uniform exploration is defined as $p(x, a_{best}) = p_{exp} + \frac{1-p_{exp}}{|A|}$. If $x \neq a_{best}$, $p(x, a) = \frac{1-p_{exp}}{|A|}$. In other words, the best action is chosen with a fixed probability p_{exp}. With probability $1 - p_{exp}$, a random action is carried out.

3.3.2. Boltzmann Exploration

The Boltzmann exploration function (Lin, 1993, Sutton, 1990) assigns the probability of doing an action a in state x as $p(x, a) = \frac{e^{\frac{U(x,a)}{T}}}{\sum_a e^{\frac{U(x,a)}{T}}}$ where T is a "temperature" parameter that controls the degree of randomness. In our experiments, the temperature T was gradually decayed from an initial fixed value using a decaying scheme similar to that described in (Barto, et al, 1995).

3.3.3. Recency-based Exploration

In recency-based exploration (Sutton, 1990), the action selected is one that maximizes the quantity $U(x, a) + \epsilon\sqrt{N(x, a)}$, where $N(x.a)$ is a recency counter and represents the last time step when action a was tried in state x. ϵ is a small constant < 1.

3.3.4. UE Exploration

Finally, the second directed exploration strategy is called *uncertainty estimation* (UE). Using this strategy, with a fixed probability p, the agent picks the action a that maximizes $U(x, a) + \frac{c}{N_f(x,a)}$, where c is a constant, and $N_f(x.a)$ represents the number of times that the action a has been tried in state x. With probability $1 - p$, the agent picks a random action.

3.4. Limit Cycles in R-learning

A key assumption underlying R-learning (and all the other methods discussed in Section 2) is that the average reward ρ is state independent. In this section, we explore the consequences of this assumption, under a sub-optimal exploration strategy that does not explore the state space sufficiently, creating non-ergodic multichains. Essentially, R-learning and Q-learning behave very differently when an exploration strategy creates a tight limit cycle such as illustrated in Figure 9. In particular, the performance of R-learning can greatly suffer. Later we will show that such limit cycles can easily arise in both the grid world domain and the simulated robot domain, using two different exploration strategies.

Consider the simple situation shown in Figure 9. In state 1, the only action is to go right (marked r), and in situation 2, the only action is to go left (marked l). Finally, the

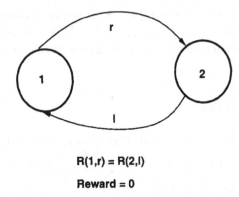

$$R(1,r) = R(2,l)$$

Reward = 0

Figure 9. A simple limit cycle situation for comparing R-learning and Q-learning.

immediate reward received for going left or right is 0. Such a limit cycle situation can easily result even when there are multiple actions possible, whenever the action values for going left or right are initially much higher than the values for the other actions. Under these conditions, the update equations for R-learning can be simplified to yield [8]

$$R(1,r) \leftarrow R(1,r) - \beta\rho$$
$$R(2,l) \leftarrow R(2,l) - \beta\rho.$$

Thus, $R(1,r)$ and $R(2,l)$ will decay over time. However, the average reward ρ is itself decaying, since under these conditions, the average reward update equation turns out to be

$$\rho \leftarrow (1 - \alpha)\rho.$$

When ρ decays to zero, the utilities $R(1,r)$ and $R(2,l)$ will stop decaying. The relative decay rate will of course depend on β and α, but one can easily show cases where the R values will decay much slower than ρ. For example, given the values $R(1,l) = 3.0$, $\rho = 0.5$, $\alpha = 0.1$, $\beta = 0.2$, after 1000 iterations, $R(1,l) = 2$, but $\rho < 10^{-46}$!

Since the $R(x,a)$ values are not changing when ρ has decayed to zero, if the agent uses the Boltzmann exploration strategy and the temperature is sufficiently low (so that the $R(x,a)$ values are primarily used to select actions), it will simply oscillate between going left and going right. We discovered this problem in a simulated robot box-pushing task where the robot had to learn to avoid obstacles (Mahadevan, 1994).

Interestingly, Q-learning will not get into the same limit cycle problem illustrated above because the Q values will always decay by a fixed amount. Under the same conditions as above, the update equation for Q-learning can be written as

$$Q(1,l) \leftarrow Q(1,l)(1 - \beta(1 - \gamma))$$
$$Q(2,r) \leftarrow Q(2,r)(1 - \beta(1 - \gamma)).$$

Now the two utilities $Q(1,r)$ and $Q(2,r)$ will continue to decay until at some point another action will be selected because its Q value will be higher.

3.4.1. Robot Task with Boltzmann Exploration

Limit cycles arise in the robot domain when the robot is learning to avoid obstacles. Consider a situation where the robot is stalled against the simulator wall, and is undecided between turning left or right. This situation is very similar to Figure 9 (in fact, we were led to the limit cycle analysis while trying to understand the lackluster performance of R-learning at learning obstacle avoidance). The limit cycles are most noticeable under Boltzmann exploration.

3.4.2. Grid World Task with Counter-based Exploration

We have observed limit cycles in the grid world domain using the UE exploration strategy. Limit cycles arise in the grid world domain at the two ends of the inverted cup membrane shape shown in Figure 7. These limit cycles involve a clockwise or counterclockwise chain of 4 states (the state left of the membrane edge, the state right of the membrane edge, and the two states below these states). Furthermore, the limit cycles arise even though the agent is getting non-zero rewards. Thus, this limit cycle is different from the idealized one shown earlier in Figure 9.

In sum, since R-learning can get into limit cycles in different tasks using different exploration strategies, this behavior is clearly not task specific or dependent on any particular exploration method. The limit cycle behavior of R-learning arises due to the fact that average reward is state independent because the underlying MDP is assumed to be unichain. An exploration method that produces multichain behavior can cause the average reward to be incorrectly estimated (for example, it can drop down to 0). This reasoning suggests that limit cycles can be avoided by using higher degrees of exploration. We show below that this is indeed the case.

3.5. Avoiding Limit Cycles by Increasing Exploration

Although limit cycles can seriously impair the performance of R learning, they can be avoided by choosing a suitable exploration strategy. The key here is to ensure that a sufficient level of exploration is carried out that will not hamper the estimate of the average reward. Figure 10 illustrates this point: here the constant c used in the UE exploration method is increased from 50 to 60. The errorbars indicate the range between the high and low values over 30 independent runs. Note that when $c = 50$, limit cycles arise (indicated by the high variance between low and high values), but disappear under a higher level of exploration ($c = 60$).

Similarly, we have found that limit cycles can be avoided in the robot domain using higher degrees of exploration. Next we demonstrate the improved performance of R-learning under these conditions.

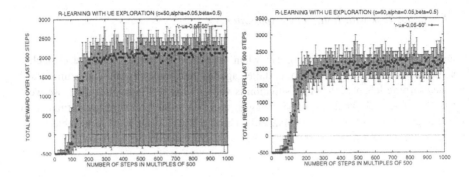

Figure 10. Comparing performance of R-learning on the grid world environment with "one-way" membranes as exploration level is increased.

3.6. Comparing R-learning and Q-learning

We now compare the performance of R-learning with Q-learning. We need to state some caveats at the outset, as such a comparison is not without some inherent difficulties. The two techniques depend on a number of different parameters, and also their performance depends on the particular exploration method used. If we optimize them across different exploration methods, then it could be argued that we are really not comparing the algorithms themselves, but the *combination* of the algorithm and the exploration method. On the other hand, using the same exploration method may be detrimental to the performance of one or both algorithms. Ideally, we would like to do both, but the number of parameter choices and alternative exploration methods can create a very large space of possibilities.

Consequently, our aim here will be more modest, and that is to provide a reasonable basis for evaluating the empirical performance of R-learning. We demonstrate that R-learning can outperform Q-learning on the robot obstacle avoidance task, even if we separately optimize the exploration method used for each technique. We also show a similar result for the grid world domain, where both techniques use the same exploration method, but with different parameter choices. We should also mention here that Tadepalli and Ok (Tadepalli & Ok, 1994) have found that R-learning outperforms Q-learning on a automated guided vehicle (AGV) task, where both techniques used the semi-uniform exploration method.

3.6.1. Simulated Obstacle Avoidance Task

Figure 11 compares the performance of R-learning with that of Q-learning on the obstacle avoidance task. Here, R-learning uses a semi-uniform exploration strategy, where the robot takes random actions 2.5% of the time, since this seemed to give the best results. Q-learning is using a recency-based exploration strategy, with $\epsilon = 0.001$ which gave the best results. Clearly, R-learning is outperforming Q-learning consistently throughout the run. Note that we are measuring the performance of both algorithms on the basis of average reward, which Q-learning is not specifically designed to optimize. However, this is consistent with many previous studies in RL, which have used average reward to judge the empirical performance of Q-learning (Lin, 1993, Mahadevan & Connell, 1992).

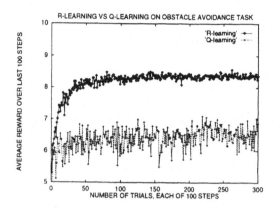

Figure 11. Comparing R and Q-learning using different exploration techniques on a simulated robot obstacle avoidance task. Here Q-learning uses a recency-based exploration method with $\epsilon = 0.001$ which gave the best results. The discount factor $\gamma = 0.99$. R-learning uses a semi-uniform exploration with random actions chosen 2.5% of the time. The curves represent median values over 30 independent runs.

3.6.2. Grid World Domain

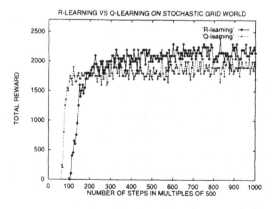

Figure 12. Comparing R vs Q-learning on a grid world environment with membranes. Here both algorithms used the UE exploration method. The UE constant for R-learning is $c = 60$, whereas for Q-learning, $c = 30$. The learning rates were set to $\beta = 0.5$ and $\alpha = 0.05$. The discount factor $\gamma = 0.995$.

Figure 12 compares the performance of R and Q-learning on the grid world domain. The exploration strategy used here is the UE counter-based strategy described above, which was separately optimized for each technique. R-learning is clearly outperforming Q-learning. Interestingly, in both domains, R-learning is slower to reach its optimal performance than Q-learning.

3.7. *Sensitivity Analysis of R-learning*

Clearly, we need to determine the sensitivity of R-learning to exploration to obtain a more complete understanding of its empirical performance. We now describe some sensitivity analysis experiments that illustrate the variation in the performance of R learning with different amounts of exploration. The experiments also illustrate the sensitivity of R-learning to the learning rate parameters α and β.

Figure 13 and Figure 14 illustrate the sensitivity of R-learning to exploration levels and learning rates for the grid world domain. Each figure shows a 3-dimensional plot of the performance of R-learning (as measured by the total cumulative reward) for different values of the two learning rates, α, for adjusting average reward ρ, and β, for adjusting the relative action values $R(x, a)$. The exploration probability parameter p was reduced gradually from an initial value of $p = 0.95$ to a final value of $p = 0.995$ in all the experiments. Each plot measures the performance for different values of the exploration constant c, and a parameter k that controls how quickly the learning rate β is decayed. β is decayed based on the number of updates of a particular $R(x, a)$ value. This state action dependent learning rate was previously used by Barto et al (Barto, et al, 1995). More precisely, the learning rate β for updating a particular $R(x, a)$ value is calculated as follows:

$$\beta(x, a) = \frac{\beta_0 k}{k + freq(x, a)},$$

where β_0 is the initial value of the β. In these experiments, the learning rate α was also decayed over time using a simpler state independent rule:

$$\alpha_{t+1} \rightarrow \alpha_t - \alpha_t \alpha_{min},$$

where α_{min} is the minimum learning rate required.

Figure 13 and Figure 14 reveal a number of interesting properties of R-learning. The first two properties can be observed by looking at each plot in isolation, whereas the second two can be observed by comparing different plots.

- *More exploration is better than less:* The degree of exploration is controlled by the parameter c. Higher values of c mean more exploration. Each plot shows that higher values of c generally produce better performance than lower values. This behavior is not surprising, given our analysis of limit cycles. Higher exploration means that R-learning will be less likely to fall into limit cycles. However, as we show below, the degree of exploration actually depends on the stochasticity of the domain, and more exploration is not always better than less.

- *Slow decay of β is better than fast decay:* Each plot also shows larger values of the parameter k (which imply that β will be decayed more slowly) produces better performance than small values of k. This behavior is similar to that of Q-learning, in that performance suffers if learning rates are decayed too quickly.

- *Low values of α are better than high values:* Another interesting pattern revealed by comparing different plots is that initializing α to smaller values (such as 0.05) clearly produces better performance as compared to larger initial values (such as 0.5). The reason for this behavior is that higher values of α cause the average reward ρ to be adjusted too frequently, causing wide fluctuations in its value over time. A low initial value of α means that ρ will be changed very slowly over time.

- *High values of β are better than low values:* Finally, comparing different plots reveals that higher initial values of β are to be preferred to lower values. The underlying reason for this behavior is not apparent, but it could be dependent on the particular grid world domain chosen for the study.

3.7.1. Degree of Exploration

The above sensitivity analysis tends to give the impression that more exploration is alway better than less. Figure 15 illustrates how the amount of exploration needed actually depends on the stochasticity of the underlying MDP. Here, the domain is a simplified grid world with no membranes. The goal is to reach the grid cell $(19, 19)$ (the bottom right corner) from any other cell in the grid world. The curves show the result of increasing semi-uniform exploration from no exploration ($p = 0.00$) to a high level of exploration ($p = 0.80$). The graph on the left shows the results for a deterministic grid world domain. Here, performance does improve as exploration is increased, but only up to a point (between $p = 0.20$ and $p = 0.40$). Beyond that, performance suffers. The graph on the right shows the same results for a stochastic grid world domain with transition probability $p = 0.75$. Here, the best performance occurs with no exploration at all, because the underlying stochasticity of the domain ensures that all states will be visited!

4. Conclusions

Based on our overview of average reward RL (Section 2) and the experimental tests on R-learning (Section 3), the main findings of this paper can be summarized as follows.

4.1. Average Reward MDP

We emphasized the distinction between gain-optimal policies that maximize average reward versus bias-optimal policies that also maximize relative values. The key finding is

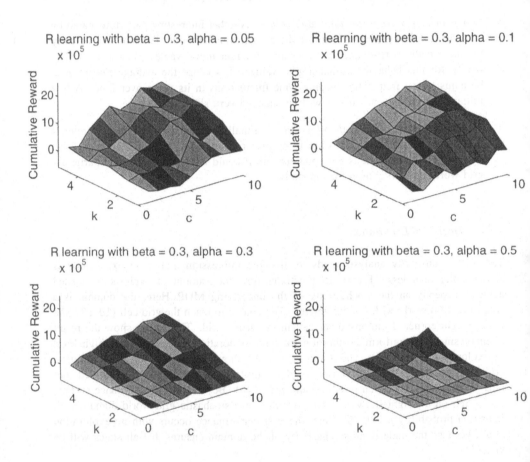

Figure 13. Performance of R-learning on a 20x20 stochastic grid world environment with one-way membranes. Each point represents the median over 30 runs of 500,000 steps. The domain was stochastic with transition probability = 0.75. The x axis represents ten different values of the constant of exploration c varied uniformly over the range [10,100]. Higher values of c mean more exploration. The y axis represents five different values of the constant k controlling the decay of the learning rate β. k was uniformly varied over the range [100,500]. Higher values of k mean more gradual decay of β.

R learning with beta = 0.5, alpha = 0.05
x 10^5

R learning with beta = 0.5, alpha = 0.1
x 10^5

R learning with beta = 0.5, alpha = 0.3
x 10^5

R learning with beta = 0.5, alpha = 0.5
x 10^5

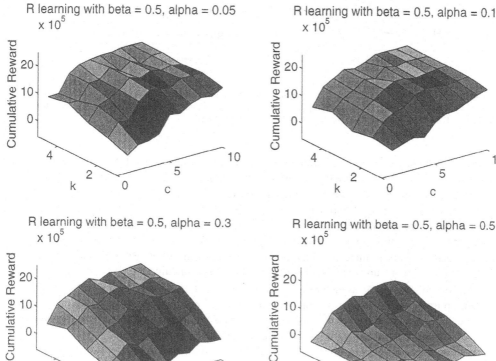

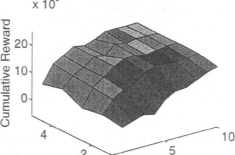

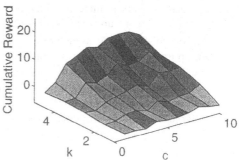

Figure 14. Performance of R-learning on a 20x20 stochastic grid world environment with one-way membranes. Each point represents the median over 30 runs of 500,000 steps. The domain was stochastic with transition probability = 0.75. The x axis plots ten different values of the constant of exploration c varied uniformly over the range [10,100]. Higher values of c mean more exploration. The y axis plots five different values of the constant k controlling the decay of the learning rate β. k was uniformly varied over the range [100,500]. Higher values of k mean more gradual decay of β.

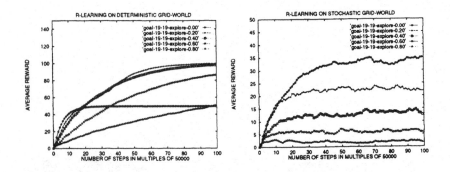

Figure 15. Performance of R-learning for varying amounts of semi-uniform exploration for a deterministic (graph on left) and stochastic (graph on right) grid world domain with no membranes. These graphs show that the best level of exploration depends on the stochasticity of the domain.

that while there are several synchronous and asynchronous algorithms for producing gain-optimal policies, none of these algorithms (including R-learning) can reliably produce bias-optimal policies for a general unichain MDP.

We presented a general optimality metric called *n-discount-optimality* (Veinott, 1969) relating discounted and undiscounted MDP. Gain optimality and bias optimality correspond to the first two terms of this optimality metric. We believe that this metric offers a general framework to understand the relationship between discounted and undiscounted RL, and also opens up some very interesting directions for future work.

We described several asynchronous algorithms that are provably guaranteed to yield gain-optimal policies. We also pointed out a key difference between these algorithms and the natural asynchronous relative value iteration method, namely that the average reward is estimated independently of the relative values, such as by averaging over sample rewards. Since R-learning shares this similarity, it brightens the prospect of proving it converges to yield gain-optimal policies.

4.2. Experimental Results on R-learning

We now discuss two key conclusions that are based on our experimental results. However, we must emphasize the tentative nature of these conclusions at this point. We are planning to undertake a more detailed study of a residual gradient (Baird, 1995) version of R-learning using larger state space problems, which will provide a further test of these hypotheses (Mahadevan & Baird).

Our first observation is that average reward methods are more sensitive to exploration than discounted methods. We showed that R learning can fall into limit cycles in two domains using two different exploration methods. We also showed that by increasing the degree of exploration, R-learning does not get into limit cycles. Finally, we presented a detailed sensitivity study of R-learning by varying the degree of exploration over different learning rates.

Our second observation is that average reward methods can be superior to discounted methods. We showed that R-learning can perform better than Q-learning in two domains, a simulated robot task and a grid world task, even if the two methods were separately optimized. More detailed tests are needed to determine if there are some generic types of domains (for example, ergodic domains) that average reward methods, such as R-learning, are inherently better at than discounted methods such as Q-learning.

Recent work by Tadepalli and Ok (Tadepalli & Ok, 1994) also supports both these conclusions. They compared a *model-based* average reward method called H learning (which is a variant of Jalali and Ferguson's B algorithm (Jalali & Ferguson, 1989) described in Section 2.5) with R learning, Q-learning, and ARTDP (Barto, et al, 1995), using an automated guided vehicle (AGV) task. In their experiments, they primarily used semi-uniform exploration. They found that H learning performs much better as the level of exploration is increased. They also found that in some cases the two average reward methods, namely H learning and R learning, outperformed the two discounted methods, namely ARTDP and Q learning.

5. Directions for Future Research

This paper suggests many promising directions for future research in average reward RL, some of which are already bearing fruit.

- *Bias-Optimal RL Algorithms:* Clearly, one of the most pressing open problems is developing a bias-optimal RL algorithm. The key difference between the algorithms described in this paper and a bias-optimal algorithm is that the latter requires solving two nested optimality equations (Puterman, 1994). Several approaches to the design of a bias-optimal algorithm have been pursued in the DP literature, ranging from policy iteration (Veinott, 1969, Puterman, 1994), linear programming (Denardo, 1970), and value iteration (Federgruen & Schweitzer, 1984). We are currently testing a model-based bias optimality algorithm, which extends Jalali and Ferguson's B algorithm (Mahadevan). We are also working on a model-free bias optimality algorithm, which extends R-learning. We expect that bias-optimal RL algorithms will scale better than bias-optimal DP algorithms.

- *Value Function Approximation in Average Reward MDP:* In this paper, we used a simple table lookup representation for representing policies. In most realistic applications of RL, such as robotics (Mahadevan & Connell, 1992) or games (Tesauro, 1992), the size of the state space rules out table-lookup representations, and necessitates using a function approximator, such as a neural net (Lin, 1993, Tesauro, 1992) or kd-trees (Moore, 1991, Salganicoff, 1993). We have recently developed some theoretical results showing the performance loss that results from using approximate value functions (Mahadevan & Baird). We are also working on a residual gradient (Baird, 1995) version of R-learning.

- *Multi-Chain Average Reward Algorithms:* In this paper we focused mainly on algorithms for unichain MDP's. This allows representing the average reward ρ by a scalar value. However, this assumption may be violated in many applications. An interesting problem is to develop RL algorithms to handle multi-chain MDP's, where different states may have different average rewards. In multi-chain MDP's, an additional optimality equation relating the gain of one state to other states has to be solved.

- *Modular Average Reward Methods:* A number of researchers have studied modular architectures for scaling RL (Dayan & Hinton, 1992, Lin, 1993, Mahadevan, 1992, Mahadevan & Connell, 1992, Singh, 1994a, Whitehead, et al., 1993). Such architectures decompose tasks into different sets of primitive subtasks. Learning primitive subtasks is easier because rewards are more frequent. Also, the solutions learned to subtasks can be transferred to yield faster task learning. All these systems have been based on Q-learning. However, an average reward method such as R-learning has significant advantages over Q-learning for multi-task learning. Average adjusted values satisfy a nice linearity property (Schwartz, 1993) (assuming a deterministic MDP): $V_y^\pi - V_x^\pi = \rho - r$. In other words, if rewards are constant, the difference in relative values across states is also constant. We have empirically found this linearity property to hold up well even in stochastic MDP's. Discounting, on the other hand, causes exponential non-linearities across states. We plan to develop an average-reward based modular architecture for multi-task learning.

Acknowledgements

This paper could not have been completed without the help of the following people, to whom I am indebted. Ronny Ashar implemented R-learning on the stochastic grid world domain. Leemon Baird provided several counter-examples for R-learning. I am especially grateful to the special issue editor, Leslie Kaelbling, for encouraging me to discuss average reward MDP. I thank Prasad Tadepalli for many discussions on average reward reinforcement learning, and for his detailed comments on this paper. This research was supported, in part, by the University of South Florida Research and Creative Scholarship Grant No. 21-08-936-RO, and by a National Science Foundation (NSF) Career Award Grant No. IRI-9501852.

Appendix: Discounted Reinforcement Learning

Most previous work in reinforcement learning has studied a formulation where agents maximize the *discounted* cumulative sum of rewards (Sutton, 1992). The discounted return of a policy π starting from a state x is defined as

$$V_\gamma^\pi(s) = \lim_{N \to \infty} E\left(\sum_{t=0}^{N-1} \gamma^t R_t^\pi(s)\right).$$

where $\gamma \leq 1$ is the discount factor, and $R_t^\pi(s)$ is the reward received at time step t starting from state s and choosing actions using policy π. An *optimal discounted policy* π^* maximizes the above value function over all states x and policies π, i.e. $V_\gamma^{\pi^*}(x) \geq V_\gamma^\pi(x)$.

The action value $Q_\gamma^\pi(x, a)$ denotes the discounted return obtained by performing action a once from state x, and thereafter following policy π

$$Q_\gamma^\pi(x, a) = r(x, a) + \gamma \sum_y P_{xy}(a) V_\gamma^\pi(y),$$

where $V_\gamma^\pi(y) = max_{a \in A} Q(y, a)$, and $r(x, a)$ is the expected reward for doing action a in state x.

5.0.1. Q-learning:

Watkins (Watkins, 1989) proposed a simple iterative method for learning Q values. All the $Q(x, a)$ values are randomly initialized to some value (say 0). At time step t, the learner either chooses the action a with the maximum $Q_t(x, a)$ value, or with some probability selects a random "exploratory" action to ensure that it does not get stuck in a local maximum. If the agent moves from state x to state y, and receives an immediate reward $r_{imm}(x, a)$, the current $Q_t(x, a)$ values are updated using the rule:

$$Q_{t+1}(x, a) \leftarrow Q_t(x, a)(1 - \beta) + \beta \left(r_{imm}(x, y) + \gamma \max_{a \in A} Q_t(y, a) \right) \qquad (2)$$

where $0 \leq \beta \leq 1$ is the learning rate controlling how quickly errors in action values are corrected. Q-learning is guaranteed to asymptotically converge to the optimal discounted policy for a finite MDP. The precise convergence conditions are given in (Tsitsiklis, 1994), but essentially every action must be tried in every state infinitely often, and the learning rate β must be slowly decayed to 0. Q-learning also converges when $\gamma = 1$, if we assume a zero reward absorbing state s which is reachable under all policies, and $Q_t(s, a) = 0$ for all actions a and time t.

Notes

1. The reason both techniques converge to a value slightly below 2 is because the robot takes random actions 5% of the time.

2. If the state space is not finite, we have to allow history dependent policies, since there may be no gain-optimal stationary policy. See (Bertsekas, 1987, Puterman, 1994, Ross, 1983) for some examples.

3. This limit is guaranteed to exist as long as the state space is countable, and we do not allow history dependent policies (Puterman, 1994). For more general policies, this limit need not exist, and two possibly different measures of average reward result from using lim sup and lim inf, respectively.

4. This limit assumes that all policies are aperiodic. For periodic policies, we need to use the Cesaro limit
$$V^\pi(s) = \lim_{N \to \infty} \frac{\sum_{k=0}^{N-1} E\left(\sum_{t=0}^k (R_t^\pi(s) - \rho^\pi) \right)}{N}.$$

5. A mapping T is a contraction mapping (w.r.t. the maximum norm) if and only if there exists a real number $0 < \delta < 1$ such that $\|T(V)(x) - T(V')(x)\| \leq \delta \|V(x) - V'(x)\|$, where $\|f(x)\| = \max_x |f(x)|$, and $V(x)$ and $V'(x)$ are any two real-valued bounded functions on S.

6. This counter-example implies that Schwartz's remark in his paper (Schwartz, 1993) – "when R-learning converges, it must produce a T-optimal policy" – is not correct for unichain MDP's.
7. Drastically compressing sensor information can easily make robot tasks non-Markovian. However, note here that the robot does not have to discriminate between boxes and non-boxes, but only avoid hitting anything and keep moving forward. Using all the sensor data would require using function approximation, which could itself create a non-Markovian problem.
8. To simplify the argument, we are assuming synchronous updating, but the results hold even for asynchronous updating.

References

Baird, L. Personal Communication.
Baird, L., (1995). Residual algorithms: Reinforcement learning with function approximation. In *Proceedings of the 12th International Conference on Machine Learning*, pages 30–37. Morgan Kaufmann.
Barto, A., Bradtke, S. & Singh, S., (1995). Learning to act using real-time dynamic programming. *Artificial Intelligence*, 72(1):81–138.
Bertsekas, D., (1982). Distributed dynamic programming. *IEEE Transactions on Automatic Control*, AC-27(3).
Bertsekas, D., (1987). *Dynamic Programming: Deterministic and Stochastic Models*. Prentice-Hall.
Blackwell, D., (1962). Discrete dynamic programming. *Annals of Mathematical Statistics*, 33:719–726.
Boutilier, C. & Puterman, M., (1995). Process-oriented planning and average-reward optimality. In *Proceedings of the Fourteenth JCAI*, pages 1096–1103. Morgan Kaufmann.
Dayan, P. & Hinton, G., (1992). Feudal reinforcement learning. In *Neural Information Processing Systems (NIPS)*, pages 271–278.
Denardo, E., (1970). Computing a bias-optimal policy in a discrete-time Markov decision problem. *Operations Research*, 18:272–289.
Dent, L., Boticario, J., McDermott, J., Mitchell, T. & Zabowski, D., (1992). A personal learning apprentice. In *Proceedings of the Tenth National Conference on Artificial Intelligence (AAAI)*, pages 96–103. MIT Press.
Engelberger, J., (1989). *Robotics in Service*. MIT Press.
Federgruen, A. & Schweitzer, P., (1984). Successive approximation methods for solving nested functional equations in Markov decision problems. *Mathematics of Operations Research*, 9:319–344.
Haviv, M. & Puterman, M., (1991) An improved algorithm for solving communicating average reward markov decision processes. *Annals of Operations Research*, 28:229–242.
Hordijk, A. & Tijms, H., (1975). A modified form of the iterative method of dynamic programming. *Annals of Statistics*, 3:203–208.
Howard, R., (1960). *Dynamic Programming and Markov Processes*. MIT Press.
Jalali, A. & Ferguson, M., (1989). Computationally efficient adaptive control algorithms for Markov chains. In *Proceedings of the 28th IEEE Conference on Decision and Control*, pages 1283–1288.
Jalali, A. & Ferguson, M., (1990). A distributed asynchronous algorithm for expected average cost dynamic programming. In *Proceedings of the 29th IEEE Conference on Decision and Control*, pages 1394–1395.
Kaelbling, L., (1993a). Hierarchical learning in stochastic domains: Preliminary results. In *Proceedings of the Tenth International Conference on Machine Learning*, pages 167–173. Morgan Kaufmann.
Kaelbling, L., (1993b) *Learning in Embedded Systems*. MIT Press.
Lin, L., (1993). *Reinforcement Learning for Robots using Neural Networks*. PhD thesis, Carnegie-Mellon Univ.
Mahadevan, S. A model-based bias-optimal reinforcement learning algorithm. In preparation.
Mahadevan, S., (1992). Enhancing transfer in reinforcement learning by building stochastic models of robot actions. In *Proceedings of the Seventh International Conference on Machine Learning*, pages 290–299. Morgan Kaufmann.
Mahadevan, S., (1994). To discount or not to discount in reinforcement learning: A case study comparing R-learning and Q-learning. In *Proceedings of the Eleventh International Conference on Machine Learning*, pages 164–172. Morgan Kaufmann.
Mahadevan, S. & Baird, L. Value function approximation in average reward reinforcement learning. In preparation.

Mahadevan, S. & Connell, J., (1992). Automatic programming of behavior-based robots using reinforcement learning. *Artificial Intelligence*, 55:311–365. Appeared originally as IBM TR RC16359, Dec 1990.

Moore, A., (1991). Variable resolution dynamic programming: Efficiently learning action maps in multivariate real-valued state spaces. In *Proceedings of the Eighth International Workshop on Machine Learning*, pages 333–337. Morgan Kaufmann.

Narendra, K. & Thathachar, M., (1989). *Learning Automata: An Introduction*. Prentice Hall.

Puterman, M., (1994). *Markov Decision Processes: Discrete Dynamic Stochastic Programming*. John Wiley.

Ross, S., (1983). *Introduction to Stochastic Dynamic Programming*. Academic Press.

Salganicoff, M., (1993). Density-adaptive learning and forgetting. In *Proceedings of the Tenth International Conference on Machine Learning*, pages 276–283. Morgan Kaufmann.

Schwartz, A., (1993). A reinforcement learning method for maximizing undiscounted rewards. In *Proceedings of the Tenth International Conference on Machine Learning*, pages 298–305. Morgan Kaufmann.

Singh, S., (1994a). *Learning to Solve Markovian Decision Processes*. PhD thesis, Univ of Massachusetts, Amherst.

Singh, S., (1994b) Reinforcement learning algorithms for average-payoff Markovian decision processes. In *Proceedings of the 12th AAAI*. MIT Press.

Sutton, R., (1988). Learning to predict by the method of temporal differences. *Machine Learning*, 3:9–44.

Sutton, R., (1990). Integrated architectures for learning, planning, and reacting based on approximating dynamic programming. In *Proceedings of the Seventh International Conference on Machine Learning*, pages 216–224. Morgan Kaufmann.

Sutton, R., editor, (1992). *Reinforcement Learning*. Kluwer Academic Press. Special Issue of Machine Learning Journal Vol 8, Nos 3-4, May 1992.

Tadepall, P. Personal Communication.

Tadepalli, P. & Ok, D., (1994). H learning: A reinforcement learning method to optimize undiscounted average reward. Technical Report 94-30-01, Oregon State Univ.

Tesauro, G., (1992). Practical issues in temporal difference learning. In R. Sutton, editor, *Reinforcement Learning*. Kluwer Academic Publishers.

Thrun, S. The role of exploration in learning control. In D. A. White and D. A. Sofge, editors, *Handbook of Intelligent Control: Neural, Fuzzy, and Adaptive Approaches*. Van Nostrand Reinhold.

Tsitsiklis, J., (1994). Asynchronous stochastic approximation and Q-learning. *Machine Learning*, 16:185–202.

Veinott, A., (1969) Discrete dynamic programming with sensitive discount optimality criteria. *Annals of Mathematical Statistics*, 40(5):1635–1660.

Watkins, C., (1989). *Learning from Delayed Rewards*. PhD thesis, King's College, Cambridge, England.

Wheeler, R. & Narendra, K., (1986). Decentralized learning in finite Markov chains. *IEEE Transactions on Automatic Control*, AC-31(6)

White, D., (1963). Dynamic programming, markov chains, and the method of successive approximations. *Journal of Mathematical Analysis and Applications*, 6:373–376.

Whitehead, S., Karlsson, J. & Tenenberg, J., (1993). Learning multiple goal behavior via task decomposition and dynamic policy merging. In J. Connell and S. Mahadevan, editors, *Robot Learning*. Kluwer Academic Publishers.

Received November 15, 1994
Accepted February 24, 1995
Final Manuscript October 4, 1995

Machine Learning, 22, 197–225 (1996)

The Loss from Imperfect Value Functions in Expectation-Based and Minimax-Based Tasks

MATTHIAS HEGER   heger@informatik.uni-bremen.de

Zentrum für Kognitionswissenschaften, Universität Bremen, FB3 Informatik, Postfach 330 440, 28334 Bremen, Germany

Editor: Leslie Pack Kaelbling

Abstract. Many reinforcement learning (RL) algorithms approximate an optimal value function. Once the function is known, it is easy to determine an optimal policy. For most real-world applications, however, the value function is too complex to be represented by lookup tables, making it necessary to use function approximators such as neural networks. In this case, convergence to the optimal value function is no longer guaranteed and it becomes important to know to which extent performance diminishes when one uses approximate value functions instead of optimal ones. This problem has recently been discussed in the context of expectation-based Markov decision problems. Our analysis generalizes this work to minimax-based Markov decision problems, yields new results for expectation-based tasks, and shows how minimax-based and expectation-based Markov decision problems relate.

Keywords: Reinforcement Learning, Dynamic Programming, Performance Bounds, Minimax Algorithms, Q-Learning

1. Introduction

Reinforcement learning (RL) is learning to solve decision problems from experience: an agent interacts with its environment (world) and receives reinforcement signals as punishments for its actions. Its task is to find a behavior that minimizes the punishment, specified as rules that tell the agent which action to choose in every possible situation. (The reinforcement signals can also be interpreted as rewards. This is equivalent to their interpretation as punishments, except that the agent now has to maximize the reward.)

The interaction of the agent with its environment is often modeled as a special kind of stochastic process, namely a Markov decision process (MDP). S is the set of states of the environment as the agent perceives it, A the set of actions that the agent has available, and C the set of scalar reinforcement signals, which we call immediate costs because they can be interpreted as the effort required to execute the actions. In this paper, we limit ourselves to finite sets S and A that are subsets of $\mathcal{N}$. We assume that the elements of $C \subset \mathcal{R}$ are countable, bounded, and nonnegative. Furthermore, the agent does not have all actions available in every state. The nonempty sets $A(i) \subseteq A$ denote the set of admissible actions in states i.

The interaction of the agent with its environment takes place in episodes, where episodes correspond to time steps: First, the agent observes the *starting state* $i \in S$. It then has to execute an *action* $a \in A(i)$, which causes a state transition from state i to a *successor state* $j \in S$. Finally, the agent receives the *reinforcement signal* $r \in C$.

MDPs make the important assumption that the probability distribution over the successor states depends only on the starting state i and the executed action a but not on t or any previous episodes. $P_S(i, a, j)$ denotes the probability that the successor state of an episode is j if action a is executed in starting state s. Similarly, MDPs assume that the probability that the immediate cost of episode t equals a given number $r \in C$ depends only on the starting state, the executed action, and the successor state of that episode, but neither on t nor past episodes. $P_C(i, a, j, r)$ denotes the probability that the immediate cost of an episode is r for a given starting state i, action a, and successor state j. Furthermore, MDPs assume that the starting state of an episode is identical to the successor state of the previous episode. If the MDP is deterministic, then the successor state j_i^a and the immediate cost c_i^a are uniquely determined by the starting state i and the executed action a.

The behavior of the agent is specified by a *policy*, which is a mapping from situations to actions. In general, the action could depend on the current time, the current state of the agent, and all previous states and actions. Furthermore, the action could be selected probabilistically. *Stationary policies* determine which action to execute based only on the current state of the agent, but prove to be very powerful.

The states and immediate costs are random variables because of the probabilistic nature of the immediate costs and state transitions. These random variables essentially depend on the policy that the agent follows. Therefore, we use the following notation: I_t^π denotes the starting state and C_t^π the immediate cost, where π and t describe the policy of the agent and the time index of the episode, respectively.

There are two problems that make it difficult to define what an optimal policy is. The first problem is often called *delayed reinforcement*, because actions may reveal their consequences only many time steps after they have been executed. Hence, the action that has the lowest immediate cost is not necessarily best from a global point of view. Therefore, one should consider all immediate costs, including the future costs, which is usually done by evaluating policies according to their return

$$R_\gamma^\pi = \sum_{\tau=0}^{\infty} \gamma^\tau C_\tau^\pi$$

where $0 \leq \gamma < 1$ is a discount factor that keeps the sum finite and has the effect that costs are weighted less if they are obtained farther into the future.

In deterministic domains, the return of a policy can already be used to measure its performance. In this case, a policy is optimal if it minimizes the return. Unfortunately, there is a second problem, since the return is usually not a real number, but a random variable. In probabilistic domains, using the same policy repeatedly in the same starting state can result in different returns. The common way to handle this problem is to use the expected value of the return as performance measure (Barto, Sutton & Anderson, 1983; Watkins, 1989). According to this so-called expected value criterion, a policy is optimal if it minimizes the expected value of the return. In operations research and decision theory, however, it is well known that it is not always reliable and can even be very misleading to use the expected value as decision criterion (e.g., Taha, 1987). In (Heger, 1994a), Heger presented dynamic programming algorithms and a RL algorithm called

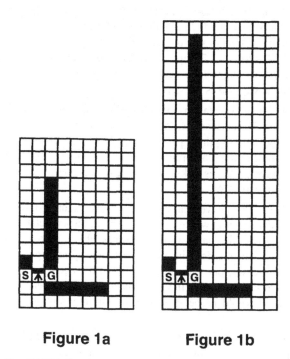

Figure 1a   Figure 1b

Figure 1. Two Grid Worlds where Expectation-Optimal and Minimax-Optimal Policies Differ

Q-hat-learning that are based on the minimax criterion. According to this criterion, a policy is optimal if it minimizes the worst-case return. The main contribution of this paper is to present new results for minimax-based tasks. The following two examples are intended to convince the reader that it is worthwhile to consider the minimax criterion as an alternative to the expected-value criterion.

In the stochastic domain of Figure 1a, the minimax-optimal policy differs from the expectation-optimal one. An agent (robot) has to navigate on the grid from location S (start) to G (goal). Dark squares represent obstacles; so does the perimeter of the grid. In each square, the agent has the four actions UP, DOWN, LEFT, and RIGHT available that take it to the corresponding adjacent square, unless the square is occupied by an obstacle, in which case the agent does not change location. Every state transition has an immediate cost of one except for state transitions from the goal state into itself, which have cost zero. The square with the arrow is a probabilistic one-way door. It can be passed through only in the direction of the arrow and behaves like an obstacle when approached from the opposite direction. If the agent is on the one-way door square, then, with probability p, it moves through the door, no matter which action it executes. With probability $1 - p$, however, the agent behaves as if it were on an empty square.

Consider the case where $p = 0.75$. When following the expectation-optimal policy π, the agent has to go RIGHT in square S and hence it moves to the one-way door square, where it has to select RIGHT again. The agent reaches its goal in the next step only with probability 0.25 because with probability 0.75 it falls through the one-way door. In the latter case, it needs 18 additional steps to move around the vertical obstacle to the goal. Hence it needs $0.25 \cdot 2 + 0.75 \cdot 20 = 15.5$ steps on average to reach the goal from the start. The minimax-optimal policy suggests to go DOWN in square S and move around the horizontal obstacle. This always takes 16 steps.

This example demonstrates that expectation-optimal policies do not necessarily have the lowest return in most cases. If, for example, the above experiment is repeated a large number of times then the expectation-optimal policy needs 25% more cost units in 75% of all cases when compared to the minimax-optimal policy. The latter policy minimizes the worst-case return and hence produces risk-avoiding behavior: the agent avoids the risk of falling through the one-way door.

In the second example (Figure 1b), the vertical obstacle is eleven squares longer than in the previous example. Consider the case where $p = 1/3$. Again, the expectation-optimal policy tells the agent to go across the one-way door because this requires on average $\frac{2 \cdot 2}{3} + \frac{42 \cdot 1}{3} = 15\frac{1}{3} < 16$ steps. In this domain, the expectation-optimal policy has the lowest return (2 steps) in most cases, but with probability $1/3$ the agent has to follow a very long path around the vertical obstacle. If the agent does not have enough energy for this long path (42 steps), it would be better to use the minimax-optimal policy (16 steps).

This example demonstrates that the expected value criterion is not reliable, especially when solving tasks with resource constraints. Resource constraints cannot be addressed by giving the agent a big punishment (i.e., high immediate cost) when its accumulated cost has exceeded the given limit, because these reinforcement signals depend on the history of the agent's interaction with its environment and, consequently, do not satisfy the Markov property – unless the state representation is augmented with information about the accumulated cost. This, however, would increase the size of the state space to an extent that makes efficient learning impossible. Fortunately, tasks for which the total cost is not allowed to exceed a given limit can be solved using the minimax criterion, because the minimax-optimal policy is guaranteed to obey this constraint if this is possible.

Recently, additional minimax-based reinforcement learning algorithms have been published. The model-based parti-game algorithm (Moore & Atkeson, 1995) uses the minimax criterion for navigation tasks in conjunction with an adaptive state-action space representation. Using the minimax criterion has the advantage that it makes the planning algorithm fast, allows it to use a simple world model, and automatically detects if a planning problem does not have a "safe" solution (in which case the partigame algorithm changes the state-action space). Littman uses the minimax-criterion for Markov games (Littman, 1994) that have two kinds of uncertainties: the transition uncertainty (as commonly represented by MDPs) and, additionally, uncertainty introduced by an opponent who does not base his decisions on well defined stationary probabilities. To deal with the latter uncertainty, Littman uses the minimax criterion as is common in game theory.

Before we formalize and analyze the problem addressed in this article, we describe the theory behind Q-learning and Q-hat-learning in Sections 3 and 4, respectively. First, however, we introduce our notation and give an overview of the commonly used RL definitions and theorems.

2. Basic Definitions and Theorems

We define the *set of admissible state-action pairs* to be

$$M := \{(i,a) \in S \times A : a \in A(i)\}$$

and call the elements of $\mathcal{R}^M$ state-action value functions or, synonymously, Q-functions. Assume that a function $Q \in \mathcal{R}^M$ and a state $i \in S$ are given. Then, we call an action

$$a \in A(i) \ \textit{greedy for i and Q} \ \text{iff} \ a = \arg \min_{b \in A(i)} Q(i,b).$$

Furthermore, we call a stationary policy

$$\pi \ \textit{greedy for Q} \ \text{iff} \ \forall i \in S : \pi(i) = \arg \min_{a \in A(i)} Q(i,a).$$

Finally, we call a function $Q \in \mathcal{R}^M$ *greedy-equivalent* to another function $Q' \in \mathcal{R}^M$ iff

$$\{a \in A(i) : a \ \text{is greedy for } i \text{ and } Q\} = \{a \in A(i) : a \ \text{is greedy for } i \text{ and } Q' \}$$

for all $i \in S$. Let $\epsilon > 0$, $i \in S$, $a \in A(i)$, and $Q \in \mathcal{R}^M$. We say that a is ϵ-*greedy with respect to i and Q* iff $|Q(i,a) - \min_{b \in A(i)} Q(i,b)| \leq \epsilon$.

We define a partial order on Q-functions as follows: $Q \leq Q'$ iff $Q(i,a) \leq Q'(i,a)$ for all $i \in S$ and $a \in A(i)$. For a given real number r and a Q-function Q we write $Q + r$ and $r \cdot Q$ to denote the Q-functions Q' and Q'' that satisfy $Q'(i,a) = Q(i,a) + r$ and $Q''(i,a) = r \cdot Q(i,a)$, respectively, for all $i \in S$ and $a \in A(i)$. We write $Q \leq r$ iff $Q(i,a) \leq r$ for all $i \in S$ and $a \in A(i)$. We use the maximum norm $\|\cdot\|$ for Q-functions, i.e.,

$$\|Q\| = \max_{i \in S, a \in A(i)} |Q(i,a)|.$$

An element of $\mathcal{R}^S$ is called a (state) value function. For a given real number r and a value function V we write $V + r$ and $r \cdot V$ to denote the value function V' and V'' that satisfy $V'(i) = V(i) + r$ and $V''(i) = r \cdot V(i)$, respectively, for all $i \in S$. We write $V \leq r$ iff $V(i) \leq r$ for all $i \in S$. We use the maximum norm $\|\cdot\|$ for value functions, i.e.,

$$\|V\| = \max_{i \in S} |V(i)|.$$

For a given Q-function Q and stationary policy π we define the value functions $V_{Q,\pi}$ and V_Q to be

$$\forall i \in S : V_{Q,\pi}(i) = Q(i, \pi(i)) \ \text{and} \ V_Q(i) = \min_{a \in A(i)} Q(i,a).$$

respectively.

In the following, we state six theorems that are used to prove the results of Section 5. The first three theorems follow immediately from the definitions of this section.

THEOREM 1 *Let Q be a Q-function, π a stationary policy, and $r \in \mathcal{R}$. Then, $V_{Q+r} = V_Q + r$ and $V_{Q+r,\pi} = V_{Q,\pi} + r$.*

THEOREM 2 *Let Q and Q' be Q-functions with $Q \leq Q'$ and let π be any stationary policy. Then, $V_Q \leq V_{Q'}$ and $V_{Q,\pi} \leq V_{Q',\pi}$.*

THEOREM 3 *Let π be a stationary policy and let Q be a Q-function. Then π is greedy for Q iff $V_{Q,\pi} = V_Q$.*

The following theorem is proven in (Williams & Baird, 1993):

THEOREM 4 *Let g_1 and g_2 be real-valued functions on a compact domain U. Then*

$$\left| \max_{u \in U} g_1(u) - \max_{u \in U} g_2(u) \right| \leq \max_{u \in U} |g_1(u) - g_2(u)|.$$

THEOREM 5 *Let g_1 and g_2 be real-valued functions on a compact domain U. Then*

$$\left| \min_{u \in U} g_1(u) - \min_{u \in U} g_2(u) \right| \leq \max_{u \in U} |g_1(u) - g_2(u)|.$$

Proof: Let $u_1 = \arg\min_{u \in U} g_1(u)$ and $u_2 = \arg\min_{u \in U} g_2(u)$. Consider first the case where $g_1(u_1) \geq g_2(u_2)$. Then

$$\left| \min_{u \in U} g_1(u) - \min_{u \in U} g_2(u) \right| = g_1(u_1) - g_2(u_2)$$
$$\leq g_1(u_2) - g_2(u_2)$$
$$\leq \max_{u \in U} |g_1(u) - g_2(u)|.$$

A symmetrical argument establishes the same result for $g_2(u_2) \geq g_1(u_1)$. ∎

THEOREM 6 *For any stationary policy π and arbitrary Q-functions Q and Q' it holds that $\|V_{Q,\pi} - V_{Q',\pi}\| \leq \|Q - Q'\|$ and $\|V_Q - V_{Q'}\| \leq \|Q - Q'\|$.*

Proof: For any state i,

$$|V_{Q,\pi}(i) - V_{Q',\pi}(i)| = |Q(i,\pi(i)) - Q'(i,\pi(i))| \leq \|Q - Q'\|.$$

Since this holds for all states, the first inequality follows. For the second inequality, consider an arbitrary state i. Applying Theorem 5 yields

$$|V_Q(i) - V_{Q'}(i)| = \left| \min_{a \in A(i)} Q(i,a) - \min_{a \in A(i)} Q'(i,a) \right|$$
$$\leq \max_{a \in A(i)} |Q(i,a) - Q'(i,a)| \leq \|Q - Q'\|.$$

and the desired result follows. ∎

3. Expectation-Based Markov Decision Tasks

For expectation-based Markov decision tasks, one uses the value function $V_\gamma^\pi \in \mathcal{R}^S$ to measure the performance of a given policy π for a given MDP. The value function is defined as follows:

$$\forall i \in S: \quad V_\gamma^\pi(i) = E\left(R_\gamma^\pi \mid I_0^\pi = i\right),$$

i.e., $V_\gamma^\pi(i)$ is the return one expects if the agent starts in state i and uses policy π. A policy is called optimal if its value function is equal to the optimal value function $V_\gamma^* \in \mathcal{R}^S$ that is defined by

$$\forall i \in S: \quad V_\gamma^*(i) = \inf_\pi V_\gamma^\pi(i).$$

3.1. Dynamic Programming Operators

A fundamental result from the theory of dynamic programming states that there always exists an optimal policy that is stationary. In dynamic programming, one approximates value functions of policies and optimal value functions by applying dynamic programming operators repeatedly. The dynamic programming operator $T_\gamma : \mathcal{R}^S \to \mathcal{R}^S$ is a mapping from value functions to value functions, defined by

$$\forall V \in \mathcal{R}^S \, \forall i \in S: \quad T_\gamma V(i) = \min_{a \in A(i)} \left[R(i, a) + \gamma \cdot \sum_{j \in S} P_S(i, a, j) V(j) \right]$$

where $R(i, a)$ denotes the expected immediate cost for a given starting state i and executed action a.[1] For convenience we write $T_\gamma V(i)$ instead of $[T_\gamma(V)](i)$ and proceed similarly for other dynamic programming operators defined below. For a given stationary policy π, the dynamic programming operator $T_\gamma^\pi : \mathcal{R}^S \to \mathcal{R}^S$ is a mapping from value functions to value functions, defined by

$$\forall V \in \mathcal{R}^S \, \forall i \in S: \quad T_\gamma^\pi V(i) = R(i, \pi(i)) + \gamma \cdot \sum_{j \in S} P_S(i, \pi(i), j) V(j).$$

A policy is called greedy for a value function V if

$$\forall i \in S: \quad \pi(i) = \arg\min_{a \in A(i)} \left[R(i, a) + \gamma \cdot \sum_{j \in S} P_S(i, a, j) \cdot V(j) \right].$$

The following three theorems are known from the theory of dynamic programming (see, e.g., Bertsekas (1987)):

THEOREM 7 *Let $V_0 \in \mathcal{R}^S$ and $V_{k+1} = T_\gamma V_k$ for all $i \in S$ and $k \in \mathbb{N}$. Then $\lim_{k \to \infty} \|V_k - V_\gamma^*\| = 0$.*

THEOREM 8 *Let $\pi \in A^S$ be a stationary policy, $V_0 \in \mathcal{R}^S$, and $V_{k+1} = T_\gamma^\pi V_k$ for all $i \in S$ and $k \in \mathbb{N}$. Then $\lim_{k \to \infty} \|V_k - V_\gamma^\pi\| = 0$.*

THEOREM 9 *A stationary policy π is optimal iff it is greedy for V_γ^*.*

Theorem 7 yields an algorithm for computing the optimal value function, and Theorem 8 yields a similar algorithm for computing the value function of a given policy. Theorem 9 shows that optimal policies can be determined easily from a given optimal state value function, which is the reason why learning algorithms often approximate the optimal value function. To be able to compute a greedy policy for a given value function, one needs to know a world model, in this case the probabilities for the state transitions and the associated expected immediate costs.

We conclude this section by defining the dynamic programming operators B_γ and B_γ^π for stationary policies π. These operators are associated with Q-functions and the expected-value criterion, and will be used in Section 5. $B_\gamma : \mathcal{R}^M \to \mathcal{R}^M$ satisfies

$$B_\gamma Q(i, a) = R(i, a) + \gamma \cdot \sum_{j \in S} P_S(i, a, j) V_Q(j)$$

for any Q-function Q, state i, and action $a \in A(i)$. $B_\gamma^\pi : \mathcal{R}^M \to \mathcal{R}^M$ is defined by

$$B_\gamma^\pi Q(i, a) = R(i, a) + \gamma \cdot \sum_{j \in S} P_S(i, a, j) V_{Q, \pi}(j)$$

for any Q-function Q, state i, and action $a \in A(i)$.

In the next section, we introduce a reinforcement learning algorithm that learns a Q-function with the property that every policy that is greedy for this Q-function is optimal according to the expected value criterion. Greedy policies can be obtained from Q-functions even if there is no world model available.

3.2. Q-Learning

For a stationary policy π, we define the Q-function Q_γ^π by

$$Q_\gamma^\pi(i, a) = R(i, a) + \gamma \cdot \sum_{j \in S} P_S(i, a, j) \cdot V_\gamma^\pi(j)$$

for all $i \in S$ and $a \in A(i)$. The Q-learning algorithm (Watkins, 1989) approximates the optimal Q-function Q_γ^* that is defined by

$$Q_\gamma^*(i, a) = R(i, a) + \gamma \cdot \sum_{j \in S} P_S(i, a, j) \cdot V_\gamma^*(j).$$

From Theorem 9, we conclude that a stationary policy is optimal iff it is greedy for Q_γ^*. The Q-learning algorithm begins with an initial estimate Q_0 of Q_γ^* and improves this estimate as the results of individual actions become apparent. After each episode

t, the Q-value associated with the starting state i and executed action a is updated as follows:

$$Q_{t+1}(i, a) = \alpha_t(i, a)(r + \gamma V_{Q_t}(j)) + (1 - \alpha_t(i, a)) Q_t(i, a)$$

where j is the successor state reached and r is the immediate cost of the episode. The learning rates $\alpha_t(i, a)$ have to satisfy

$$\sum_{t=0}^{\infty} \alpha_t(i, a) = \infty; \quad \sum_{t=0}^{\infty} \alpha_t(i, a)^2 < \infty.$$

In (Watkins & Dayan, 1992) and recently in (Tsitsiklis, 1994) it is proven that the Q-learning algorithm converges to Q_γ^* if every action that is admissible in a state is executed infinitely often in that state.

4. Minimax-Based Markov Decision Tasks

The theory behind minimax-based Markov decision tasks is very similar to the one behind expectation-based tasks, as will become apparent in the following. We therefore use the same notation where appropriate, which also simplifies our presentation of a unified analysis of both kinds of Markov decision tasks in a later section. From the context, it is always obvious which kind of task we are referring to.

The definitions in this section do not depend on the earlier definitions for expectation-based Markov decision tasks, which avoids cyclic definitions. For minimax-based Markov decision tasks, we use the value function $V_\gamma^\pi \in \mathcal{R}^S$ to measure the performance of a given policy π for a given MDP. The value function is defined as follows:

$$\forall i \in S : V_\gamma^\pi(i) = \sup \left\{ r \in \mathcal{R} : P\left(R_\gamma^\pi > r \mid I_0^\pi = i\right) > 0 \right\},$$

i.e., $V_\gamma^\pi(i)$ is the worst-case return that can possibly occur if the agent starts in state i and uses policy π. A policy is called optimal if its value function is equal to the optimal value function $V_\gamma^* \in \mathcal{R}^S$ that is defined by

$$\forall i \in S : \quad V_\gamma^*(i) = \inf_\pi V_\gamma^\pi(i).$$

4.1. Dynamic Programming Operators

From (Heger, 1994b) we obtain the result that there always exists an optimal policy that is stationary. Let

$$c(i, a, j) = \sup \{r \in C : P_C(i, a, j, r) > 0\}$$

denote the worst-case immediate cost that can be obtained for the state transition from i to j under action a, and let

$$N(i, a) = \{j \in S : P_S(i, a, j) > 0\}$$

be the set of possible successor states when executing action a in state i. The dynamic programming operator $T_\gamma : \mathcal{R}^S \to \mathcal{R}^S$ is a mapping from value functions to value functions, defined by

$$\forall V \in \mathcal{R}^S \ \forall i \in S : \ T_\gamma V(i) = \min_{a \in A(i)} \max_{j \in N(i,a)} [c(i,a,j) + \gamma \cdot V(j)]. \tag{1}$$

For a given stationary policy π, the dynamic programming operator $T_\gamma^\pi : \mathcal{R}^S \to \mathcal{R}^S$ is a mapping from value functions to value functions, defined by

$$\forall V \in \mathcal{R}^S \ \forall i \in S : \ T_\gamma^\pi V(i) = \max_{j \in N(i,\pi(i))} [c(i,\pi(i),j) + \gamma \cdot V(j)]. \tag{2}$$

A policy is called greedy for a value function V if

$$\forall i \in S : \ \pi(i) = \arg \min_{a \in A(i)} \left[\max_{j \in N(i,a)} (c(i,a,j) + \gamma \cdot V(j)) \right].$$

The following three theorems are proven in (Heger, 1994b):

THEOREM 10 *Let $\pi \in A^S$ be a stationary policy, $V_0 \in \mathcal{R}^S$, and $V_{k+1} = T_\gamma V_k$ for all $i \in S$ and $k \in \mathbb{N}$. Then $\lim_{k \to \infty} \|V_k - V_\gamma^*\| = 0$.*

THEOREM 11 *Let $\pi \in A^S$ be a stationary policy, $V_0 \in \mathcal{R}^S$, and $V_{k+1} = T_\gamma^\pi V_k$ for all $i \in S$ and $k \in \mathbb{N}$. Then $\lim_{k \to \infty} \|V_k - V_\gamma^\pi\| = 0$.*

THEOREM 12 *A stationary policy π is optimal iff it is greedy for V_γ^*.*

The first two theorems yield algorithms for computing the optimal value function and the value function of a given policy, respectively. Theorem 12 shows that optimal policies can be determined easily from a given optimal value function. To be able to compute a greedy policy for a given value function, one needs to know a world model, in this case the sets of possible successor states for all state-action pairs and the worst immediate costs which the action executions can result in. Note that this world model is, in general, much simpler than the corresponding world model needed for expectation-based Markov decision tasks.

We conclude this section by defining the dynamic programming operators B_γ and B_γ^π for stationary policies π. These operators are associated with Q-functions and the minimax criterion, and will be used in Section 5. $B_\gamma : \mathcal{R}^M \to \mathcal{R}^M$ satisfies

$$B_\gamma Q(i,a) = \max_{j \in N(i,a)} [c(i,a,j) + \gamma \cdot V_Q(j)]$$

for any Q-function Q, state i, and action $a \in A(i)$. $B_\gamma^\pi : \mathcal{R}^M \to \mathcal{R}^M$ is defined by

$$B_\gamma^\pi Q(i,a) = \max_{j \in N(i,a)} [c(i,a,j) + \gamma \cdot V_{Q,\pi}(j)]$$

for any Q-function Q, state i, and action $a \in A(i)$.

In the next section, we introduce a reinforcement learning algorithm that learns a Q-function with the property that every policy that is greedy for this Q-function is optimal according to the minimax criterion. Its advantage is, as in the case of expectation-based Markov decision tasks, that one does not need to know a world model to compute a policy that is greedy for a given Q-function.

4.2. Q-Hat-Learning

For a stationary policy π, we define the Q-function Q_γ^π by

$$Q_\gamma^\pi(i,a) = \max_{j \in N(i,a)} \left[c(i,a,j) + \gamma \cdot V_\gamma^\pi(j) \right]$$

for all $i \in S$ and $a \in A(i)$. The Q-hat-learning algorithm (Heger, 1994a) approximates the optimal Q-function Q_γ^* that is defined by

$$Q_\gamma^*(i,a) = \max_{j \in N(i,a)} \left[c(i,a,j) + \gamma \cdot V_\gamma^*(j) \right].$$

From Theorem 12, we conclude that a stationary policy is optimal iff it is greedy for Q_γ^*. The Q-hat-learning algorithm begins with an initial estimate $Q_0 \le Q_\gamma^*$ and improves this estimate as the results of individual actions become apparent. After each episode t, the Q-value associated with the starting state i and executed action a is updated as follows:

$$Q_{t+1}(i,a) = \max \left\{ Q_t(i,a), \; r + \gamma V_{Q_t}(j) \right\}$$

where j is the successor state reached and r is the immediate cost of the episode. Note that the Q-hat-learning algorithm has no need for a learning rate, and that the Q-values are monotonely increasing in time. In (Heger, 1995) it is proven that the Q-hat-learning algorithm converges to Q_γ^* if every action that is admissible in a state is executed infinitely often in that state.

An advantage of the Q-hat-learning algorithm is that the exploration-exploitation problem (e.g., Thrun, 1992) for Q-hat-learning is not as severe as the one for Q-learning: Let $\epsilon > 0$ be a constant and assume that every state becomes infinitely often a starting state. Assume further that the agent always selects an action from the set of actions that are ϵ-greedy with respect to its current Q-function and starting state, and that it selects an action from this set with uniform probability. Then, the Q-hat-learning algorithm converges, with probability one, to a Q-function $\hat{Q}$ that is greedy-equivalent to Q_γ^*, and $\hat{Q}(i,a) = Q_\gamma^*(i,a)$ with probability one for every state $i \in S$ and action $a \in A(i)$ that is greedy for i and Q_γ^*.

5. The Loss from Imperfect Value Functions

In the previous two sections, we have seen that expectation-based and minimax-based tasks have in common that optimal policies can be determined by finding policies that are greedy for the optimal value function. This is the reason why many dynamic programming approaches, such as the famous Q-learning algorithm and its counterpart for the minimax criterion, the Q-hat-learning algorithm, are only concerned with finding an optimal value function.

The question remains to which extent the performance of the agent differs from its optimal performance if it uses a policy that is greedy for an *approximation* of the optimal value function instead of a policy that is greedy for the optimal value function itself;

see for example (Singh & Yee, 1994) and (Williams & Baird, 1993). Knowing the magnitude of this loss in performance is important for at least two reasons: First, dynamic programming and RL algorithms iteratively improve an approximation of the optimal value function. The value function obtained after a finite number of iterations is, in general, only an approximation of the optimal value function. Second, one cannot use lookup tables to represent value functions in complex domains, because there are too many states. This makes it necessary to use parametric function approximators, such as neural networks, that have the ability to generalize. Since function approximators cannot guarantee that they generalize correctly, they can only be expected to approximate the optimal value function.

In practice, one can therefore only expect to obtain approximations of optimal value functions. If small deviations of these approximations from the optimal value functions resulted in an arbitrarily bad performance of the agent, this would raise significant concerns about the use of function approximators in dynamic programming-based learning.

5.1. Results for Value Functions

The maximum norm distance was used in (Singh & Yee, 1994) to measure the approximation quality of a value function; in (Williams & Baird, 1993), the so-called Bellman error magnitude was used for the same purpose. In both cases, the loss in performance turned out to be at most proportional to the approximation error for expectation-based Markov decision tasks.

We generalize this work by giving an abstract analysis that holds for both expectation-based tasks and minimax-based tasks. We only use properties that dynamic programming operators of both tasks have in common. In the following, we present these common features as "assumptions" and defer their proof to Appendix A. Based on these assumptions, we present in this and the following section our results about the loss from imperfect value functions and Q-functions, respectively. The significance of these results will be discussed in the conclusion of this article, and the corresponding proofs can be found in Appendix B.

Assume V_γ^* and V_γ^π (for some stationary policy π) are value functions, and T_γ and $T_\gamma^\pi : \mathcal{R}^S \to \mathcal{R}^S$ are operators. These value functions and operators have to satisfy the following eight assumptions.

The first assumption guarantees that the value function V_γ^π (V_γ^*) is the only fixed point of the operator T_γ^π (T_γ).

Assumption 1 *Let π be a stationary policy and V be a value function. Then $T_\gamma^\pi V = V$ iff $V = V_\gamma^\pi$, and $T_\gamma V = V$ iff $V = V_\gamma^*$.*

Assume that we apply operator T_γ or T_γ^π to a value function that is the sum of a real value r and a value function V. The next assumption states that we obtain $\gamma \cdot r$ plus the value function that is obtained by applying the same operator to V.

Assumption 2 *Let π be a stationary policy, V be a value function, and $r \in \mathcal{R}$. Then $T_\gamma (V + r) = T_\gamma V + \gamma \cdot r$ and $T_\gamma^\pi (V + r) = T_\gamma^\pi V + \gamma \cdot r$.*

Third, we assume that both operators maintain the partial relation $\leq$ that was defined earlier for value functions.

Assumption 3 *Let U and V be value functions with $U \leq V$ and π be a stationary policy. Then $T_\gamma U \leq T_\gamma V$ and $T_\gamma^\pi U \leq T_\gamma^\pi V$.*

Furthermore, we assume that the maximum norm distance between the images of any two value functions U and V is at most γ times the maximum norm distance between U and V, no matter which of the two operators is used.

Assumption 4 *Let π be a stationary policy and U and V be value functions. Then $\|T_\gamma U - T_\gamma V\| \leq \gamma \cdot \|U - V\|$ and $\|T_\gamma^\pi U - T_\gamma^\pi V\| \leq \gamma \cdot \|U - V\|$.*

The next assumption defines greedy policies in an abstract way that does not depend on details of either expectation-based or minimax-based decision tasks. It also shows the fundamental relationship between the operators T_γ^π and T_γ.

Assumption 5 *Let π be any stationary policy and V be any value function. Then π is greedy for V iff $T_\gamma^\pi V = T_\gamma V$.*

The following assumption shows the fundamental relationship between the value functions V_γ^* and V_γ^π.

Assumption 6 *Let i be any state and Π_S be the set of stationary policies. Then $V_\gamma^*(i) = \min_{\pi \in \Pi_S} V_\gamma^\pi(i)$.*

For deterministic MDPs, the starting state i and the executed action a determine uniquely the successor state j_i^a reached and the immediate cost c_i^a obtained. We assume that for deterministic MDPs, the operators T_γ and T_γ^π are defined in the following way.

Assumption 7 *Let π be a stationary policy, i be a state, and V be a value function. Then $T_\gamma V(i) = \min_{a \in A(i)} [c_i^a + \gamma \cdot V(j_i^a)]$ and $T_\gamma^\pi V(i) = c_i^{\pi(i)} + \gamma \cdot V\left(j_i^{\pi(i)}\right)$ if the MDP is deterministic.*

Consider a deterministic MDP, any stationary policy π, and a given starting state i at time $t = 0$. In this case the return, call it $r_\gamma^\pi(i)$, is uniquely determined. The eighth and final assumption states that V_γ^π is equal to this value.[2]

Assumption 8 *$V_\gamma^\pi(i) = r_\gamma^\pi(i)$ for any deterministic Markov decision process, stationary policy π, and state i.*

Before we can analyze the loss in performance due to imperfect value functions we have to define what we precisely mean by "imperfect." We measure the distance between a value function V and the optimal value function V_γ^* in two different ways:

$$d_M(V) := \|V - V_\gamma^*\| = \max_{i \in S} |V(i) - V_\gamma^*(i)| ; \tag{3}$$

$$d_B(V) := \|T_\gamma V - V\| = \max_{i \in S} |T_\gamma V(i) - V(i)| . \tag{4}$$

We call $d_M(V)$ the *maximum norm distance* between V and V^*_γ and, following Williams and Baird, $d_B(V)$ the *Bellman error magnitude*. From its definition it is not obvious that the Bellman error magnitude is a reasonable measure for the distance between V and V^*_γ. The next two theorems show that this is indeed the case.

THEOREM 13 *Let V be any value function and π be any stationary policy. Then*

$$(1 - \gamma) \cdot \|V - V^*_\gamma\| \ \leq \|V - T_\gamma V\| \ \leq (1 + \gamma) \cdot \|V - V^*_\gamma\|;$$
$$(1 - \gamma) \cdot \|V - V^\pi_\gamma\| \ \leq \|V - T^\pi_\gamma V\| \ \leq (1 + \gamma) \cdot \|V - V^\pi_\gamma\|.$$

THEOREM 14 *Let V be a value function that satisfies $V \leq V^*_\gamma$ and π be any stationary policy. Then*

$$(1 - \gamma) \cdot \|V - V^*_\gamma\| \ \leq \|V - T_\gamma V\| \ \leq \|V - V^*_\gamma\|;$$
$$(1 - \gamma) \cdot \|V - V^\pi_\gamma\| \ \leq \|V - T^\pi_\gamma V\| \ \leq \|V - V^\pi_\gamma\|.$$

Theorem 13 ensures that $(1 - \gamma) \cdot d_M(V) \leq d_B(V) \leq (1 + \gamma) \cdot d_M(V)$. This tight relationship between the Bellman error magnitude and the maximum norm distance justifies the use of the Bellman error magnitude as a measure for the distance between V and V^*_γ. Its advantage over the maximum norm distance is that it can be computed easily even if V^*_γ is unknown.

We measure the loss in performance due to a policy π by $\|V^\pi_\gamma - V^*_\gamma\|$. The following two theorems guarantee that the loss in performance is small if dynamic programming based learning approaches are used that satisfy the assumptions above, good approximations of the optimal value functions are achieved, and then a greedy policy is followed – provided that the discount factor γ is not too close to 1.0.

THEOREM 15 *Let V be a value function and π be a stationary policy that is greedy for V. Then $\|V^\pi_\gamma - V^*_\gamma\| \leq \frac{2\gamma}{1-\gamma} \cdot d_B(V)$. Furthermore, this bound is tight, i.e., there exists an example with $\|V^\pi_\gamma - V^*_\gamma\| = \frac{2\gamma}{1-\gamma} \cdot d_B(V)$.*

THEOREM 16 *Let V be a value function and π be a stationary policy that is greedy for V. Then $\|V^\pi_\gamma - V^*_\gamma\| \leq \frac{2\gamma}{1-\gamma} \cdot d_M(V)$. Furthermore, this bound is tight.*

The next theorem and corollary guarantee that the bound for the loss in performance can be reduced by a factor of 2 if the approximation V underestimates V^*_γ.

THEOREM 17 *Let V be a value function with $V \leq V^*_\gamma$ and π be a stationary policy that is greedy for V. Then $\|V^\pi_\gamma - V^*_\gamma\| \leq \frac{\gamma}{1-\gamma} \cdot d_B(V)$. Furthermore, this bound is tight.*

COROLLARY 1 *Let V be a value function with $V \leq V^*_\gamma$ and π be a stationary policy that is greedy for V. Then $\|V^\pi_\gamma - V^*_\gamma\| \leq \frac{\gamma}{1-\gamma} \cdot d_M(V)$. Furthermore, this bound is tight.*

It might not always be easy to verify that $V \leq V^*_\gamma$ if V^*_γ is unknown, but the following two corollaries provide a sufficient condition for $V \leq V^*_\gamma$ that depends only on the dynamic programming operator T_γ and is easy to verify.

COROLLARY 2 *Let V be a value function and π be a greedy policy for V. If $V \le T_\gamma V$, then $\left\| V_\gamma^\pi - V_\gamma^* \right\| \le \frac{\gamma}{1-\gamma} \cdot d_B(V)$. Furthermore, this bound is tight.*

COROLLARY 3 *Let V be a value function and π be a greedy policy for V. If $V \le T_\gamma V$, then $\left\| V_\gamma^\pi - V_\gamma^* \right\| \le \frac{\gamma}{1-\gamma} \cdot d_M(V)$. Furthermore, this bound is tight.*

5.2. Results for Q-Functions

This section proceeds similarly to the last section, except that our results now apply to Q-functions instead of value functions: We first present features of dynamic programming operators that hold for both expectation-based and minimax-based Markov decision tasks and then derive theorems based on these assumptions.

Assume that V_γ^* and V_γ^π (for some stationary policy π) are value functions, Q_γ^* and Q_γ^π are Q-functions, and $B_\gamma : \mathbb{R}^M \to \mathcal{R}^M$ and $B_\gamma^\pi : \mathcal{R}^M \to \mathcal{R}^M$ are operators. These value functions, Q-functions, and operators have to satisfy the following assumptions in addition to the eight assumptions stated in the previous section. We show in Appendix A that all of these assumptions are satisfied for both expectation-based and minimax-based Markov decision tasks.

Assumption 9 relates the value functions V_γ^* and V_γ^π to their corresponding Q-functions Q_γ^* and Q_γ^π, respectively.

Assumption 9 *Let π be a stationary policy. Then $V_{Q_\gamma^*} = V_\gamma^*$ and $V_{Q_\gamma^\pi, \pi} = V_\gamma^\pi$.*

Assumption 10 guarantees that the Q-function Q_γ^π (Q_γ^*) is the only fixed point of the operator B_γ^π (B_γ).

Assumption 10 *Let π be a stationary policy and Q be a Q-function. Then $B_\gamma Q = Q$ iff $Q = Q_\gamma^*$, and $B_\gamma^\pi Q = Q$ iff $Q = Q_\gamma^\pi$.*

Assume that we apply operator B_γ or B_γ^π to a Q-function that is the sum of a real value r and a Q-function Q. Assumption 11 states that we obtain $\gamma \cdot r$ plus the Q-function that is obtained by applying the same operator to Q.

Assumption 11 *Let π be a stationary policy, Q be a Q-function, and $r \in \mathcal{R}$. Then $B_\gamma(Q + r) = B_\gamma Q + \gamma \cdot r$ and $B_\gamma^\pi(Q + r) = B_\gamma^\pi Q + \gamma \cdot r$.*

Assumption 12 states that both operators maintain the partial relation $\le$ for Q-functions.

Assumption 12 *Let Q and Q' be Q-functions with $Q \le Q'$ and π be a stationary policy. Then $B_\gamma Q \le B_\gamma Q'$ and $B_\gamma^\pi Q \le B_\gamma^\pi Q'$.*

We assume that the maximum norm distance between the images of any two Q-functions Q and Q' is at most γ times the maximum norm distance between Q and Q', no matter which of the two operators is used.

Assumption 13 *Let π be a stationary policy and Q and Q' be Q-functions. Then $\left\| B_\gamma Q - B_\gamma Q' \right\| \le \gamma \cdot \left\| Q - Q' \right\|$ and $\left\| B_\gamma^\pi Q - B_\gamma^\pi Q' \right\| \le \gamma \cdot \left\| Q - Q' \right\|$.*

Assumption 14 shows the fundamental relationship between the operators B_γ^π and B_γ.

Assumption 14 *Let Q be a Q-function and π be a stationary policy that is greedy for Q. Then $B_\gamma Q = B_\gamma^\pi Q$.*

Assumption 15 shows the fundamental relationship between the Q-functions Q_γ^* and Q_γ^π.

Assumption 15 *Let i be a state, $a \in A(i)$, and Π_S be the set of stationary policies. Then $Q_\gamma^*(i, a) = \min_{\pi \in \Pi_S} Q_\gamma^\pi(i, a)$.*

The following assumption gives the definitions of the operators B_γ and B_γ^π for deterministic MDPs.

Assumption 16 *Let π be a stationary policy, i be a state, $a \in A(i)$ be an action, and Q be a Q-function. Then $B_\gamma Q(i, a) = c_i^a + \gamma \cdot V_Q(j_i^a)$ and similarly, $B_\gamma^\pi Q(i, a) = c_i^a + \gamma \cdot V_{Q,\pi}(j_i^a)$ if the MDP is deterministic.*

The final assumption relates the definition of an optimal policy to the Q-function Q_γ^*.

Assumption 17 *A stationary policy is optimal iff it is greedy for Q_γ^*.*

Before we can analyze the loss in performance due to imperfect Q-functions we have to define what we precisely mean by "imperfect." We measure the distance between a Q-function Q and the optimal Q-function Q_γ^* in two different ways:

$$d_M(Q) := \left\| Q - Q_\gamma^* \right\| = \max_{i \in S, a \in A(i)} \left| Q(i, a) - Q_\gamma^*(i, a) \right|; \tag{5}$$

$$d_B(Q) := \left\| B_\gamma Q - Q \right\| = \max_{i \in S, a \in A(i)} \left| B_\gamma Q(i, a) - Q(i, a) \right|. \tag{6}$$

We call $d_M(Q)$ the *maximum norm distance* between Q and Q_γ^* and, following Williams and Baird (Williams & Baird, 1993), $d_B(Q)$ the *Bellman error magnitude*.

THEOREM 18 *Let Q be a Q-function and π be a stationary policy. Then*

$$(1 - \gamma) \cdot \left\| Q - Q_\gamma^* \right\| \leq \left\| Q - B_\gamma Q \right\| \leq (1 + \gamma) \cdot \left\| Q - Q_\gamma^* \right\|;$$
$$(1 - \gamma) \cdot \left\| Q - Q_\gamma^\pi \right\| \leq \left\| Q - B_\gamma^\pi Q \right\| \leq (1 + \gamma) \cdot \left\| Q - Q_\gamma^\pi \right\|.$$

THEOREM 19 *Let Q be a Q-function with $Q \leq Q_\gamma^*$ and π be a stationary policy. Then*

$$(1 - \gamma) \cdot \left\| Q - Q_\gamma^* \right\| \leq \left\| Q - B_\gamma Q \right\| \leq \left\| Q - Q_\gamma^* \right\|;$$
$$(1 - \gamma) \cdot \left\| Q - Q_\gamma^\pi \right\| \leq \left\| Q - B_\pi^\pi Q \right\| \leq \left\| Q - Q_\gamma^\pi \right\|.$$

Theorem 18 ensures that $(1 - \gamma) \cdot d_M(Q) \leq d_B(Q) \leq (1 + \gamma) \cdot d_M(Q)$. This relationship justifies to use the Bellman error magnitude as a measure for the distance between Q and Q_γ^*. Its advantage over the maximum norm distance is that it can be computed easily even if Q_γ^* is unknown.

The following two theorems guarantee that the loss in performance is small if dynamic programming based learning approaches are used that satisfy the assumptions above, have achieved good approximations of the optimal Q-function, and then follow a greedy policy – provided that the discount factor γ is not too close to 1.0.

THEOREM 20 *Let Q be a Q-function and π be a stationary policy that is greedy for Q. Then $\left\|V_\gamma^\pi - V_\gamma^*\right\| \leq \frac{2}{1-\gamma} \cdot d_B(Q)$. Furthermore, this bound is tight.*

THEOREM 21 *Let Q be a Q-function and π be a stationary policy that is greedy for Q. Then $\left\|V_\gamma^\pi - V_\gamma^*\right\| \leq \frac{2}{1-\gamma} \cdot d_M(Q)$. Furthermore, this bound is tight.*

The next theorem and corollary guarantee that the bound for the loss in performance can be reduced by a factor of 2 if the approximation Q underestimates Q_γ^*. These results apply especially to Q-hat-learning because this algorithm operates with underestimations of the minimax-optimal Q-function.

THEOREM 22 *Let Q be a Q-function with $Q \leq Q_\gamma^*$ and π be a stationary policy that is greedy for Q. Then $\left\|V_\gamma^\pi - V_\gamma^*\right\| \leq \frac{1}{1-\gamma} \cdot d_B(Q)$. Furthermore, this bound is tight.*

COROLLARY 4 *Let Q be a Q-function with $Q \leq Q_\gamma^*$ and π be a stationary policy that is greedy for Q. Then $\left\|V_\gamma^\pi - V_\gamma^*\right\| \leq \frac{1}{1-\gamma} \cdot d_M(Q)$. Furthermore, this bound is tight.*

It might not always be easy to verify that $Q \leq Q_\gamma^*$ if Q_γ^* is unknown, but the following two corollaries provide a sufficient condition for $Q \leq Q_\gamma^*$ that depends only on the dynamic programming operator B_γ and is easy to verify.

COROLLARY 5 *Let Q be a Q-function with $Q \leq B_\gamma Q$ and π be a stationary policy that is greedy for Q. Then $\left\|V_\gamma^\pi - V_\gamma^*\right\| \leq \frac{1}{1-\gamma} \cdot d_B(Q)$. Furthermore, this bound is tight.*

COROLLARY 6 *Let Q be a Q-function with $Q \leq B_\gamma Q$ and π be a stationary policy that is greedy for Q. Then $\left\|V_\gamma^\pi - V_\gamma^*\right\| \leq \frac{1}{1-\gamma} \cdot d_M(Q)$. Furthermore, this bound is tight.*

The following theorem and corollary show that policies that are greedy for Q-functions are optimal if the Q-functions approximate the optimal Q-function closely.

THEOREM 23 *There exists an $\epsilon > 0$ with the following property: Every stationary policy is optimal if it is greedy for a Q-function Q with $d_M(Q) < \epsilon$.*

COROLLARY 7 *There exists an $\epsilon > 0$ with the following property: Every stationary policy is optimal if it is greedy for a Q-function Q with $d_B(Q) < \epsilon$.*

However, Q-functions that do not approximate the optimal Q-function well do not necessarily prevent one from obtaining optimal policies, as the following two results show.

THEOREM 24 *For every real number r there exists a Q-function Q with $d_M(Q) > r$ that has the following property: Every stationary policy that is greedy for Q is optimal.*

COROLLARY 8 *For every real number r there exists a Q-function Q with $d_B(Q) > r$ that has the following property: Every stationary policy that is greedy for Q is optimal.*

6. Conclusions

In this paper, we have studied both minimax-based and expectation-based Markov decision tasks and quantified how much performance is lost if one uses value functions that approximate the optimal value function instead of the optimal value function itself. In particular, we described the properties that both Markov decision tasks have in common (namely Assumptions 1 to 17) and based our analysis solely on these properties. Using this framework, we were able to transfer the results of (Williams & Baird, 1993) and (Singh & Yee, 1994) to minimax-based Markov decision tasks. To the best of our knowledge, our analysis is the first one that studies approximate value functions for minimax-based Markov decision tasks and the resulting loss in performance.

Our main result is that this loss in performance is at most proportional to the approximation error (Theorems 15, 16, 20 and 21). The factor of proportionality, however, depends on the discount factor γ and approaches infinity if γ approaches one. Unfortunately, there are some Markov decision tasks with delayed reinforcement for which it is not reliable to choose γ freely (Schwartz, 1993; see also McDonald & Hingson, 1994).

We obtained previously unknown results that hold for both the expectation-based tasks and the minimax-based tasks. The right hand side inequalities of the Theorems 13, 14, 18, and 19 we proved show that the Bellman error magnitude (used in (Williams & Baird, 1993) in order to measure the quality of approximation) is more related to the maximum norm distance (used in (Singh & Yee, 1994)) than was previously known. The results were useful to derive new bounds. Furthermore, on the one hand, we showed that the bounds given in (Singh & Yee, 1994) are tight, i.e., there are examples where these bounds are attained (Theorem 16 and 21). On the other hand if one assumes $V \leq V_\gamma^*$ or $Q \leq Q_\gamma^*$ then the corresponding bounds can be reduced by a factor of two. These new bounds were also proven to be tight (Corollaries 1 and 4).

In the Q-hat-learning algorithm, the Q-function Q always satisfies $Q \leq Q_\gamma^*$ with probability one. This is generally not true in Q-learning. Hence the results of Theorem 22 and Corollary 4 are especially important in minimax-based tasks.

We showed that small non-zero approximation errors in Q-functions are already sufficient to obtain optimal policies (Theorem 23 and Corollary 7). As a consequence, every algorithm that iteratively produces a sequence of Q-functions that converges in infinity to the optimal Q-function, is able to determine optimal policies after a finite time. On the other hand we also showed that small approximation errors are not necessary to obtain optimal policies (Theorem 24 and Corollary 8).

It is still an open problem why expectation-based and minimax-based Markov decision tasks share so many properties. Another related open problem is whether there are other types of Markov decision tasks that satisfy our assumptions. Answers to these questions would probably lead to a better understanding of the existing reinforcement learning algorithms and might lead to the development of algorithms that use decision criteria that are different from the ones we use today.

We have seen that the loss in performance is small when acting greedily on imperfect value functions if the value functions approximate the optimal value function well. A related issue is whether the reinforcement learning algorithms that were designed to be used in conjunction with lookup tables work equally well with function approximators. Theoretical results can be found in (Thrun & Schwartz, 1993), (Baird, 1995), and (Gordon, 1995). Practical results in using RL algorithms with function approximators are published in (Tesauro, 1992), (Lin, 1993), (Crites & Barto, 1995), (Boyan & Moore, 1995), (Zhang & Dietterich, 1995), and (Sutton, 1995). RL algorithms that are specifically designed to be used with function approximators were recently published in (Baird, 1995) and (Boyan & Moore 1995).

Acknowledgments

Many special thanks to Sven Koenig for his substantial help with the final editing of this article.

Appendix A

Proofs of the Assumptions

It is easy to see that the seventeen assumptions hold for expectation-based Markov decision tasks. They are either standard results from dynamic programming (see e.g., Bertsekas, 1987; Ross, 1970) or follow immediately from Section 3. Assumptions 9 and 10 were proven in (Williams & Baird, 1993).

The seventeen assumptions also hold for minimax-based Markov decision tasks. Assumption 1 was proven in (Heger, 1994b). We prove Assumptions 4, 9, 10, and 13 in the following. The remaining eleven assumptions then follow immediately from the results of Section 4.

Proof of Assumption 4 for Minimax-Based Markov Decision Tasks:

Assume $T_\gamma U(i^*) \geq T_\gamma V(i^*)$ for $i^* = \arg\max_{i \in S} |T_\gamma U(i) - T_\gamma V(i)|$, and let $a^* = \arg\min_{a \in A(i^*)} \left[\max_{j \in N(i^*,a)} [c(i^*, a, j) + \gamma V(j)] \right]$. Then, according to Theorem 4,

$$
\begin{aligned}
\|T_\gamma U - T_\gamma V\| &= |T_\gamma U(i^*) - T_\gamma V(i^*)| \\
&= T_\gamma U(i^*) - T_\gamma V(i^*) \\
&= \min_{a \in A(i^*)} \max_{j \in N(i^*,a)} [c(i^*, a, j) + \gamma U(j)] \\
&\quad - \min_{a \in A(i^*)} \max_{j \in N(i^*,a)} [c(i^*, a, j) + \gamma V(j)] \\
&\leq \max_{j \in N(i^*,a^*)} [c(i^*, a^*, j) + \gamma U(j)] \\
&\quad - \max_{j \in N(i^*,a^*)} [c(i^*, a^*, j) + \gamma V(j)] \\
&\leq \max_{j \in N(i^*,a^*)} |c(i^*, a^*, j) + \gamma U(j) - (c(i^*, a^*, j) + \gamma V(j))| \\
&= \gamma \cdot \max_{j \in N(i^*,a^*)} |U(j) - V(j)| \leq \gamma \cdot \|U - V\|.
\end{aligned}
$$

A symmetrical arguments holds if $T_\gamma V(i^*) \geq T_\gamma U(i^*)$. This proves the first inequality.

To prove the second inequality, first consider the special case where $A(i) = \{\pi(i)\}$ for all $i \in S$. In this case, $T_\gamma = T_\gamma^\pi$ (see (1) and (2) at page 206) and the second inequality follows from the first one. To show that the second inequality holds in general, notice that it does not depend on any assumptions about the sets $A(i)$ other than $\pi(i) \in A(i)$, which is trivially true.   ∎

Proof of Assumption 9 for Minimax-Based Markov Decision Tasks: According to Assumption 1, which was proven in (Heger 1994b), it holds for any state i that

$$V_{Q_\gamma^*}(i) = \min_{a \in A(i)} Q_\gamma^*(i, a) = \min_{a \in A(i)} \max_{j \in N(i,a)} \left[c(i, a, j) + \gamma \cdot V_\gamma^*(j) \right]$$
$$= T_\gamma V_\gamma^*(i) = V_\gamma^*(i)$$

which proves the first equation. Similarly, it follows from Assumption 1 that

$$V_{Q_\gamma^\pi, \pi}(i) = Q_\gamma^\pi(i, \pi(i)) = \max_{j \in N(i, \pi(i))} \left[c(i, \pi(i), j) + \gamma \cdot V_\gamma^\pi(j) \right]$$
$$= T_\gamma^\pi V_\gamma^\pi(i) = V_\gamma^\pi(i)$$

which proves the second equation.   ∎

Proof of Assumption 10 for Minimax-Based Markov Decision Tasks: According to Assumption 9, it holds for any state i and action $a \in A(i)$ that

$$B_\gamma Q_\gamma^*(i, a) = \max_{j \in N(i,a)} \left[c(i, a, j) + \gamma \cdot V_{Q_\gamma^*}(j) \right]$$
$$= \max_{j \in N(i,a)} \left[c(i, a, j) + \gamma \cdot V_\gamma^*(j) \right] = Q_\gamma^*(i, a)$$

which proves the first equation. Similarly, it follows from Assumption 9 that

$$B_\gamma^\pi Q_\gamma^\pi(i, a) = \max_{j \in N(i,a)} \left[c(i, a, j) + \gamma \cdot V_{Q_\gamma^\pi, \pi}(j) \right]$$
$$= \max_{j \in N(i,a)} \left[c(i, a, j) + \gamma \cdot V_\gamma^\pi(j) \right] = Q_\gamma^\pi(i, a)$$

which proves the second equation.   ∎

Proof of Assumption 13 for Minimax-Based Markov Decision Tasks: Let $i \in S$ and $a \in A(i)$. From Theorems 4 and 6 we can conclude

$$\| B_\gamma^\pi Q - B_\gamma^\pi Q' \|$$
$$\leq \left| B_\gamma^\pi Q(i, a) - B_\gamma^\pi Q'(i, a) \right|$$
$$= \left| \max_{j \in N(i,a)} \left[c(i, a, j) + \gamma \cdot V_{Q, \pi}(j) \right] - \max_{j \in N(i,a)} \left[c(i, a, j) + \gamma \cdot V_{Q', \pi}(j) \right] \right|$$

$$\leq \max_{j \in N(i,a)} |c(i,a,j) + \gamma \cdot V_{Q,\pi}(j) - (c(i,a,j) + \gamma \cdot V_{Q',\pi}(j))|$$
$$= \gamma \cdot \max_{j \in N(i,a)} |V_{Q,\pi}(j) - V_{Q',\pi}(j)|$$
$$\leq \gamma \cdot \|V_{Q,\pi} - V_{Q',\pi}\|$$
$$\leq \gamma \cdot \|Q - Q'\|.$$

This proves the second inequality. The proof of the first inequality is obtained from this proof by substituting B_γ, V_Q, and $V_{Q'}$ for B_γ^π, $V_{Q,\pi}$, and $V_{Q',\pi}$, respectively.

∎

Appendix B

Proofs of the Results from Section 5

Proof of Theorem 13: The following four inequalities follow from Assumptions 1 and 4 in conjunction with the triangle inequality.

$$\|V - V_\gamma^*\| \leq \|V - T_\gamma V\| + \|T_\gamma V - V_\gamma^*\| \leq \|V - T_\gamma V\| + \gamma \cdot \|V - V_\gamma^*\|; \quad \text{(B.1)}$$
$$\|V - T_\gamma V\| \leq \|V - V_\gamma^*\| + \|V_\gamma^* - T_\gamma V\| \leq \|V - V_\gamma^*\| + \gamma \cdot \|V_\gamma^* - V\|; \quad \text{(B.2)}$$
$$\|V - V_\gamma^\pi\| \leq \|V - T_\gamma^\pi V\| + \|T_\gamma^\pi V - V_\gamma^\pi\| \leq \|V - T_\gamma^\pi V\| + \gamma \cdot \|V - V_\gamma^\pi\|; \quad \text{(B.3)}$$
$$\|V - T_\gamma^\pi V\| \leq \|V - V_\gamma^\pi\| + \|V_\gamma^\pi - T_\gamma^\pi V\| \leq \|V - V_\gamma^\pi\| + \gamma \cdot \|V_\gamma^\pi - V\|. \quad \text{(B.4)}$$

Inequalities (B.1), (B.2), (B.3), and (B.4) imply the first, second, third, and fourth inequality of this theorem, respectively.

∎

Proof of Theorem 14: The first and third inequality have already been proven in the context of Theorem 13. From Assumptions 3, 1, and 6 we obtain

$$T_\gamma V - \|V - V_\gamma^*\| \leq T_\gamma V_\gamma^* - \|V - V_\gamma^*\| = V_\gamma^* - \|V - V_\gamma^*\| \leq V; \quad \text{(B.5)}$$
$$T_\gamma^\pi V - \|V - V_\gamma^\pi\| \leq T_\gamma^\pi V_\gamma^\pi - \|V - V_\gamma^\pi\| = V_\gamma^\pi - \|V - V_\gamma^\pi\| \leq V. \quad \text{(B.6)}$$

Assumptions 1 and 4 imply

$$\|V_\gamma^* - T_\gamma V\| = \|T_\gamma V_\gamma^* - T_\gamma V\| \leq \gamma \cdot \|V_\gamma^* - V\| \leq \|V - V_\gamma^*\|;$$
$$\|V_\gamma^\pi - T_\gamma^\pi V\| = \|T_\gamma^\pi V_\gamma^\pi - T_\gamma^\pi V\| \leq \gamma \cdot \|V_\gamma^\pi - V\| \leq \|V - V_\gamma^\pi\|.$$

and therefore, in conjunction with Assumption 6,

$$V - T_\gamma V \leq V_\gamma^* - T_\gamma V \leq \|V_\gamma^* - T_\gamma V\| \leq \|V - V_\gamma^*\|; \quad \text{(B.7)}$$
$$V - T_\gamma^\pi V \leq V_\gamma^\pi - T_\gamma^\pi V \leq \|V_\gamma^\pi - T_\gamma^\pi V\| \leq \|V - V_\gamma^\pi\|. \quad \text{(B.8)}$$

Inequalities (B.5) and (B.7) imply $-\|V - V_\gamma^*\| \leq V - T_\gamma V \leq \|V - V_\gamma^*\|$ and thus the second inequality of the theorem. Inequalities (B.6) and (B.8) imply $-\|V - V_\gamma^\pi\| \leq V - T_\gamma^\pi V \leq \|V - V_\gamma^\pi\|$ and therefore the third inequality of the theorem. ∎

Proof of Theorem 15: We prove the desired bound by applying, in turn, the triangle inequality, Assumptions 1, 5, and 4, Theorem 13, and finally Assumption 5 a second time:

$$
\begin{aligned}
\|V_\gamma^\pi - V_\gamma^*\| &\leq \|V_\gamma^\pi - T_\gamma V\| + \|T_\gamma V - V_\gamma^*\| \\
&= \|T_\gamma^\pi V_\gamma^\pi - T_\gamma^\pi V\| + \|T_\gamma V - T_\gamma V_\gamma^*\| \\
&\leq \gamma \cdot \|V_\gamma^\pi - V\| + \gamma \cdot \|V - V_\gamma^*\| \\
&\leq \frac{\gamma}{1-\gamma} \cdot \|V - T_\gamma^\pi V\| + \frac{\gamma}{1-\gamma} \cdot \|V - T_\gamma V\| \\
&= \frac{\gamma}{1-\gamma} \cdot \|V - T_\gamma V\| + \frac{\gamma}{1-\gamma} \cdot \|V - T_\gamma V\| = \frac{2\gamma}{1-\gamma} \cdot d_B(V).
\end{aligned}
$$

To see that this bound cannot be tightened any further, consider a deterministic MDP with two states, 1 and 2, and two actions, 1 and 2. Action 1 causes a state transition to state 1 no matter which state it is executed in. Similarly, action 2 always causes a state transition to state 2. All immediate costs are two, except for the execution of action 2 in state 2, which costs nothing. Now consider the value function V that is defined by $V(1) = V(2) = \frac{1}{1-\gamma}$. From Assumption 7 it follows that $T_\gamma V(1) - V(1) = 2 + \gamma V(1) - V(1) = 1$ and $T_\gamma V(2) - V(2) = 0 + \gamma V(2) - V(2) = -1$. Therefore $d_B(V) = \|T_\gamma V - V\| = 1$. The stationary policy π with $\pi(1) = 1$ and $\pi(2) = 2$ is greedy for V according to Assumptions 5 and 7. Assumptions 6 and 8 imply $V_\gamma^*(1) = 2$, $V_\gamma^\pi(1) = \frac{2}{1-\gamma}$, $V_\gamma^*(2) = 0$, and $V_\gamma^\pi(2) = 0$. It follows that $\|V_\gamma^\pi - V_\gamma^*\| = \frac{2\gamma}{1-\gamma} = \frac{2\gamma}{1-\gamma} \cdot d_B(V)$. ∎

Proof of Theorem 16: We prove the desired bound by applying, in turn, the triangle inequality, Assumptions 1, 5, and 4, Theorem 13, Assumption 5 again, and finally Theorem 13 a second time:

$$
\begin{aligned}
\|V_\gamma^\pi - V_\gamma^*\| &\leq \|V_\gamma^\pi - T_\gamma^\pi V\| + \|T_\gamma^\pi V - V_\gamma^*\| \\
&= \|T_\gamma^\pi V_\gamma^\pi - T_\gamma^\pi V\| + \|T_\gamma V - T_\gamma V_\gamma^*\| \\
&\leq \gamma \cdot \|V_\gamma^\pi - V\| + \gamma \cdot d_M(V) \\
&\leq \gamma \cdot \frac{\|V - T_\gamma^\pi V\|}{1-\gamma} + \gamma \cdot d_M(V) \\
&\leq \gamma \cdot \frac{(1+\gamma) \cdot \|V - V_\gamma^*\|}{1-\gamma} + \gamma \cdot d_M(V) \\
&= \gamma \cdot \frac{(1+\gamma) \cdot d_M(V)}{1-\gamma} + \gamma \cdot d_M(V)
\end{aligned}
$$

$$= \frac{2\gamma}{1-\gamma} \cdot d_M(V).$$

To see that this bound cannot be tightened any further, consider again the Markov decision problem that we used in the proof of Theorem 15, but this time in conjunction with the value function V that is defined by $V(1) = V(2) = 1$. Assumptions 6 and 8 imply that $V_\gamma^*(1) = 2$ and $V_\gamma^*(2) = 0$. Since $V_\gamma^*(1) - V(1) = 2 - 1 = 1$ and $V_\gamma^*(2) - V(2) = 0 - 1 = -1$, it follows that $d_M(V) = \|V_\gamma^* - V\| = 1$. The stationary policy π with $\pi(1) = 1$ and $\pi(2) = 2$ is greedy for V according to Assumptions 5 and 7. Assumption 8 implies $V_\gamma^\pi(1) = \frac{2}{1-\gamma}$, $V_\gamma^\pi(2) = 0$, and, consequently, $\|V_\gamma^\pi - V_\gamma^*\| = \frac{2\gamma}{1-\gamma} = \frac{2\gamma}{1-\gamma} \cdot d_M(V)$. ■

Proof of Theorem 17: From Theorem 13 and Assumption 5 we obtain $\|V - V_\gamma^\pi\| \leq \|V - T_\gamma V\| \cdot \frac{1}{1-\gamma} = d_B(V) \cdot \frac{1}{1-\gamma}$. Therefore,

$$V \geq V_\gamma^\pi - d_B(V) \cdot \frac{1}{1-\gamma}. \tag{B.9}$$

We can now prove the desired bound by applying, in turn, Assumptions 1, 3, and 5, Inequality (B.9), and finally Assumptions 3, 2, and 1.

$$V_\gamma^* = T_\gamma V_\gamma^* \geq T_\gamma V = T_\gamma^\pi V$$

$$\geq T_\gamma^\pi \left(V_\gamma^\pi - \frac{d_B(V)}{1-\gamma}\right) = T_\gamma^\pi V_\gamma^\pi - \frac{\gamma}{1-\gamma} \cdot d_B(V)$$

$$= V_\gamma^\pi - \frac{\gamma}{1-\gamma} \cdot d_B(V).$$

Together with Assumption 6 we obtain $0 \leq V_\gamma^\pi - V_\gamma^* \leq \frac{\gamma}{1-\gamma} \cdot d_B(V)$ and, therefore, $\|V_\gamma^\pi - V_\gamma^*\| \leq \frac{\gamma}{1-\gamma} \cdot d_B(V)$.

To see that this bound cannot be tightened any further, consider again the MDP that we used in the proof of Theorem 15, but this time in conjunction with the value function V that is defined by $V(1) = V(2) = 0$. Assumptions 6 and 8 imply that $V_\gamma^*(1) = 2$ and $V_\gamma^*(2) = 0$ and, therefore, $V_\gamma^*(1) = 2 \geq V(1)$ and $V_\gamma^*(2) = 0 \geq V(2)$. From Assumption 7 it follows that

$$T_\gamma V(1) - V(1) = 2 - 0 = 2; \tag{B.10}$$

$$T_\gamma V(2) - V(2) = 0 - 0 = 0. \tag{B.11}$$

Therefore $d_B(V) = \|T_\gamma V - V\| = 2$. The stationary policy π with $\pi(1) = 1$ and $\pi(2) = 2$ is greedy for V according to Assumptions 5 and 7, and Assumption 8 implies $V_\gamma^\pi(1) = \frac{2}{1-\gamma}$ and $V_\gamma^\pi(2) = 0$. We can conclude that $\|V_\gamma^\pi - V_\gamma^*\| = \frac{2\gamma}{1-\gamma} = \frac{\gamma}{1-\gamma} \cdot d_B(V)$. ■

Proof of Corollary 1: The desired bound is an immediate consequence of Theorem 17 in conjunction with the second inequality of Theorem 14. The example used to prove

Theorem 17 proves that the bound is tight, because it satisfies $d_B(V) = d_M(V)$. ■

LEMMA 1 *Let V be a value function that satisfies $V \le T_\gamma V$. Then $V \le V_\gamma^*$.*

Proof: Consider mappings $T_{\gamma,n}$ from $\mathcal{R}^S$ to $\mathcal{R}^S$ that are inductively defined as follows:

$$\forall n \in \mathbb{N} \,\, \forall V \in \mathcal{R}^S : T_{\gamma,0}V = V, \quad T_{\gamma,n+1}V = T_\gamma(T_{\gamma,n}V).$$

It follows by induction from Assumption 3 that $V \le T_{\gamma,n}V$ for all n if $V \le T_\gamma V$. Similarly, it follows by induction from Assumption 1 that $T_{\gamma,n}V_\gamma^* = V_\gamma^*$ and from Assumption 4 that $\left\| T_{\gamma,n}V_\gamma^* - T_{\gamma,n}V \right\| \le \gamma^n \left\| V_\gamma^* - V \right\|$, for all n. Put together, we can conclude that $\left\| V_\gamma^* - T_{\gamma,n}V \right\| \le \gamma^n \left\| V_\gamma^* - V \right\|$ for all n which further implies $\lim_{n\to\infty} T_{\gamma,n}V = V_\gamma^*$ and, finally, $V \le V_\gamma^*$. ■

Proof of Corollary 2: The desired bound is an immediate consequence of Lemma 1 and Theorem 17. The example used to prove Theorem 17 proves that the bound is tight, because it satisfies $V \le T_\gamma V$ according to Equations (B.10) and (B.11). ■

Proof of Corollary 3: The desired bound follows immediately from Lemma 1 and Corollary 1. The example used to prove Theorem 17 proves that the bound is tight, because it satisfies $V \le T_\gamma V$ according to Equations (B.10) and (B.11) and, furthermore, $d_B(V) = d_M(V)$. ■

Proof of Theorem 18: The following four inequalities follow from Assumptions 10 and 13 in conjunction with the triangle inequality.

$$\left\| Q - Q_\gamma^* \right\| \le \left\| Q - B_\gamma Q \right\| + \left\| B_\gamma Q - Q_\gamma^* \right\| \le \left\| Q - B_\gamma Q \right\| + \gamma \cdot \left\| Q - Q_\gamma^* \right\|; \quad \text{(B.12)}$$

$$\left\| Q - B_\gamma Q \right\| \le \left\| Q - Q_\gamma^* \right\| + \left\| Q_\gamma^* - B_\gamma Q \right\| \le \left\| Q - Q_\gamma^* \right\| + \gamma \cdot \left\| Q_\gamma^* - Q \right\|; \quad \text{(B.13)}$$

$$\left\| Q - Q_\gamma^\pi \right\| \le \left\| Q - B_\gamma^\pi Q \right\| + \left\| B_\gamma^\pi Q - Q_\gamma^\pi \right\| \le \left\| Q - B_\gamma^\pi Q \right\| + \gamma \cdot \left\| Q - Q_\gamma^\pi \right\|; \quad \text{(B.14)}$$

$$\left\| Q - B_\gamma^\pi V \right\| \le \left\| Q - Q_\gamma^\pi \right\| + \left\| Q_\gamma^\pi - B_\gamma^\pi Q \right\| \le \left\| Q - Q_\gamma^\pi \right\| + \gamma \cdot \left\| Q_\gamma^\pi - Q \right\|. \quad \text{(B.15)}$$

Inequalities (B.12), (B.13), (B.14), and (B.15) imply the first, second, third, and fourth inequality of this theorem, respectively. ■

Proof of Theorem 19: The first and third inequality have already been proven in the context of Theorem 18. From Assumptions 12 and 10, we obtain

$$B_\gamma Q - \left\| Q - Q_\gamma^* \right\| \le B_\gamma Q_\gamma^* - \left\| Q - Q_\gamma^* \right\| = Q_\gamma^* - \left\| Q - Q_\gamma^* \right\| \le Q: \quad \text{(B.16)}$$

$$B_\gamma^\pi Q - \left\| Q - Q_\gamma^\pi \right\| \le B_\gamma^\pi Q_\gamma^\pi - \left\| Q - Q_\gamma^\pi \right\| = Q_\gamma^\pi - \left\| Q - Q_\gamma^\pi \right\| \le Q. \quad \text{(B.17)}$$

Assumptions 10 and 13 imply

$$\|Q_\gamma^* - B_\gamma Q\| = \|B_\gamma Q_\gamma^* - B_\gamma Q\| \le \gamma \cdot \|Q_\gamma^* - Q\| \le \|Q - Q_\gamma^*\|,$$
$$\|Q_\gamma^\pi - B_\gamma^\pi Q\| = \|B_\gamma^\pi Q_\gamma^\pi - B_\gamma^\pi Q\| \le \gamma \cdot \|Q_\gamma^\pi - Q\| \le \|Q - Q_\gamma^\pi\|,$$

and therefore, in conjunction with Assumption 15

$$Q - B_\gamma Q \le Q_\gamma^* - B_\gamma Q \le \|Q_\gamma^* - B_\gamma Q\| \le \|Q - Q_\gamma^*\|; \quad \text{(B.18)}$$
$$Q - B_\gamma^\pi Q \le Q_\gamma^* - B_\gamma^\pi Q \le Q_\gamma^\pi - B_\gamma^\pi Q \le \|Q_\gamma^\pi - B_\gamma^\pi Q\| \le \|Q - Q_\gamma^\pi\|. \quad \text{(B.19)}$$

Inequalities (B.16) and (B.18) imply $- \|Q - Q_\gamma^*\| \le Q - B_\gamma Q \le \|Q - Q_\gamma^*\|$ and thus the second inequality of the theorem. Inequalities (B.17) and (B.19) imply $- \|Q - Q_\gamma^\pi\| \le Q - B_\gamma^\pi Q \le \|Q - Q_\gamma^\pi\|$ and therefore the third inequality of the theorem. ∎

Proof of Theorem 20: We prove the desired bound by applying, in turn, the triangle inequality, Assumption 9, Theorems 3, 6 and 18, and finally Assumption 14:

$$\begin{aligned}
\|V_\gamma^\pi - V_\gamma^*\| &\le \|V_\gamma^\pi - V_Q\| + \|V_Q - V_\gamma^*\| \\
&= \|V_{Q_\gamma^\pi,\pi} - V_{Q,\pi}\| + \|V_Q - V_{Q_\gamma^*}\| \\
&\le \|Q_\gamma^\pi - Q\| + \|Q - Q_\gamma^*\| \\
&\le \frac{\|Q - B_\gamma^\pi Q\|}{1 - \gamma} + \frac{\|Q - B_\gamma Q\|}{1 - \gamma} \\
&= \frac{2\|Q - B_\gamma Q\|}{1 - \gamma} = \frac{2}{1 - \gamma} \cdot d_B(Q).
\end{aligned}$$

To see that this bound cannot be tightened any further, consider a deterministic MDP with a single state 1 and two actions, 1 and 2, that both cause self transitions and have immediate costs 2 and 0, respectively. Now consider the Q-function Q that is defined by $Q(1,1) = Q(1,2) = 1/(1 - \gamma)$. From Assumption 16 it follows that $B_\gamma Q(1,1) - Q(1,1) = 2 + \gamma \cdot V_Q(1) - Q(1,1) = 2 + \frac{\gamma}{1-\gamma} - \frac{1}{1-\gamma} = 1$ and $B_\gamma Q(1.2) - Q(1,2) = 0 + \gamma \cdot V_Q(1) - Q(1.2) = \frac{\gamma}{1-\gamma} - \frac{1}{1-\gamma} = -1$. Therefore $d_B(Q) = \|B_\gamma Q - Q\| = 1$. The stationary policy π with $\pi(1) = 1$ is greedy for Q, and Assumptions 6 and 8 imply $V_\gamma^\pi(1) = 2/(1 - \gamma)$ and $V_\gamma^*(1) = 0$. It follows that $\|V_\gamma^\pi - V_\gamma^*\| = 2/(1 - \gamma) = 2d_B(Q)/(1 - \gamma)$. ∎

Proof of Theorem 21: We prove the desired bound by applying, in turn, the triangle inequality, Assumption 9, Theorems 3, 6 and 18, Assumption 14, and finally Theorem 18 a second time:

$$\begin{aligned}
\|V_\gamma^\pi - V_\gamma^*\| &\le \|V_\gamma^\pi - V_{Q,\pi}\| + \|V_Q - V_\gamma^*\| \\
&= \|V_{Q_\gamma^\pi,\pi} - V_Q\| + \|V_Q - V_{Q_\gamma^*}\|
\end{aligned}$$

$$\leq \|Q_\gamma^\pi - Q\| + \|Q - Q_\gamma^*\|$$

$$\leq \frac{\|Q - B_\gamma^\pi Q\|}{1 - \gamma} + \|Q - Q_\gamma^*\|$$

$$= \frac{\|Q - B_\gamma Q\|}{1 - \gamma} + d_M(Q)$$

$$\leq \frac{(1 + \gamma) \cdot d_M(Q)}{1 - \gamma} + d_M(Q)$$

$$= \frac{2}{1 - \gamma} \cdot d_M(Q).$$

To see that this bound cannot be tightened any further, consider again the MDP that we used in the proof of Theorem 20, but this time in conjunction with the Q-function Q that is defined by $Q(1, 1) = Q(1, 2) = 1$. The stationary policy π defined by $\pi(1) = 1$ is greedy for Q and Assumptions 6 and 8 imply that $V_\gamma^\pi(1) = 2/(1 - \gamma)$ and $V_\gamma^*(1) = 0$. Assumptions 9, 10, and 16 imply $Q_\gamma^*(i, a) = c_i^a + \gamma \cdot V_{Q_\gamma^*}(j_i^a) = c_i^a + \gamma \cdot V_\gamma^*(j_i^a)$ for all deterministic MDPs, states i, and actions $a \in A(i)$. Hence, $Q_\gamma^*(1, 1) = 2$ and $Q_\gamma^*(1, 2) = 0$ for the example. It follows that $d_M(Q) = \|Q_\gamma^* - Q\| = 1$ and, finally, $\|V_\gamma^\pi - V_\gamma^*\| = 2/(1 - \gamma) = 2 d_M(Q)/(1 - \gamma)$. ∎

Proof of Theorem 22: Let i be an arbitrary state. We prove the desired bound by applying, in turn, Assumptions 6 and 9, Theorems 2, 3, 6, and 18, and finally Assumption 14.

$$\|V_\gamma^\pi - V_\gamma^*\| \leq V_\gamma^\pi(i) - V_\gamma^*(i)$$

$$= V_{Q_\gamma^\pi, \pi}(i) - V_{Q_\gamma^*}(i) \leq V_{Q_\gamma^\pi, \pi}(i) - V_Q(i)$$

$$= V_{Q_\gamma^\pi, \pi}(i) - V_{Q, \pi}(i) \leq \|V_{Q_\gamma^\pi, \pi} - V_{Q, \pi}\|$$

$$\leq \|Q_\gamma^\pi - Q\| \leq \frac{\|Q - B_\gamma^\pi Q\|}{1 - \gamma}$$

$$= \frac{\|Q - B_\gamma Q\|}{1 - \gamma} = \frac{d_B(Q)}{1 - \gamma}.$$

To see that this bound cannot be tightened any further, consider again the MDP that we used in the proof of Theorems 20 and 21, but this time in conjunction with the Q-function Q that is defined by $Q(1, 1) = Q(1, 2) = 0$. In the proof of Theorem 21, we showed that $V_\gamma^*(1) = 0$, $Q_\gamma^*(1, 1) = 2$, and $Q_\gamma^*(1, 0) = 0$. Therefore, $Q \leq Q_\gamma^*$. Assumption 16 implies that $B_\gamma Q(1, 1) - Q(1, 1) = 2 + \gamma \cdot V_Q(1) - Q(1.1) = 2$, $B_\gamma Q(1, 2) - Q(1, 2) = 0 + \gamma \cdot V_Q(1) - Q(1, 2) = 0$, and therefore $d_B(Q) = \|B_\gamma Q - Q\| = 2$. The stationary policy π with $\pi(1) = 1$ is greedy for Q. Assumptions 6 and 8 imply $V_\gamma^\pi(1) = 2/(1 - \gamma)$, and, therefore, $\|V_\gamma^\pi - V_\gamma^*\| = 2/(1 - \gamma) = d_B(Q)/(1 - \gamma)$. ∎

Proof of Corollary 4: The desired bound is an immediate consequence of Theorems 22 and 19. The example used to prove Theorem 22 proves that the bound is tight, because it satisfies $Q_\gamma^*(1,1) = 2$, $Q_\gamma^*(1,2) = 0$, and therefore $d_B(Q) = d_M(Q)$. ∎

LEMMA 2 *Let Q be a Q-function that satisfies $Q \leq B_\gamma Q$. Then $Q \leq Q_\gamma^*$.*

Proof: Consider mappings $B_{\gamma,n}$ from $\mathcal{R}^M$ to $\mathcal{R}^M$ that are inductively defined as follows:

$$\forall n \in \mathbb{N} \, \forall Q \in \mathcal{R}^M : B_{\gamma,0}Q = Q, \quad B_{\gamma,n+1}Q = B_\gamma(B_{\gamma,n}Q).$$

It follows by induction from Assumption 12 that $Q \leq B_{\gamma,n}Q$ for all n if $Q \leq B_\gamma Q$. Similarly, it follows by induction from Assumption 10 that $B_{\gamma,n}Q_\gamma^* = Q_\gamma^*$ and from Assumption 13 that $\|B_{\gamma,n}Q_\gamma^* - B_{\gamma,n}Q\| \leq \gamma^n \|Q_\gamma^* - Q\|$, for all n. Put together, we can conclude that $\|Q_\gamma^* - B_{\gamma,n}Q\| \leq \gamma^n \|Q_\gamma^* - Q\|$ for all n which further implies $\lim_{n \to \infty} B_{\gamma,n}Q = Q_\gamma^*$ and, finally, $Q \leq Q_\gamma^*$. ∎

Proof of corollary 5: Lemma 2 implies $Q \leq Q_\gamma^*$. Thus, Theorem 22 can be applied and gives the desired result. ∎

Proof of Corollary 6: Lemma 2 implies $Q \leq Q_\gamma^*$. Thus, Corollary 4 can be applied and gives the desired result. ∎

Proof of Theorem 23: Let Q be an arbitrary Q-function. We will show that for every state i there exists a real value $\epsilon(i) > 0$ with the following property: If $\max_{a \in A(i)} |Q(i,a) - Q_\gamma^*(i,a)| < \epsilon(i)$ then every action that is greedy for Q and i is also greedy for Q_γ^* and i. If such values $\epsilon(i)$ exist, then $\epsilon := \min_{i \in S} \epsilon(i)$ gives the desired result.

Let i be any state. If every action in $A(i)$ is greedy for Q_γ^* and i, then $\epsilon(i)$ can be set to an arbitrary non-zero value, for example one. Otherwise, set

$$\epsilon(i) := \frac{\min\left\{r \in \mathcal{R} : \quad \exists a \in A(i) : r = Q_\gamma^*(i,a) \neq V_{Q_\gamma^*}(i)\right\}}{2}.$$

The desired result follows using Assumption 17. ∎

Proof of Corollary 7: The corollary follows immediately from Theorems 23 and 18. ∎

Proof of Theorem 24: From Assumption 17 we conclude that the Q-function $Q_\gamma^* + 2r$ has the desired property. ∎

Proof of Corollary 8: The corollary follows immediately from Theorems 24 and 18.

∎

Notes

1. In the dynamic programming literature, it is often assumed that the immediate cost of a state transition depends only on the starting state and the executed action, but not on the successor state reached. This is the assumption that we make in this section. We could easily extend the theory to the more general case, but are afraid that the resulting dynamic programming operators might look unfamiliar to the reader.
2. This assumption is made for convenience. It follows already from Assumptions 1 and 7.

References

Baird, L., (1995). "Residual Algorithms: Reinforcement Learning with Function Approximation," *Machine Learning, Proceedings of the 12th International Conference*, Morgan Kaufmann Publishers, Inc., San Francisco, CA.

Barto, A. G., Sutton, R. S., & Anderson, C. W., (1983)."Neuronlike adaptive elements that can solve difficult learning control problems," *IEEE Transactions on Systems, Man, and Cybernetics 13*, pp. 834-846.

Bertsekas, D.P., (1987). *Dynamic Programming: Deterministc and Stochstic Models*, Englewood Cliffs, NJ: Prentice Hall.

Boyan, J. & Moore, A., (1995). "Generalization in Reinforcement Learning: Safely Approximating the Value Function," G. Tesauro, D. S. Touretzky and T. K. Lean, editors, *Advances in Neural Information Processing Systems 7*, MIT Press, Cambridge, MA.

Crites, R. H. & Barto, A. G., (1995, unpublished). "Improving Elevator Performance Using Reinforcement Learning."

Gordon, G., (1995). "Stable function approximation in Dynamic Programming." *Machine Learning, Proceedings of the 12th International Conference*, Morgan Kaufmann Publishers, Inc., San Francisco, CA.

Heger, M. (1994a), "Consideration of Risk in Reinforcement Learning," *Machine Learning, Proceedings of the 11th International Conference*, pp. 105-111, Morgan Kaufmann Publishers, Inc., San Francisco, CA.

Heger, M. (1994b), "Risk and Reinforcement Learning: Concepts and Dynamic Programming," *Technical report No. 8/94*, Zentrum für Kognitionswissenschaften, Universität Bremen, Fachbereich 3 Informatik, Germany.

Heger, M. (1995), "Risk and Reinforcement Learning: Algorithms." *Technical report (forthcomming)*, Zentrum für Kognitionswissenschaften, Universität Bremen, Fachbereich 3 Informatik, Germany.

Lin, L.-J., (1993). "Scaling Up Reinforcement Learning for Robot Control." *Machine Learning, Proceedings of the 10th International Conference*, pp. 182-189, Morgan Kaufmann Publishers, Inc., San Francisco, CA.

Littman, M. L., (1994). "Markov games as a framework for multi-agent reinforcement learning," *Machine Learning, Proceedings of the 11th International Conference*. pp. 157-111. Morgan Kaufmann Publishers, Inc., San Francisco, CA.

McDonald, M. A. F. & Hingson P., (1994). "Approximate Discounted Dynamic Programming Is Unreliable," *Technical report 94/6*, University of Western Australia.

Moore, A.W. & Atkeson, C. G., (to appear). "The Parti-game Algorithm for Variable Resolution Reinforcement Learning in Multidimensional State-spaces," *Machine Learning*.

Ross, S. M., (1970). *Applied Probability Models with Optimization Applications*, San Francisco, California: Holden Day.

Schwartz, A., (1993). "A reinforcement learning method for maximizing undiscounted rewards," *Machine Learning, Proceedings of the 10th International Conference*. pp. 298-305. Morgan Kaufmann Publishers, Inc., San Francisco, CA.

Singh, P. S. & Yee, R. C., (1994). "An Upper Bound on the Loss from Approximate Optimal-Value Functions," *Machine Learning*, No 16, pp. 227-233, Kluwer Academic Publishers, Boston.

Sutton, R. S., (unpublished). "Generalization in Reinforcement Learning: Successful Examples Using Sparse Coarse Coding."

Taha, H. A., (1987). *Operations Research: An Introduction, Fourth Edition*, Macmillan Publishing Company, New York.

Tesauro, G. J., (1992). "Practical Issues in Temporal Difference Learning," *Machine Learning, 8(3/4)*, pp. 257-277.

Thrun, S. B., (1992) "Efficient Exploration in Reinforcement Learning," *Technical report, CMU-CS-92-102*, School of Computer Science, Carnegie Mellon University, Pittsburgh, PA.

Thrun, S. B. & Schwartz, A., (1993). "Issues in Using Function Approximation for Reinforcement Learning," *Proceedings of the Fourth Connectionist Models Summer School*, Lawrence Erlbaum Publishers, Hillsdale, NJ.

Tsitsiklis, J. N., (1994). "Asynchronous Stochastic Approximation and Q-Learning," *Machine Learning*, No 16, pp. 185–202, Kluwer Academic Publishers, Boston.

Watkins, C. J. C. H., (1989). *Learning from Delayed Rewards*, PhD thesis, Cambridge University, England.

Watkins, C. J. C. H. & Dayan, P., (1992). "Q-Learning," *Machine Learning*, No 8, pp. 279-292.

Williams, R.J. & Baird, III, L.C., (1993). 'Tight Performance Bounds on Greedy Policis Based on Imperfect Value Functions," *Technical Report NU-CCS-93-14*, Northeastern University.

Zhang, W. & Dietterich, T. G., (1995). "A Reinforcement Learning Approach to Job-shop Scheduling," *Proceedings of the 14th International Joint Conference on Artificial Intelligence*, Morgan Kaufmann Publishers, Inc., San Francisco, CA.

Received November 8, 1994
Accepted February 24, 1995
Final Manuscript October 6, 1995

Machine Learning, 22, 227–250 (1996)

The Effect of Representation and Knowledge on Goal-Directed Exploration with Reinforcement-Learning Algorithms*

SVEN KOENIG   skoenig@cs.cmu.edu

REID G. SIMMONS   reids@cs.cmu.edu

School of Computer Science, Carnegie Mellon University, Pittsburgh, PA 15213-3890, USA

Editor: Leslie Pack Kaelbling

Abstract. We analyze the complexity of on-line reinforcement-learning algorithms applied to goal-directed exploration tasks. Previous work had concluded that, even in deterministic state spaces, initially uninformed reinforcement learning was at least exponential for such problems, or that it was of polynomial worst-case time-complexity only if the learning methods were augmented. We prove that, to the contrary, the algorithms are tractable with only a simple change in the reward structure ("penalizing the agent for action executions") or in the initialization of the values that they maintain. In particular, we provide tight complexity bounds for both Watkins' Q-learning and Heger's Q-hat-learning and show how their complexity depends on properties of the state spaces. We also demonstrate how one can decrease the complexity even further by either learning action models or utilizing prior knowledge of the topology of the state spaces. Our results provide guidance for empirical reinforcement-learning researchers on how to distinguish hard reinforcement-learning problems from easy ones and how to represent them in a way that allows them to be solved efficiently.

Keywords: action models, admissible and consistent heuristics, action-penalty representation, complexity, goal-directed exploration, goal-reward representation, on-line reinforcement learning, prior knowledge, reward structure, Q-hat-learning, Q-learning

1. Introduction

A **goal-directed reinforcement-learning problem** can often be stated as: an agent has to learn an optimal policy for reaching a goal state in an initially unknown state space. One necessary step towards a solution to this problem is to locate a goal state. We therefore analyze the complexity of on-line reinforcement-learning methods (measured in action executions) until the agent reaches a goal state for the first time. If a reinforcement-learning method is terminated when the agent reaches a goal state, then it solves the following **goal-directed exploration problem**: the agent has to find some path to a goal state in an initially unknown or only partially known state space, but it does not need to find a shortest path. Studying goal-directed exploration problems provides insight into the corresponding reinforcement-learning problems. Whitehead (1991a), for example, proved that goal-directed exploration with reinforcement-learning methods can be intractable, which demonstrates that solving reinforcement-learning problems can be intractable as

* This research was supported in part by NASA under contract NAGW-1175 . The views and conclusions contained in this document are those of the authors and should not be interpreted as representing the official policies, either expressed or implied, of NASA or the U.S. government.

well. In particular, he showed that the behavior of an agent that is controlled by an initially uninformed (i.e. tabula rasa) reinforcement-learning algorithm can degenerate to a random walk until it reaches a goal state. In this case, it might have to execute a number of actions that is, on average, exponential in the size of the state space and even speed-up techniques such as Lin's action-replay technique (Lin, 1992) do not improve its performance. This theoretical result contrasts sharply with experimental observations of Kaelbling (1990), Whitehead (1991b), Peng and Williams (1992), and Moore and Atkeson (1993b), who reported good performance results.

These results motivated us to study how the performance of reinforcement-learning algorithms in deterministic and non-deterministic state spaces is affected by different representations, where a **representation** determines both the immediate rewards that the reinforcement-learning algorithms receive and the initialization of the values that they maintain. We study versions of both Watkins' Q-learning and Heger's Q̂-learning that perform only a minimal amount of computation between action executions, choosing only which action to execute next, and basing this decision only on information local to the current state of the agent. If these inefficient algorithms have a low complexity, then we can expect other reinforcement-learning algorithms to have a low complexity as well. One main result of our analysis is that the choice of representation can have a tremendous impact on the performance of these reinforcement-learning algorithms. If a poor representation is chosen, they are intractable. However, if a good reward structure (such as "penalizing the agent for action executions") or suitable initialization is chosen, the complexity of the same algorithms is only a small polynomial in the size of the state space. Consequently, reinforcement-learning algorithms can be tractable without any need for augmentation. One can decrease their complexity even further by either learning action models or utilizing prior knowledge of the topology of the state space. These results, the proofs of which are found in (Koenig and Simmons, 1995a), provide guidance for empirical reinforcement-learning researchers on how to distinguish hard reinforcement-learning problems from easy ones and how to represent them in a way that allows them to be solved efficiently. They also provide the theoretical underpinnings for the experimental results mentioned above.

2. Notation and Assumptions

We use the following notation: S denotes the finite set of states of the state space, $s_{start} \in S$ is the start state, and G (with $\emptyset \neq G \subseteq S$) is the set of goal states. $A(s)$ is the finite set of actions that can be executed in state $s \in S$. Executing action $a \in A(s)$ causes a (potentially nondeterministic) state transition into one of the states $succ(s, a)$ (with $\emptyset \neq succ(s, a) \subseteq S$). The **size** of the state space is $n := |S|$, and the total number of state-action pairs (loosely called **actions**) is $e := \sum_{s \in S} |A(s)|$. We assume that the state space does not change over time and is completely observable: the agent can always determine its current state with certainty. It also knows at every point in time which actions it can execute in its current state and whether its current state is a goal state.

A state space is **deterministic** iff the cardinality of $succ(s, a)$ equals one for all $s \in S$ and $a \in A(s)$. For deterministic state spaces, we use $succ(s. a)$ to denote both the set of successor states and the single element of this set. We call a state space **non-deterministic** if we want to stress that we do not require it to be deterministic. In non-deterministic state spaces, we view reinforcement learning as a two-player game: The reinforcement-learning algorithm selects the action to execute and is only constrained by having to choose an action that is applicable in the current state of the agent. The action determines the possible successor states, from which some mechanism, which we call **nature**, chooses one. We do not impose any restrictions on how nature makes its decisions (its **strategy**). In particular, its choice can depend on the state-action history (we do not make the Markov assumption). Nature might, for example, select a successor state randomly or deliberately, in the latter case either to help or hurt the agent. Reinforcement learning in deterministic state spaces is simply a special case of reinforcement learning in non-deterministic state spaces where every action execution results in a unique successor state and nature has no choice which one to select.

The **distance** $d(s, s') \in [0, \infty]$ between $s \in S$ and $s' \in S$ is defined to be the (unique) solution of the following set of equations

$$d(s, s') = \begin{cases} 0 & \text{for all } s, s' \in S \text{ with } s = s' \\ 1 + \min_{a \in A(s)} \max_{s'' \in succ(s,a)} d(s'', s') & \text{for all } s, s' \in S \text{ with } s \neq s'. \end{cases}$$

This means that an agent that knows the state space and acts optimally can reach s' from s with at most $d(s, s')$ action executions no matter which strategy nature uses. The **goal distance** $gd(s)$ of $s \in S$ is defined to be $gd(s) := \min_{s' \in G} d(s, s')$. If the agent had to traverse a state more than once in order to reach a goal state, then nature could force it to traverse this cycle infinitely often, which would imply $gd(s) = \infty$. Thus, $gd(s) \leq n - 1$ if $gd(s)$ is finite. We define the **diameter** (depth) of the state space with respect to G as $d := \max_{s \in S} gd(s)$. If d is finite, then $d \leq n - 1$. These definitions correspond to a worst-case scenario, in which nature is an opponent of the agent. For deterministic state spaces, the definitions of $d(s, s')$ and $gd(s)$ simplify to the standard definitions of distance and goal distance, respectively.

We call a reinforcement-learning algorithm **uninformed** if it initially does not know which successor states an action can lead to. By its **complexity** we mean an upper bound on the number of actions that an agent controlled by the uninformed reinforcement-learning algorithm can execute in state spaces of a given size until it reaches a goal state for the first time. This bound has to hold for all possible topologies of the state space, start and goal states, tie breaking rules among indistinguishable actions, and strategies of nature. If $gd(s_{start}) = \infty$, then even completely informed reinforcement-learning algorithms have infinite complexity, since there exists a strategy of nature that prevents the agent from reaching any goal state. This problem could be solved by requiring $gd(s_{start}) < \infty$, but this is not a sufficient condition to guarantee the existence of uninformed reinforcement-learning algorithms with finite complexity. We call a state s **lost** iff $gd(s) = \infty$. If the agent enters a lost state during exploration, then nature can prevent it from reaching a goal state. Thus, the complexity can only be finite if no states are lost (i.e. if $d < \infty$).[1] We call state spaces with this property **safely explorable** and limit our analysis to such state spaces. Moore and Atkeson's parti-game algorithm (Moore and Atkeson, 1993a), for example, learns safely explorable abstractions of spatial

Initially, the agent is in state $s_{start} \in S$.

1. Set $s :=$ the current state.

2. If $s \in G$, then stop.

3. Set $a := argmax_{a' \in A(s)} Q(s, a')$.
   (Read: "Choose an action $a \in A(s)$ with the largest $Q(s, a)$ value.")

4. Execute action a.
   (As a consequence, nature selects a state $s' \in succ(s, a)$, the agent receives the immediate reward $r(s, a, s')$, and its new state becomes s'.)

5. Update $Q(s, a)$ using $r(s, a, s')$ and $U(s')$.

6. Go to 1.

where $U(s) := \max_{a' \in A(s)} Q(s, a')$ (i.e. the value of a state is the largest value of its actions).

Figure 1. A framework for 1-step Q-learning

state spaces. Other examples of safely explorable state spaces include two-player zero-sum games where one player can force a win and the game is restarted if this player loses. For deterministic state spaces, researchers such as Whitehead (1991a) and Ishida (1992) usually make the more restrictive assumption that the state spaces are safely explorable for all $s - start \in S$ and $G \subseteq S$, not just the $s - start$ and G given (i.e. that they are strongly connected).

3. A General Q-Learning Framework

Reinforcement learning is learning from positive and negative rewards (costs). When the agent executes action a in state s and successor state s' results, it receives the immediate reward $r(s, a, s') \in \mathcal{R}$. If the agent receives immediate reward r_t when it executes the $(t + 1)$st action, then the **total reward** that it receives over its lifetime for this particular behavior is $\sum_{t=0}^{\infty} \gamma^t r_t$. The **discount factor** $\gamma \in (0.1]$ specifies the relative value of a reward received after t action executions compared to the same reward received one action execution earlier. We say that **discounting** is used if $\gamma < 1$, otherwise no discounting is used. Reinforcement learning algorithms solve goal-directed reinforcement-learning problems by determining a behavior for the agent that maximizes its total reward. They specify such behaviors as state-action rules, called **policies**.

We analyze reinforcement-learning methods that are variants of Q-learning, probably the most popular reinforcement-learning method. In particular, we study 1-step versions of on-line Q-learning. They perform only a minimal amount of computation between action executions, choosing only which action to execute next, and basing this decision only on information local to the current state of the agent. A general framework for 1-step Q-learning (Figure 1) consists of a **termination-checking step** (line 2), an **action-selection step** (line 3), an **action-execution step** (line 4), and a **value-update step** (line 5). The termination-checking step stops the agent when it reaches a goal state.

This is possible, because we have limited ourselves to studying goal-directed exploration problems. The action-selection step determines which action to execute next. This decision is based on values that store information about the relative goodness of the actions, one **Q-value** $Q(s, a) \in \mathcal{R}$ for each action a that can be executed in state s. $Q(s, a)$ approximates the total reward that the agent receives if it starts in s, executes a, and then behaves optimally. The action-selection strategy is greedy: it chooses the action with the largest Q-value in the curent state of the agent. (If several actions tie, an arbitrary one of the equally good actions can be selected.) Because the action-selection step uses only those Q-values that are local to the current state of the agent, there is no need to predict $succ(s, a)$ for any $a \in A(s)$, which means that 1-step Q-learning does not need to learn an **action model** of the state space. After the action-execution step has directed the agent to execute the desired action a in its current state s, nature selects the successor state s' from the states $succ(s, a)$ and the agent receives the immediate reward $r(s, a, s')$. 1-step Q-learning then has temporary access to the Q-values of the agent's former and new state at the same time, and the value-update step adjusts $Q(s, a)$ using $r(s, a, s')$ and the values $Q(s', a')$ for $a' \in A(s')$. This is done because a 1-step look-ahead value of the total reward is more accurate than, and should therefore replace, the current value of $Q(s, a)$. For now, we leave the initial Q-values unspecified.

Two well-known Q-learning algorithms fit this general Q-learning framework:

- A 1-step version (Whitehead, 1991a) of Watkins' **Q-learning algorithm** (Watkins, 1989) can be obtained from the Q-learning framework by making the value-update step

  "Set $Q(s, a) := (1 - \alpha)Q(s, a) + \alpha(r(s, a, s') + \gamma U(s'))$,"

  where $\alpha \in (0, 1]$ is called the **learning rate**. This Q-learning algorithm assumes that nature always selects the successor state $s' \in succ(s, a)$ with some time-invariant (but unknown) probability $p(s, a, s')$. Thus, nature selects the successor states probabilistically, and the agent plans for the average (i.e. risk-neutral) case: it tries to maximize its *expected total reward*. Convergence results for Q-learning are given in (Watkins and Dayan, 1992).

  The learning rate determines how much $Q(s, a)$ changes with every update. In order for the Q-values to converge in non-deterministic state spaces to the desired values, it has to approach zero asymptotically – in a manner described in (Watkins and Dayan, 1992). Q-learning needs the learning rate to be able to average the values $r(s, a, s') + \gamma U(s')$ for all successor states $s' \in succ(s, a)$. In deterministic state spaces, however, there is only one successor state and, consequently, averaging is not necessary. Thus, the learning rate can be set to one.

- Heger's **Q̂-learning algorithm** (Heger, 1994) (pronounced "Q-hat-learning") can be obtained from the Q-learning framework by making the value-update step

  "Set $Q(s, a) := \min(Q(s, a), r(s, a, s') + \gamma U(s'))$."

and initializing the Q-values optimistically. $\hat{Q}$-learning assumes that nature always selects the worst successor state for the agent. Thus, nature is an opponent of the agent, and the agent tries to maximize the *total reward that it can receive in the worst case*. Even if the successor states are determined probabilistically, the agent could be extremely risk-averse (in the sense of utility theory; see (Koenig and Simmons, 1994)) and believe in Murphy's law ("anything that can go wrong, will indeed go wrong"). It then assumes that an imaginary opponent exists that always makes the worst possible successor state occur that its action can result in. In this case, $\hat{Q}$-learning can be used to learn a behavior that reflects a completely risk-averse attitude and is optimal for the agent provided that it accepts the axioms of utility theory.

$\hat{Q}$-learning needs less information about the state space than Q-learning and converges faster. In particular, it does not need to execute each action $a \in A(s)$ in state s often enough to get a representative distribution over the successor states $succ(s, a)$ and has no need for a learning rate; for details see (Heger, 1996). Convergence results for $\hat{Q}$-learning are given in (Heger, 1994).

If the agent wants to learn an optimal policy, its risk attitude determines which Q-learning algorithm it should use. In the following sections, we analyze the complexity of Q-learning and $\hat{Q}$-learning for different representations of goal-directed exploration problems.

4. Representations

When modeling a goal-directed reinforcement-learning or exploration problem, one has to decide on both the immediate rewards and the initial Q-values. So far, we have left these values unspecified. In this section, we introduce possible reward structures and initializations, and discuss their properties. All of these representations have been used in the experimental reinforcement-learning literature to represent goal-directed reinforcement-learning problems, i.e. for learning shortest paths to a goal state. Since we have restricted our analysis to goal-directed exploration problems, we have to decide on the following values: For the reward structure, we have to decide on the values $r(s, a, s') \in \mathcal{R}$ for $s \in S \setminus G := S \cap \overline{G}$, $a \in A(s)$, and $s' \in succ(s, a)$. For the initial Q-values, we have to decide on the values $Q(s, a)$ for $s \in S \setminus G$ and $a \in A(s)$. In both cases, the values for $s \in G$ do not matter, because the reinforcement-learning methods terminate when the agent reaches a goal state. For goal-directed reinforcement learning problems, we define $Q(s, a) = 0$ for $s \in G$ and a $a \in A(s)$.)

4.1. Reward Structures

Developing appropriate reward structures can, in general, be a complex engineering problem (Matarić, 1994). Since goal-directed reinforcement-learning methods determine policies that maximize the total reward, a reward structure for learning shortest paths must guarantee that a state with a smaller goal distance also has a larger optimal total reward. Two different reward structures have been used in the literature. Neither of them encodes any information about the topology of the state space.

- In the **goal-reward representation**, the agent is rewarded for entering a goal state, but not rewarded or penalized otherwise. This reward structure has been used by Sutton (1990), Whitehead (1991a), Peng and Williams (1992), and Thrun (1992b), among others. Formally,

$$r(s, a, s') := \begin{cases} 1 & \text{for } s \in S \setminus G, \ a \in A(s), \text{ and } s' \in G \\ 0 & \text{for } s \in S \setminus G, \ a \in A(s), \text{ and } s' \in S \setminus G. \end{cases}$$

Discounting is necessary for learning shortest paths with this reward structure. If no discounting were used, all behaviors that lead the agent eventually to a goal state would have a total reward of one and the reinforcement-learning algorithms could not distinguish between paths of different lengths. If discounting is used, then the goal reward gets discounted with every action execution, and the agent tries to reach a goal state with as few action executions as possible in order to maximize the portion of the goal reward that it receives.

- In the **action-penalty representation**, the agent is penalized for every action that it executes. This reward structure is **denser** than the goal-reward representation (the agent receives non-zero rewards more often) if goals are relatively sparse. It has been used by Barto, Sutton, and Watkins (1989), Koenig (1991), and Barto, Bradtke, and Singh (1995), among others. Formally,

$$r(s, a, s') := -1 \qquad \text{for } s \in S \setminus G, \ a \in A(s), \text{ and } s' \in S.$$

For learning shortest paths, discounting can be used, but is not necessary. In both cases, the agent tries to reach a goal state with as few action executions as possible in order to minimize the amount of penalty that it receives. The action-penalty representation can be generalized to non-uniform immediate rewards; see Section 7.

4.2. Initial Q-Values

The complexity of Q- or Q̂-learning depends on properties of the initial Q-values. An important special case of initial Q-values are those that do not convey any information about the state space, such as uniformly initialized Q-values.

Definition. Q- or Q̂-learning is **uniformly initialized** with $q \in \mathcal{R}$ (or, synonymously, q-initialized), iff initially

$$Q(s, a) = \begin{cases} 0 & \text{for all } s \in G \text{ and } a \in A(s) \\ q & \text{for all } s \in S \setminus G \text{ and } a \in A(s). \end{cases}$$

If Q- or Q̂-learning is q-initialized with $q \neq 0$, then goal states are initialized differently from non-goal states. It is important to understand that this particular initialization does not convey any information, since the agent is able to distinguish whether its current

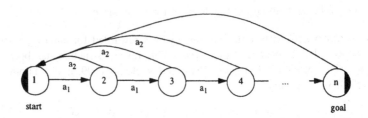

Figure 2. A state space ("reset state space") for which random walks need $3 \times 2^{n-2} - 2$ action executions on average to reach the goal state (for $n \geq 2$)

state is a goal state or not and can therefore initialize the Q-values differently for goal and non-goal states. Consequently, all uniformly initialized Q-learning algorithms are uninformed. This includes both zero-initialized and (minus one) -initialized Q-learning algorithms.

The following definitions characterize properties of Q-values for undiscounted Q- or $\hat{Q}$-learning with action-penalty representation.

Definition. Q-values are **consistent** for undiscounted Q- or $\hat{Q}$-learning with action-penalty representation iff

$$\left. \begin{array}{c} 0 \\ -1 + \min_{s' \in succ(s,a)} U(s') \end{array} \right\} \leq Q(s,a) \leq 0 \quad \left\{ \begin{array}{l} \text{for all } s \in G \text{ and } a \in A(s) \\ \text{for all } s \in S \setminus G \text{ and } a \in A(s). \end{array} \right.$$

Consistency means that the triangle inequality holds. It corresponds to the concept of consistency or monotonicity in heuristic search. Both zero-initialized and (minus one)-initialized Q-values are consistent, for example.

Definition. Q-values are **admissible** for undiscounted Q- or $\hat{Q}$-learning with action-penalty representation iff

$$\left. \begin{array}{c} 0 \\ -1 - \max_{s' \in succ(s,a)} gd(s') \end{array} \right\} \leq Q(s,a) \leq 0 \quad \left\{ \begin{array}{l} \text{for all } s \in G \text{ and } a \in A(s) \\ \text{for all } s \in S \setminus G \text{ and } a \in A(s). \end{array} \right.$$

Admissibility means that $-Q(s,a)$ never overestimates the number of action executions that the agent needs for reaching a goal state if it knows the state space, starts in state s, executes action a, and then behaves optimally, but nature tries to keep it away from a goal state for as long as possible. This corresponds to the concept of admissibility in heuristic search. All consistent Q-values are admissible.

5. An Intractable Representation

In this section, we assume that Q-learning operates on the goal-reward representation and is zero-initialized. We show that this representation makes the exploration problem intractable.

The agent receives a non-zero immediate reward only if an action execution results in a goal state. Thus, during the search for a goal state, all Q-values remain zero, and the action-selection step has no information on which to base the decision which action to execute next.[2] If Q-learning had a systematic bias for actions, for example if it always chose the smallest action according to some predetermined ordering, then the agent could cycle in the state space forever. We therefore assume that it selects among the available actions with uniform probability, in which case the agent performs a random walk. Although random walks reach a goal state with probability one in safely explorable state spaces, the required number of action executions can exceed every given bound. This implies that the complexity of Q-learning is infinite. In deterministic state spaces, the expected number of action executions x_s that the agent needs to reach a goal state from state s can be calculated by solving the following set of linear equations:

$$x_s = \begin{cases} 0 & \text{for } s \in G \\ 1 + \dfrac{1}{|A(s)|} \displaystyle\sum_{a \in A(s)} x_{succ(s,a)} & \text{for } s \in S \setminus G. \end{cases}$$

Although $x_{s_{start}}$ is finite, it can scale exponentially with n. Consider for example the (deterministic) reset state space shown in Figure 2. A **reset state space** is one in which all states (except for the start state) have an action that leads back to the start state. It corresponds for example to the task of stacking n blocks if the agent can, in every state, either stack another block or scramble the stack. Solving the linear equations $x_1 = 1 + x_2$, $x_s = 1 + 0.5x_1 + 0.5x_{s+1}$ for all $s \in \{2, 3, \ldots, n-1\}$, and $x_n = 0$ yields $x_{s_{start}} = x_1 = 3 \times 2^{n-2} - 2$. Since the complexity of Q-learning cannot be lower in non-deterministic state spaces (a superset of deterministic state spaces), the following result holds:

THEOREM 1 *The expected number of action executions that zero-initialized Q-learning with goal-reward representation needs to reach a goal state in deterministic or non-deterministic state spaces can be exponential in n, the size of the state space.*

Whitehead (1991a) made the same observation for the behavior of Q-learning in deterministic state spaces that have the following property: in every state (except for the states on the perimeter of the state space), the probability of choosing an action that leads away from the only goal state is larger than the probability of choosing an action that leads closer to the goal state. This observation motivated him to explore cooperative reinforcement-learning methods in order to decrease the complexity. Subsequently, Thrun (1992a) showed that even non-cooperative reinforcement-learning algorithms can have polynomial complexity if they are extended with a directed exploration mechanism that he calls "counter-based exploration." Counter-based Q-learning, for example, maintains, in addition to the Q-values, a second kind of state-action values, called "counters." These values are used exclusively during learning (exploration) and are meaningless afterwards. Thrun was able to specify action-selection and value-update rules that use these counters and achieve polynomial complexity. There are other techniques that can improve the performance of reinforcement-learning algorithms, such as Lin's action-replay technique (Lin, 1992). However, this method does not change the Q-values before a goal state has been reached for the first time when it is applied to zero-initialized Q-learning with goal-reward representation and, thus, cannot be used to reduce its complexity.

6. Tractable Representations

In the following, we show that one does not need to augment Q-learning algorithms to reduce their complexity. They are tractable if one uses either the action-penalty representation or initial Q-values different from zero. The intuitive explanation is that, in both cases, the Q-values change immediately, starting with the first action execution. This way, the agent remembers the effects of previous action executions. Our analysis, that formally shows by how much the complexity is reduced, also explains experimental findings by Kaelbling (1990), Whitehead (1991b), Peng and Williams (1992), and Moore and Atkeson (1993b), who reported good performance results for Q-learning when using similar representations.

6.1. Deterministic State Spaces

We first analyze two representations that make Q- or $\hat{Q}$-learning tractable in deterministic state spaces and then discuss how their complexity can be reduced even further. Our analysis of deterministic state spaces is very detailed, because we can then transfer its results to $\hat{Q}$-learning in arbitrary non-deterministic state spaces.

6.1.1. Zero-Initialized Q-Values with Action-Penalty Representation

In this section, we analyze zero-initialized Q- or $\hat{Q}$-learning that operates on the action-penalty representation. Since the agent receives a non-zero immediate reward for every action execution, the Q-values change immediately,

6.1.1.1. Complexity Analysis

We first define admissible Q- or $\hat{Q}$-learning, show that undiscounted admissible Q- or $\hat{Q}$-learning is tractable in deterministic state spaces, and state how its complexity depends on properties of the state space. Then, we show how the analysis can be applied to discounted Q- or $\hat{Q}$-learning.

Definition. Undiscounted Q- or $\hat{Q}$-learning with action-penalty representation is **admissible** in deterministic state spaces[3] iff either

1. its initial Q-values are consistent and its value-update step[4] is "Set $Q(s,a) := -1 + U(s')$," or

2. its initial Q-values are admissible and its value-update step is "Set $Q(s,a) := \min(Q(s,a), -1 + U(s'))$."

The value-update step in the first part of the definition is the one of Q-learning with learning rate one. The value-update step in the second part of the definition is the one of Q̂-learning. If the Q-values are consistent, then either value-update step can be used, since consistent Q-values are always admissible. The most important property of admissible Q- or Q̂-learning is that initially consistent or admissible Q-values remain consistent or admissible, respectively, after every action execution and are monotonically non-increasing.

The correctness of admissible Q- or Q̂-learning for goal-directed exploration is easy to show. The argument that it reaches a goal state eventually parallels a similar argument for RTA*-type search algorithms (Korf, 1990) (Russell and Wefald, 1991) and is by contradiction: If the agent did not reach a goal state eventually, it would execute actions forever. In this case, there is a time t from which on the agent only executes those actions that it executes infinitely often. Every time the agent has executed all of these actions at least once, the largest Q-value of these actions has decreased by at least one. Eventually, the Q-values of all these actions drop below every bound. In particular, they drop below the Q-value of an action that Q- or Q̂-learning considers infinitely often for execution, but never executes after time t. Such an action exists, since the state space is safely explorable and thus the agent can always reach a goal state. Then, however, Q- or Q̂-learning is forced to execute this action after time t, which is a contradiction.

To understand intuitively why admissible Q- or Q̂-learning is tractable, assume that it is zero-initialized and consider the set $X := \{s \in S : U(s) = 0\} \supseteq G$. X is always the set of states in which the agent has not yet executed all of the available actions at least once. At every point in time, the following relationship holds:

$$-1 - \min_{s' \in X} d(succ(s, a), s') \leq Q(s, a) \leq 0 \qquad \text{for all } s \in S \setminus G \text{ and } a \in A(s).$$

The action-selection step can be interpreted as using $Q(s, a)$ to approximate $-1 - \min_{s' \in X} d(succ(s, a), s')$. Since it always executes the action with the largest Q-value, it tries (sometimes unsuccessfully) to direct the agent with as few action executions as possible from its current state to the closest state that has at least one **unexplored action** (an action that it has never executed before) and make it then take the unexplored action. Since any unexplored action can potentially lead to a goal state, Q- or Q̂-learning always executes the action that appears to be best according to its local view of the state space. This suggests that admissible Q- or Q̂-learning might have a low complexity. We therefore conduct a formal complexity analysis. It is centered around the invariant shown in Lemma 1. This lemma states that the number of executed actions is always bounded by an expression that depends only on the initial and current Q-values and, moreover, that "every action execution decreases the sum of all Q-values by one, except for a bounded number of action executions that leave the sum unchanged" (this paraphrase is somewhat simplified). A time superscript of t in Lemmas 1 and 2 refers to the values of the variables immediately before the agent executes the $(t+1)$st action (e.g. $s^0 = s_{start}$).

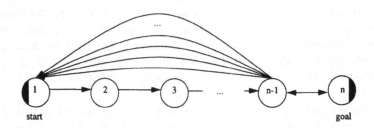

Figure 3. A state space for which uninformed Q- or Q̂-learning can need at least $(e - n + 1)(n - 1)$ action executions to reach the goal state (for $n \geq 2$ and $e \geq n$)

LEMMA 1 *For all times $t \in \mathcal{N}_0$ (until termination) of undiscounted admissible Q- or Q̂-learning with action-penalty representation in deterministic state spaces, it holds that*

$$U^t(s^t) + \sum_{s \in S} \sum_{a \in A(s)} Q^0(s, a) - t \geq \sum_{s \in S} \sum_{a \in A(s)} Q^t(s, a) + U^0(s^0) - loop^t$$

and

$$loop^t \leq \sum_{s \in S} \sum_{a \in A(s)} Q^0(s, a) - \sum_{s \in S} \sum_{a \in A(s)} Q^t(s, a),$$

where $loop^t := |\{t' \in \{0, \ldots, t-1\} : s^{t'} = s^{t'+1}\}|$ (the number of actions executed before t that did not change the state).

Lemma 2 uses Lemma 1 to derive a bound on t. This is possible, because "the sum of all Q-values decreases by one for every executed action, ..." (according to the invariant), but is bounded from below. That each of the e different Q-values is bounded from below by an expression that depends only on the goal distances follows directly from the definition of consistent or admissible Q-values and the fact that the Q-values maintain this property after every action execution.

LEMMA 2 *Undiscounted admissible Q- or Q̂-learning with action-penalty representation reaches a goal state after at most*

$$2 \sum_{s \in S \setminus G} \sum_{a \in A(s)} [Q^0(s, a) + gd(succ(s, a)) + 1] - U^0(s^0)$$

action executions in deterministic state spaces.

Theorem 2 uses Lemma 2 and the fact that $gd(s) \leq d$ for all $s \in S$ to state how the complexity of Q- or Q̂-learning depends on e and d. Theorem 2 also guarantees that admissible Q- or Q̂-learning terminates in safely explorable state spaces, because d is finite for such state spaces.

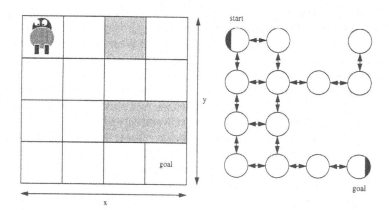

Figure 4. An example of a rectangular gridworld

THEOREM 2 *Undiscounted admissible Q- or Q̂-learning with action-penalty represen-*
tation reaches a goal state after at most $O(ed)$ action executions in deterministic state
spaces.

Theorem 2 provides an upper bound on the number of action executions that an agent
controlled by Q- or Q̂-learning needs to reach a goal state. (Such a worst-case bound is,
of course, also a bound on its average-case performance.) To demonstrate that $O(ed)$ is
a tight bound for uninformed Q- or Q̂-learning, we show that it is also a lower bound.
Lower bounds can be proven by example, one of which is depicted in Figure 3. In
this state space, uninformed Q- or Q̂-learning and every other uninformed exploration
algorithm can make the agent traverse a supersequence of the following state sequence
if ties are broken in favor of actions that lead to states with smaller numbers: $e - n + 1$
times the sequence $123 \ldots n - 1$, and finally n. Basically, all of the $O(e)$ actions in
state $n - 1$ are executed once. All of these (but one) lead the agent back to state 1
and therefore force it to execute another $O(d)$ actions before it can execute the next
unexplored action, resulting in a total of $O(ed)$ action executions before it reaches the
goal state. Thus, the following corollary holds:

COROLLARY 1 *Zero-initialized (or (minus one)-initialized) undiscounted Q- or Q̂-*
learning with action-penalty representation has a tight complexity of $O(ed)$ action exe-
cutions for reaching a goal state in deterministic state spaces.

This complexity can be expressed as a function of the size of the state space: $O(ed) \leq$
$O(en)$, since $d \leq n-1$ in all safely explorable state spaces. A state space has no duplicate
actions iff either $a = a'$ or $succ(s.a) \neq succ(s.a')$ for all $s \in S$ and $a, a' \in A(s)$.
$O(en) \leq O(n^3)$ for such state spaces, since $e \leq n^2$. The complexity of uninformed Q-
or Q̂-learning is even lower in many state spaces that are typically used by reinforcement-
learning researchers, because (a) their number of edges often increases only linearly with

the number of states, and/or (b) their diameter increases only sublinearly in n. Consider for example deterministic gridworlds with discrete states. They have often been used in studying reinforcement learning; see (Barto, Sutton, and Watkins, 1989), (Sutton, 1990), (Peng and Williams, 1992), (Singh, 1992), (Thrun, 1992b), or (Whitehead, 1992). In the state space shown in Figure 4, for example, the agent can move from any square to each of its four neighboring squares as long as it stays on the grid and the target square does not contain an obstacle. Thus, the number of actions increases at most linearly with the number of states. A rectangular obstacle-free gridworld of size $x \times y$ has $n = xy$ states, $e = 4xy - 2x - 2y$ state-action pairs, and diameter $d = x + y - 2$. If the gridworld is square ($x = y$), then its diameter increases sublinearly in n, since $d = 2\sqrt{n} - 2$, and $O(ed) = O(x^3) = O(n^{3/2})$. Even for safely-explorable rectangular gridworlds that contain arbitrary obstacles, the complexity of Q- or Q̂-learning cannot be larger than $O(n^2)$ action executions, since $e \leq 4n$, $d \leq n - 1$, and consequently $ed \leq 4n^2 - 4n$. Therefore, Q- or Q̂-learning actually has a very low complexity for finding goal states in unknown gridworlds.

6.1.1.2. Generalization of the Complexity Analysis

We can reduce the analysis of discounted Q- or Q̂-learning with action-penalty representation to the one of undiscounted Q- or Q̂-learning with the same reward structure. Consider the following strictly monotonically increasing bijection from the Q-values of a discounted Q- or Q̂-learning algorithm to the Q-values of the same algorithm with discount rate one: map a Q-value $Q_1^t(s,a) \in (1/(\gamma - 1), 0]$ of the former algorithm to the Q-value $Q_2^t(s,a) = -\log_\gamma(1 + (1 - \gamma)Q_1^t(s,a)) \in \mathcal{R}_0^-$ of the latter algorithm. If this relationship holds initially for the Q-values of the two algorithms, it continues to hold: if the execution of action a in state s results in successor state s', then either none of the Q-values changes or only the Q-values $Q_1(s,a)$ and $Q_2(s,a)$ change, in which case

$$
\begin{aligned}
Q_2^{t+1}(s,a) &= -1 + U_2^t(s') \\
&= -1 + \max_{a' \in A(s')} Q_2^t(s',a') \\
&= -1 + \max_{a' \in A(s')} (-\log_\gamma(1 + (1 - \gamma)Q_1^t(s',a'))) \\
&= -\log_\gamma(1 + (1 - \gamma)(-1 + \gamma \max_{a' \in A(s')} Q_1^t(s',a'))) \\
&= -\log_\gamma(1 + (1 - \gamma)(-1 + \gamma U_1^t(s'))) \\
&= -\log_\gamma(1 + (1 - \gamma)Q_1^{t+1}(s,a)).
\end{aligned}
$$

Because both algorithms always execute the action with the largest Q-value in the current state, they always choose the same action for execution (if ties are broken in the same way). Thus, they behave identically and the complexity analysis of the latter algorithm also applies to the former algorithm. This means that one can proceed as follows: Given initial Q-values $Q_1^0(s,a)$ for Q- or Q̂-learning with discounting, one first determines the corresponding Q-values $Q_2^0(s, a)$ according to the above formula. These Q-values can then be tested for their consistency or admissibility and finally be used in the formulas of Lemmas 1 and 2 to determine the complexity of discounted Q- or $\hat{Q}$-learning with initial Q-values $Q_1^0(s, a)$. If the values $Q_1^0(s, a)$ are zero-initialized, then the corresponding undiscounted Q- or $\hat{Q}$-learning algorithm has the same initialization, which implies that it has a tight complexity of $O(ed)$ action executions. Consequently, the following corollary holds:

COROLLARY 2 *Zero-initialized discounted Q- or $\hat{Q}$-learning with action-penalty representation has a tight complexity of $O(ed)$ action executions for reaching a goal state in deterministic state spaces.*

The other results from Section 6.1.1.1. can be transferred in the same way.

6.1.2. One-Initialized Q-Values with Goal-Reward Representation

We have shown that uninformed Q- or $\hat{Q}$-learning with action-penalty representation is tractable in deterministic state spaces, because the Q-values change immediately. Since the value-update step of Q- or $\hat{Q}$-learning updates the Q-values using both the immediate reward and the Q-values of the successor state, this result suggests that one can achieve tractability not only by changing the reward structure, but also by initializing the Q-values differently. In this section we show that, indeed, Q- or $\hat{Q}$-learning is also tractable if goal-reward representation is used (which implies that discounting has to be used as well, i.e. $\gamma < 1$) and the Q-values are $1/\gamma$- or one-initialized. The latter initialization has the advantage that it does not depend on any parameters of Q- or $\hat{Q}$-learning.

The complexity analysis of $1/\gamma$-initialized Q- or $\hat{Q}$-learning with goal-reward representation can be reduced to the one of zero-initialized Q- or $\hat{Q}$-learning with action-penalty representation. Similarly, the complexity analysis of one-initialized Q- or $\hat{Q}$-learning with goal-reward representation can be reduced to the one of (minus one)-initialized Q- or $\hat{Q}$-learning with action-penalty representation. In both cases, we can proceed in a way similar to the method of Section 6.1.1.2. This time, we consider the following strictly monotonically increasing bijection from the Q-values of a discounted Q- or $\hat{Q}$-learning algorithm with goal-reward representation to the Q-values of an undiscounted Q- or $\hat{Q}$-learning algorithm with action-penalty representation: map a Q-value $Q_1^t(s, a) \in (0, 1/\gamma]$ (or zero) of the former algorithm to the Q-value $Q_2^t(s, a) = -1 - \log_\gamma Q_1^t(s, a) \in \mathcal{R}^-$ (or zero, respectively) of the latter algorithm.[5] Similarly to the proof sketch in the previous section, one can now easily show that this relationship continues to hold if it holds for the initial Q-values. If the values $Q_1^0(s, a)$ are $1/\gamma$- or one-initialized, then the corresponding Q- or $\hat{Q}$-learning algorithm with action-penalty representation is zero- or (minus one)-initialized, which implies that it has a tight complexity of $O(ed)$ action executions. Consequently, the following corollary holds:

COROLLARY 3 *One-initialized discounted Q- or $\hat{Q}$-learning with goal-reward representation has a tight complexity of $O(ed)$ action executions for reaching a goal state in deterministic state spaces.*

The other results from Section 6.1.1.1. can be transferred in the same way.

6.1.3. Decreasing the Complexity Further

We have shown that finding a goal state with uninformed Q- or Q̂-learning is tractable if an appropriate representation is chosen. In particular, we identified both undiscounted or discounted zero-initialized Q-values with action-penalty representation and discounted one-initialized Q-values with goal-reward representation as representations that make these algorithms tractable (**tractable representations**). In this section, we investigate how their complexity can be decreased even further by utilizing prior knowledge of the state space or by learning and using an action model.

6.1.3.1. Using Prior Knowledge

The larger the absolute values of consistent or admissible initial Q-values are, the lower the complexity of Q- or Q̂-learning is (if we assume without loss of generality that action-penalty representation and no discounting is used), as can be seen from Lemma 2. For example, in the **completely informed** case, the Q-values are initialized as follows:

$$Q(s,a) = \begin{cases} 0 & \text{for } s \in G \text{ and } a \in A(s) \\ -1 - gd(succ(s,a)) & \text{for } s \in S \setminus G \text{ and } a \in A(s). \end{cases}$$

Lemma 2 predicts in this case correctly that the agent needs at most $-U(s) = gd(s)$ action executions to reach a goal state from any given $s \in S$. Often, however, the agent will only have partial prior knowledge of the state space. Heuristic functions that approximate the goal distances and are consistent or admissible for A*-search (Nilsson, 1971) can be used to encode such prior knowledge. Consistent or admissible heuristics are known for many deterministic state spaces, for example for path planning problems in spatial domains (Pearl, 1984). If a heuristic h (with $h(s) \geq 0$ for all $s \in S$) is consistent or admissible for A*-search, then the following Q-values are consistent or admissible, respectively, as well:

$$Q(s,a) = \begin{cases} 0 & \text{for } s \in G \text{ and } a \in A(s) \\ -1 - h(succ(s,a)) & \text{for } s \in S \setminus G \text{ and } a \in A(s). \end{cases}$$

Thus, consistent or admissible heuristics can be used to initialize the Q-values, which makes Q- or Q̂-learning better informed and lowers its complexity.

6.1.3.2. Using Action Models

Although the complexity of uninformed Q- or Q̂-learning is a small polynomial in the size of the state space if it operates on a tractable representation, it often directs the agent to execute actions that are suboptimal when judged according to the knowledge that the agent could have acquired had it memorized all of its experience. In particular, the agent

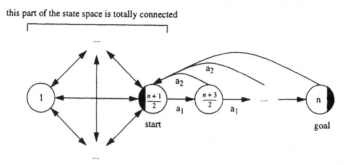

Figure 5. A state space for which uninformed Q- or Q̂-learning can need at least $1/16n^3 - 3/16n^2 - 1/16n + 3/16$ action executions to reach the goal state (even if it uses a tractable representation), but Q_{map}-learning needs only at most $3/8n^2 + 3/2n - 23/8$ action executions (for odd $n \geq 3$)

can move around for a long time in parts of the state space that it has already explored. This inefficiency can be avoided by increasing the amount of planning performed between action executions, because it enables the agent to make more informed decisions about which action to execute next. To be able to plan between action executions, the agent has to learn an action model. In this section, we show that learning and using an action model can decrease the complexity of Q- or Q̂-learning.

Sutton's **DYNA architecture** (Sutton, 1990 and 1991), for example, implements planning in the framework of Q- or Q̂-learning. The learned action model is used to simulate action executions and thereby to create experiences that are indistinguishable from the execution of actions in the world. This way, the world and its model can interchangeably be used to provide input for Q- or Q̂-learning. Actions are executed in the world mainly to refine (and, in non-stationary environments, to update) the model. Using the model, the agent can optimize its behavior according to its current knowledge without having to execute actions in the world. In particular, the model allows the agent to simulate any action at any time, whereas the world constrains the agent to execute only actions that are available in its current state. Various researchers, for example Peng and Williams (1992), Yee (1992), and Moore and Atkeson (1993b), have devised strategies for choosing which actions to simulate in order to speed up planning.

Q_{map}**-learning** is an algorithm that fits the DYNA framework. It remembers the action model of the part of the state space that it has already explored and executes unexplored actions in the same order as Q- or Q̂-learning, but always chooses the shortest known path to the next unexplored action: Q_{map}-learning uses its current (incomplete) action model to simulate the behavior of Q- or Q̂-learning until it would execute an unexplored action. Then, it uses the same action model to find the shortest known action sequence that leads from its current state in the world to the state in which it can execute this action, directs the agent to execute the action sequence and the unexplored action, and repeats the cycle. Per construction, Q_{map}-learning cannot perform worse than Q- or Q̂-learning if ties are broken in the same way, but it can be more efficient: Consider, for example, the state sequence that uninformed Q- or $\hat{Q}$-learning on a tractable representation traverses in a reset state space (shown in Figure 2) of size $n = 6$ if ties are broken in favor of actions that lead to states with smaller numbers: 121231234123451212 3456. First, Q- or $\hat{Q}$-learning finds out about the effect of action a_1 in state 1 and then about a_2 in 2, a_1 in 2, a_2 in 3, a_1 in 3, a_2 in 4, a_1 in 4, a_2 in 5, and a_1 in 5, in this order. Q_{map}-learning explores the actions in the same order. However, after it has executed action a_2 in state 5 for the first time and, as a consequence, got reset into state 1, it reaches state 5 again faster than Q- or $\hat{Q}$-learning: it goes from state 1 through states 2, 3, and 4, to state 5, whereas Q- or $\hat{Q}$-learning goes through states 2, 1, 2, 3, and 4. Thus, Q_{map}-learning traverses the state sequence 12123123412345123456, and is two action executions faster than Q- or $\hat{Q}$-learning. Figure 5 gives an example of a state space for which the big-O complexities of the two algorithms differ: there is a tie-breaking rule that causes Q- or $\hat{Q}$-learning to need $O(n^3)$ action executions to reach the goal state (even if it uses a tractable representation), whereas Q_{map}-learning needs at most $O(n^2)$ action executions no matter how ties are broken; see (Koenig and Simmons, 1992) for the proof. This shows that it is possible to augment Q- or $\hat{Q}$-learning with a component that learns action models such that the resulting algorithm never performs worse than Q- or $\hat{Q}$-learning, but often performs better. In some state spaces, one can reduce the number of action executions even by more than a constant factor. There are "hard" state spaces, however, for which no uninformed algorithm can have a lower big-O complexity than uninformed Q- or $\hat{Q}$-learning with a tractable representation (an example is the state space shown in Figure 3).

Note that we have not included the planning time in the complexity measure. When learning an action model and using it for planning, the agent has to keep more information around and perform more computations between action executions. However, chances are that the increased deliberation time reduces the number of action executions that are needed for reaching a goal state. Consequently, using an action model can be advantageous if executing actions in the world is slow (and expensive), but simulating the execution of actions in a model of the world is fast (and inexpensive). If planning time is not negligible compared to execution time, then there is a trade-off: simulating actions allows the agent to utilize its current knowledge better, whereas executing actions in the world increases its knowledge and allows it to stumble across a goal. Also, the fewer actions are simulated between action executions, the less opportunity a non-stationary state space has to change and the closer the model still reflects reality. Planning approaches that take these kinds of trade-offs into account have, for example, been investigated by Boddy and Dean (1989), Russell and Wefald (1991), Goodwin (1994), and Zilberstein (1993).

6.2. Non-Deterministic State Spaces

So far, we have studied representations that make Q- or $\hat{Q}$-learning tractable in deterministic state spaces. We now show that our analysis can easily be generalized to $\hat{Q}$-learning in non-deterministic state spaces. In non-deterministic state spaces, admissible $\hat{Q}$-learning is defined as follows:

Definition. Undiscounted $\hat{Q}$-learning with action-penalty representation is **admissible** in non-deterministic state spaces iff its initial Q-values are admissible and its value-update step is "Set $Q(s,a) := \min(Q(s,a), -1 + U(s'))$."

This definition uses the value-update step of undiscounted $\hat{Q}$-learning and is identical to the second part of the definition of admissible Q- or $\hat{Q}$-learning in the deterministic case (Page 236). It allows us to transfer the complexity analysis from deterministic to non-deterministic state spaces. In particular, our complexity analysis continues to hold if we replace $gd(succ(s,a))$ with $\max_{s' \in succ(s,a)} gd(s')$, which reflects that we perform a worst-case analysis over all strategies of nature; and undiscounted admissible $\hat{Q}$-learning with action-penalty representation reaches a goal state after at most

$$2 \sum_{s \in S \setminus G} \sum_{a \in A(s)} [Q^0(s,a) + \max_{s' \in succ(s,a)} gd(s') + 1] - U^0(s^0)$$

action executions in non-deterministic state spaces no matter which strategy nature uses. Therefore, its complexity is at most $O(ed)$ action executions, just like in the deterministic case. (This is due to our definition of d that assumes that nature is an opponent of the agent and encompasses deterministic state spaces as a special case.) The complexity is tight for uninformed $\hat{Q}$-learning, because it is tight for deterministic state spaces and cannot be lower in non-deterministic state spaces. We can use the transformations of the Q-values that we used in Sections 6.1.1.2. and 6.1.2 to show that the same complexity result holds for discounted $\hat{Q}$-learning in non-deterministic state spaces if either zero-initialized Q-values and action-penalty representation or one-initialized Q-values and goal-reward representation is used. Consequently, the following corollary holds:

COROLLARY 4 *Zero-initialized undiscounted or discounted $\hat{Q}$-learning with action-penalty representation and one-initialized discounted $\hat{Q}$-learning with goal-reward representation have a tight complexity of $O(ed)$ action executions for reaching a goal state in non-deterministic state spaces.*

This complexity result is fairly general, since it provides an upper bound on the number of action executions that holds for all strategies of nature. For example, it holds for both nature's strategy to select successor states randomly (the assumption underlying Q-learning) and its strategy to select successor states deliberately so that it hurts the agent most (the assumption underlying $\hat{Q}$-learning). But it also applies to scenarios where the agent cannot make assumptions about nature's strategy. Assume, for example, a deterministic world which the agent can only model with low granularity. Then, it might not be able to identify its current state uniquely, and actions can appear to have non-deterministic effects: sometimes the execution of an action results in one successor state, sometimes in another. The agent has no way of predicting which of the potential successor states will occur and could attribute this to nature having a strategy that is unknown to the agent. Our complexity bound holds even in this case.

7. Extensions

Space limitations forced us to limit our presentation to an analysis of the goal-directed exploration behavior of two reinforcement-learning algorithms (Q-learning and $\hat{Q}$-learning) for two different reward structures (action-penalty and goal-reward representation). We have generalized this analysis in three orthogonal directions:

- *Reward Structure*: In our analysis, we have assumed that one can choose an appropriate reward structure when representing a reinforcement-learning or exploration problem. Although this is usually true, sometimes the reward structure is given. In this case, even if the reward structure is dense, the immediate rewards might not all be uniformly minus one, as assumed by our analysis of Q- or $\hat{Q}$-learning with action-penalty representation. The results presented in this article have been generalized to cover dense reward structures with non-uniform costs. The complexity of zero-initialized Q- or $\hat{Q}$-learning with a non-uniform action-penalty representation, for example, is tight at $O(ed)$ action executions, where d is now the weighted diameter of the state space divided by the smallest absolute value of all immediate costs; see (Koenig and Simmons, 1992).

- *Exploration Algorithm*: We have restricted our analysis to two reinforcement-learning algorithms, Q-learning and $\hat{Q}$-learning. Q-learning behaves very similarly to value-iteration (Bellman, 1957), an algorithm that does not use Q-values, but rather the U-values directly. 1-step on-line value-iteration with action-penalty representation behaves in deterministic state spaces identically to Korf's Learning Real-Time A* (LRTA*) search algorithm (Korf, 1990) with search horizon one. Korf showed that LRTA* is guaranteed to reach a goal state and, if it is repeatedly reset into the start state when it reaches a goal state, eventually determines a shortest path from the start state to a goal state. Subsequently, Barto, Bradtke, and Singh (1995) generalized these results to probabilistic state spaces. Since on-line value-iteration and, consequently, LRTA* behave like 1-step Q-learning in a modified state space (Koenig and Simmons, 1992), we have been able to transfer our complexity results to LRTA* and, in the process, generalized previous complexity results for LRTA* by Ishida and Korf (1991); see (Koenig and Simmons, 1995b). The complexity of zero-initialized LRTA* and uninformed value-iteration that operates on a tractable representation is tight at $O(nd)$ action executions; see (Koenig and Simmons, 1995b).

- *Task*: We have analyzed goal-directed exploration problems, because these tasks need to be solved if one wants to solve goal-directed reinforcement-learning problems in unknown state spaces. Our analysis generalizes to finding optimal policies, because this problem can be solved either by repeatedly resetting the agent into its start state when it reaches a goal state (if the task is to find an optimal behavior from the start state) or, if such a reset action is not available, by iteratively executing two independent exploration algorithms: one Q- or $\hat{Q}$-learning algorithm that finds a goal state, and another one that finds a state for which the optimal action assignment has not yet been determined. (The latter method determines optimal behaviors from all states in strongly connected state spaces.) In both cases, we have been able to show that the complexity of uninformed Q- or Q̂-learning that operates on a tractable representation remains tight at $O(ed)$; see (Koenig and Simmons, 1993). (For non-deterministic state spaces, one has to make assumptions about how long nature is allowed to "trick" the agent by never choosing a bad action outcome and thereby hiding its existence.)

8. Conclusion

In this article, we have analyzed how 1-step reinforcement-learning methods solve goal-directed reinforcement-learning problems in safely explorable state spaces – we have studied their behavior until they reach a goal state for the first time. In particular, we studied how the complexity of Q-learning methods (measured in action executions), such as Watkins' Q-learning or Heger's Q̂-learning, depends on the number of states of the state space (n), the total number of state-action pairs (e), and the largest goal distance (d). When formulating a goal-directed reinforcement-learning problem, one has to decide on an appropriate representation, which consists of both the immediate rewards that the reinforcement-learning algorithms receive and their initial Q-values. We showed that the choice of representation can have a tremendous impact on the performance of Q- or Q̂-learning.

We considered two reward structures that have been used in the literature to learn optimal policies, the goal-reward representation and the action-penalty representation. In the action-penalty representation, the agent is penalized for every action that it executes. In the goal-reward representation, it is rewarded for entering a goal state, but not rewarded or penalized otherwise. Zero-initialized Q-learning with goal-reward representation provides only sparse rewards. Even in deterministic state spaces, it performs a random walk. Although a random walk reaches a goal state with probability one, its complexity is infinite and even its average number of action executions can be exponential in n. Furthermore, speed-up techniques such as Lin's action-replay technique do not improve its performance. This provides motivation for making the reward structure dense. Since the value-update step of Q- or Q̂-learning updates the Q-values using both the immediate reward and the Q-values of the successor state, this can be achieved by either changing the reward structure or initializing the Q-values differently. And indeed, we showed that both (undiscounted and discounted) zero-initialized Q- or Q̂-learning with action-penalty representation and (discounted) one-initialized Q- or Q̂-learning with goal-reward representation are tractable. For the proof, we developed conditions on the initial Q-values, called consistency and admissibility, and – for initial Q-values with these properties – a time-invariant relationship between the number of executed actions, the initial Q-values, and the current Q-values. This relationship allowed us to express how the complexity of Q- or Q̂-learning depends on the initial Q-values and properties of the state space. Our analysis shows that, if a tractable representation is used, the greedy action-selection strategy of Q- or Q̂-learning always executes the action that locally appears to be best, and even uninformed Q- or Q̂-learning has a complexity of at most $O(ed)$ action executions in deterministic state spaces. The same result holds for

$\hat{Q}$-learning in non-deterministic state spaces, in which we viewed reinforcement learning as a game where the reinforcement-learning algorithm selects the actions and "nature," a fictitious second agent, selects their outcomes. (The bound holds for all possible outcome selection strategies of nature.)

If a safely explorable state space has no duplicate actions, then $O(ed) \leq O(n^3)$. The complexity of Q- or $\hat{Q}$-learning is even lower in many state spaces, since e often grows only linearly in n and/or d grows only sublinearly in n. Examples of such state spaces are gridworlds, which are popular reinforcement-learning domains. The complexity can be reduced further by using prior knowledge of the state space and by learning action models. Prior knowledge can be encoded in the initial Q-values by utilizing heuristics that are consistent or admissible for A*-search. Action models predict what would happen if a particular action were executed. We showed how to augment Q- or $\hat{Q}$-learning with a component that learns action models such that the resulting algorithm, which we called Q_{map}-learning, never performs worse than Q- or $\hat{Q}$-learning, but reduces the complexity by at least a factor of n (i.e. by more than a constant factor) in some state spaces.

To summarize, reinforcement learning algorithms are tractable if a suitable representation is chosen. Our complexity results provide guidance for empirical reinforcement-learning researchers on how to model reinforcement learning problems in a way that allows them to be solved efficiently – even for reinforcement-learning tasks that cannot be reformulated as goal-directed reinforcement-learning or exploration problems in safely explorable state spaces: the performance can be improved by making the reward structure dense. Our results also characterize which properties of state spaces make them easy to solve with reinforcement-learning methods, which helps empirical reinforcement-learning researchers to choose appropriate state spaces for their experiments.

Acknowledgments

Avrim Blum, Lonnie Chrisman, Matthias Heger, Long-Ji Lin, Michael Littman, Andrew Moore, Martha Pollack, and Sebastian Thrun provided helpful comments on the ideas presented in this article. Special thanks to Sebastian Thrun for stimulating discussions and to Lonnie Chrisman also for taking the time to check the proofs.

Notes

1. To be precise: it does not matter whether states that the agent cannot possibly reach from its start state are lost. Furthermore, for goal-directed exploration problems, we can disregard all states that the agent can only reach by passing through a goal state, since the agent can never occupy those states. We assume without loss of generality that all such states have been removed from the state space.

2. A similar statement also holds for $\hat{Q}$-learning: it never changes a Q-value. However, since the Q-values are not initialized optimistically, zero-initialized $\hat{Q}$-learning with goal-reward representation cannot be used to learn shortest paths. Studying the goal-directed exploration problem for $\hat{Q}$-learning with this representation therefore does not provide any insight into the corresponding reinforcement-learning problem.

3. Our analysis can be generalized to goal-directed reinforcement-learning problems. For goal-directed exploration problems, the definition of admissible Q- or Q̂-learning can be broadened, since one can add the same constant to all Q-values without changing the behavior of Q- or Q̂-learning until it reaches a goal state for the first time. Given initial Q-values for a goal-directed exploration problem, one can therefore add a constant to all Q-values before determining whether they are consistent or admissible.

4. In other words: the value-update step is "Set $Q(s, a) := (1 - \alpha)Q(s, a) + \alpha(r(s, a, s') + \gamma U(s'))$," where $r(s, a, s') = -1$ and $\alpha = \gamma = 1$.

5. Our analysis can be generalized to goal-directed reinforcement-learning problems. For goal-directed exploration problems, the statement can be broadened as follows: According to Note 3, one can add the same constant to all values $Q_2^0(s, a)$ without changing the behavior of undiscounted Q- or Q̂-learning with action-penalty representation until it reaches a goal state for the first time. Since $Q_2^t(s, a) - \log_\gamma c = (-1 - \log_\gamma Q_1^t(s, a)) - \log_\gamma c = -1 - \log_\gamma cQ_1^t(s, a)$, one can multiply all values $Q_1^0(s, a)$ with some positive constant c without changing the behavior of discounted Q- or Q̂-learning with goal-reward representation until it reaches a goal state for the first time.

References

Barto, A.G., S.J. Bradtke, and S.P. Singh. (1995). Learning to act using real-time dynamic programming. *Artificial Intelligence*, 73(1):81–138.

Barto, A.G., R.S. Sutton, and C.J. Watkins. (1989). Learning and sequential decision making. Technical Report 89–95, Department of Computer Science, University of Massachusetts at Amherst.

Bellman, R. (1957). *Dynamic Programming*. Princeton University Press, Princeton (New Jersey).

Boddy, M. and T. Dean. (1989). Solving time-dependent planning problems. In *Proceedings of the IJCAI*, pages 979–984.

Goodwin, R. (1994). Reasoning about when to start acting. In *Proceedings of the International Conference on Artificial Intelligence Planning Systems*, pages 86–91.

Heger, M. (1996). The loss from imperfect value functions in expectation-based and minimax-based tasks. *Machine Learning*, pages 197-225.

Heger, M. (1994). Consideration of risk in reinforcement learning. In *Proceedings of the International Conference on Machine Learning*, pages 105–111.

Ishida, T. (1992). Moving target search with intelligence. In *Proceedings of the AAAI*, pages 525-532.

Ishida, T. and R.E. Korf. (1991). Moving target search. In *Proceedings of the IJCAI*, pages 204–210.

Kaelbling, L.P. (1990). *Learning in Embedded Systems*. MIT Press, Cambridge (Massachusetts).

Koenig, S. (1991). Optimal probabilistic and decision-theoretic planning using Markovian decision theory. Master's thesis, Computer Science Department, University of California at Berkeley. (Available as Technical Report UCB/CSD 92/685).

Koenig, S. and R.G. Simmons. (1992). Complexity analysis of real-time reinforcement learning applied to finding shortest paths in deterministic domains. Technical Report CMU–CS–93-106, School of Computer Science, Carnegie Mellon University.

Koenig, S. and R.G. Simmons. (1993). Complexity analysis of real-time reinforcement learning. In *Proceedings of the AAAI*, pages 99–105.

Koenig, S. and R.G. Simmons. (1994). How to make reactive planners risk-sensitive. In *Proceedings of the International Conference on Artificial Intelligence Planning Systems*, pages 293–298.

Koenig, S. and R.G. Simmons. (1995a). The effect of representation and knowledge on goal-directed exploration with reinforcement learning algorithms: The proofs. Technical Report CMU–CS–95-177, School of Computer Science, Carnegie Mellon University.

Koenig, S. and R.G. Simmons. (1995b). Real-time search in non-deterministic domains. In *Proceedings of the IJCAI*, pages 1660-1667.

Korf, R.E. (1990). Real-time heuristic search. *Artificial Intelligence*, 42(2-3):189–211.

Lin, L.-J. (1992). Self-improving reactive agents based on reinforcement learning, planning, and teaching. *Machine Learning*, 8:293–321.

Matarić, M. (1994). *Interaction and Intelligent Behavior*. PhD thesis, Department of Electrical Engineering and Computer Science, Massachusetts Institute of Technology.

Moore,A.W. and C.G. Atkeson. (1993a). The parti-game algorithm for variable resolution reinforcement learning in multidimensional state-spaces. In *Proceedings of the NIPS*.

Moore,A.W. and C.G. Atkeson. (1993b). Prioritized sweeping: Reinforcement learning with less data and less time. *Machine Learning*, 13:103–130.

Nilsson,N.J. (1971). *Problem-Solving Methods in Artificial Intelligence*. McGraw-Hill, New York (New York).

Pearl,J. (1984). *Heuristics: Intelligent Search Strategies for Computer Problem Solving*. Addison-Wesley, Menlo Park (California).

Peng,J. and R.J. Williams. (1992). Efficient learning and planning within the DYNA framework. In *Proceedings of the International Conference on Simulation of Adaptive Behavior: From Animals to Animats*, pages 281–290.

Russell,S. and E. Wefald. (1991). *Do the Right Thing – Studies in Limited Rationality*. MIT Press, Cambridge (Massachusetts).

Singh,S.P. (1992). Reinforcement learning with a hierarchy of abstract models. In *Proceedings of the AAAI*, pages 202–207.

Sutton,R.S. (1990). Integrated architectures for learning, planning, and reacting based on approximating dynamic programming. In *Proceedings of the International Conference on Machine Learning*, pages 216–224.

Sutton,R.S. (1991). DYNA, an integrated architecture for learning, planning, and reacting. *SIGART Bulletin*, 2(4):160–163.

Thrun,S.B. (1992a). Efficient exploration in reinforcement learning. Technical Report CMU-CS-92-102, School of Computer Science, Carnegie Mellon University.

Thrun,S.B. (1992b). The role of exploration in learning control with neural networks. In D.A. White and D.A. Sofge, editors, *Handbook of Intelligent Control: Neural, Fuzzy and Adaptive Approaches*, pages 527–559. Van Nostrand Reinhold, New York (New York).

Watkins,C.J. (1989). *Learning from Delayed Rewards*. PhD thesis, King's College, Cambridge University.

Watkins,C.J. and P. Dayan. (1992). Q-learning. *Machine Learning*, 8(3-4):279–292.

Whitehead,S.D. (1991a). A complexity analysis of cooperative mechanisms in reinforcement learning. In *Proceedings of the AAAI*, pages 607–613.

Whitehead,S.D. (1991b). A study of cooperative mechanisms for faster reinforcement learning. Technical Report 365, Computer Science Department, University of Rochester.

Whitehead,S.D. (1992). *Reinforcement Learning for the Adaptive Control of Perception and Action*. PhD thesis, Computer Science Department, University of Rochester.

Yee, R. (1992). Abstraction in control learning. Technical Report 92–16, Department of Computer Science, University of Massachusetts at Amherst.

Zilberstein, S. (1993). *Operational Rationality through Compilation of Anytime Algorithms*. PhD thesis, Computer Science Department, University of California at Berkeley.

Received November 3, 1994
Accepted March 10, 1995
Final Manuscript October 17, 1995

Machine Learning, 22, 251–281 (1996)

Creating Advice-Taking Reinforcement Learners

RICHARD MACLIN AND JUDE W. SHAVLIK   maclin@cs.wisc.edu and shavlik@cs.wisc.edu

Computer Sciences Dept., University of Wisconsin, 1210 W. Dayton St., Madison, WI 53706

Editor: Leslie Pack Kaelbling

Abstract. Learning from reinforcements is a promising approach for creating intelligent agents. However, reinforcement learning usually requires a large number of training episodes. We present and evaluate a design that addresses this shortcoming by allowing a connectionist Q-learner to accept advice given, at any time and in a natural manner, by an external observer. In our approach, the advice-giver watches the learner and occasionally makes suggestions, expressed as instructions in a simple imperative programming language. Based on techniques from knowledge-based neural networks, we insert these programs directly into the agent's utility function. Subsequent reinforcement learning further integrates and refines the advice. We present empirical evidence that investigates several aspects of our approach and shows that, given good advice, a learner can achieve statistically significant gains in expected reward. A second experiment shows that advice improves the expected reward regardless of the stage of training at which it is given, while another study demonstrates that subsequent advice can result in further gains in reward. Finally, we present experimental results that indicate our method is more powerful than a naive technique for making use of advice.

Keywords: Reinforcement learning, advice-giving, neural networks, Q-learning, learning from instruction, theory refinement, knowledge-based neural networks, adaptive agents

1. Introduction

A successful and increasingly popular method for creating intelligent agents is to have them learn from reinforcements (Barto, Sutton, & Watkins, 1990; Lin, 1992; Mahadevan & Connell, 1992; Tesauro, 1992; Watkins, 1989). However, these approaches suffer from their need for large numbers of training episodes. Several methods for speeding up reinforcement learning have been proposed; one promising approach is to design a learner that can also accept *advice* from an external observer (Clouse & Utgoff, 1992; Gordon & Subramanian, 1994; Lin, 1992; Maclin & Shavlik, 1994). Figure 1 shows the general structure of a reinforcement learner, augmented (in bold) with an observer that provides advice. We present and evaluate a connectionist approach in which agents learn from both experience and instruction. Our approach produces agents that significantly outperform agents that only learn from reinforcements.

To illustrate the general idea of advice-taking, imagine that you are watching an agent learning to play some video game. Assume you notice that frequently the agent loses because it goes into a "box canyon" in search of food and then gets trapped by its opponents. One would like to give the learner broad advice such as "do not go into box canyons when opponents are in sight." This approach is more appealing than the current alternative: repeatedly place the learner in similar circumstances and expect it to learn this advice from direct experience, while not forgetting what it previously learned.

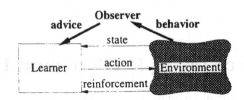

Figure 1. In basic reinforcement learning the learner receives a description of the current environment (the state), selects an action to choose, and receives a reinforcement as a consequence of selecting that action. We augment this with a process that allows an observer to watch the learner and suggest advice based on the learner's behavior.

Recognition of the value of advice-taking has a long history in AI. The general idea of a program accepting advice was first proposed nearly 40 years ago by McCarthy (1958). Over a decade ago, Mostow (1982) developed a program that accepted and "operational-ized" high-level advice about how to better play the card game Hearts. Recently, after a decade-long lull, there has been a growing amount of research on advice-taking (Gordon & Subramanian, 1994; Huffman & Laird, 1993; Maclin & Shavlik, 1994; Noelle & Cottrell, 1994). For example, Gordon and Subramanian (1994) created a system that de-ductively compiles high-level advice into concrete actions, which are then refined using genetic algorithms.

Several characteristics of our approach to providing advice are particularly interesting. One, we allow the advisor to provide instruction in a quasi-natural language using terms about the specific task domain; the advisor does not have to be aware of the internal representations and algorithms used by the learner in order to provide useful advice. Two, the advice need not be precisely specified; vague terms such as "big," "near," and "old" are acceptable. Three, the learner does not follow the advice blindly; rather, the learner judges the usefulness of the advice and is capable of altering the advice based on subsequent experience.

In Section 2 we present a framework for using advice with reinforcement learners, and in Section 3 we outline an implemented system that instantiates this framework. The fourth section describes experiments that investigate the value of our approach. Finally, we discuss possible extensions to our research, relate our work to other research, and present some conclusions.

2. A General Framework for Advice-Taking

In this section we present our design for a reinforcement learning (RL) advice-taker, following the five-step framework for advice-taking developed by Hayes-Roth, Klahr, and Mostow (1981). In Section 3 we present specific details of our implemented system, named RATLE, which concretizes the design described below.

Step 1. Request/receive the advice. To begin the process of advice-taking, a decision must be made that advice is needed. Often, approaches to advice-taking focus on having the learner ask for advice when it needs help (Clouse & Utgoff, 1992; Whitehead, 1991). Rather than having the learner request advice, we allow the external observer to provide advice whenever the observer feels it is appropriate. There are two reasons to allow the observer to determine when advice is needed: (i) it places less of a burden on the observer; and (ii) it is an open question how to create the best mechanism for having an agent recognize (and express) its need for general advice. Other RL methods (Clouse & Utgoff, 1992; Whitehead, 1991) focus on having the observer provide information about the action to take in a specific state. However, this can require a lot of interaction between the human advisor and computer learner, and also means that the learner must induce the generality of the advice.

Step 2. Convert the advice into an internal representation. Once the observer has created a piece of advice, the agent must try to understand the advice. Due to the complexities of natural language processing, we require that the external observer express its advice using a simple programming language and a list of task-specific terms. We then parse the advice, using traditional methods from the programming-languages literature (Levine, Mason, & Brown, 1992).

Table 1 shows some sample advice that the observer could provide to an agent learning to play a video game. The left column contains the advice as expressed in our programming language, the center column shows the advice in English, and the right column illustrates the advice. (In Section 3 we use these samples to illustrate our algorithm for integrating advice.)

Step 3. Convert the advice into a usable form. After the advice has been parsed, the system transforms the general advice into terms that it can directly understand. Using techniques from *knowledge compilation* (Dietterich, 1991), a learner can convert ("operationalize") high-level advice into a (usually larger) collection of directly interpretable statements (Gordon & Subramanian, 1994; Kaelbling & Rosenschein, 1990; Nilsson, 1994). We only address a limited form of operationalization, namely the concretization of imprecise terms such as "near" and "many." Terms such as these allow the advice-giver to provide natural, yet partially vague, instructions, and eliminate the need for the advisor to fully understand the learner's sensors.

Step 4. Integrate the reformulated advice into the agent's knowledge base. In this work we employ a connectionist approach to RL (Anderson, 1987; Barto, Sutton, & Anderson, 1983; Lin, 1992). Hence, to incorporate the observer's advice, the agent's neural network must be updated. We use ideas from *knowledge-based neural networks* (Fu, 1989; Omlin & Giles, 1992; Shavlik & Towell, 1989) to directly install the advice into the agent. In one approach to knowledge-based neural networks, KBANN (Towell, Shavlik, & Noordewier, 1990; Towell & Shavlik, 1994), a set of propositional rules is re-represented as a neural network. KBANN converts a ruleset into a network by mapping the "target concepts" of the ruleset to output units and creating hidden units that represent the intermediate conclusions (for details, see Section 3). We extend the KBANN method to accommodate our advice-giving language.

Figure 2 illustrates our basic approach for adding advice into the reinforcement learner's action-choosing network. This network computes a function from sensations to the utility

Table 1. Samples of advice in our advice language (left column).

Advice	English Version	Pictorial Version

IF An Enemy IS (Near ∧ West) ∧
   An Obstacle IS (Near ∧ North)
THEN
   MULTIACTION
     MoveEast
     MoveNorth
   END
END;
	If an enemy is near and west and an obstacle is near and north, hide behind the obstacle.	

WHEN Surrounded ∧
   OKtoPushEast ∧
   An Enemy IS Near
REPEAT
   MULTIACTION
     PushEast
     MoveEast
   END
UNTIL ¬ OKtoPushEast ∨
   ¬ Surrounded
END;
	When the agent is surrounded, pushing east is possible, and an enemy is near, then keep pushing (moving the obstacle out of the way) and moving east until there is nothing more to push or the agent is no longer surrounded.	

IF An Enemy IS (Near ∧ East) THEN DO_NOT MoveEast END;	Do not move toward a nearby enemy.	

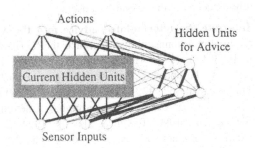

Figure 2. Adding advice to the RL agent's neural network by creating new hidden units that represent the advice. The thick links on the right capture the semantics of the advice. The added thin links initially have near-zero weight; during subsequent backpropagation training the magnitude of their weights can change, thereby refining the original advice. Details and an example appear in Section 3.

of actions. Incorporating advice involves adding to the existing neural network new hidden units that represent the advice.

Step 5. Judge the value of the advice. The final step of the advice-taking process is to evaluate the advice. We view this process from two perspectives: (i) the learner's, who must decide if the advice is useful; and (ii) the advisor's, who must decide if the advice had the desired effect on the behavior of the learner. Our learner evaluates advice by continued operation in its environment; the feedback provided by the environment offers a crude measure of the advice's quality. (One can also envision that in some circumstances – such as a game-learner that can play against itself (Tesauro, 1992) or when an agent builds an internal world model (Sutton, 1991) – it would be possible to quickly estimate whether the advice improves performance.) The advisor judges the value of his or her advice similarly (i.e., by watching the learner's post-advice behavior). This may lead to the advisor giving further advice – thereby restarting the advice-taking process.

3. The RATLE System

Figure 3 summarizes the approach we discussed in the previous section. We implemented the RATLE (**R**einforcement and **A**dvice-**T**aking **L**earning **E**nvironment) system as a mechanism for evaluating this framework. In order to explain RATLE, we first review connectionist Q-learning (Sutton, 1988; Watkins, 1989), the form of reinforcement learning that we use in our implementation, and then KBANN (Towell & Shavlik, 1994), a technique for incorporating knowledge in the form of rules into a neural network. We then discuss our extensions to these techniques by showing how we implement each of the five steps described in the previous section.

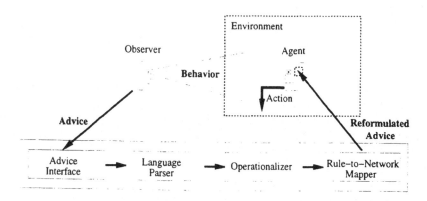

Figure 3. Interaction of the observer, agent, and our advice-taking system. The process is a cycle: the observer watches the agent's behavior to determine what advice to give, the advice-taking system processes the advice and inserts it into the agent, which changes the agent's behavior, thus possibly causing the observer to provide more advice. The agent operates as a normal Q-learning agent when not presented with advice.

Background – Connectionist Q-Learning

In standard RL, the learner senses the current world state, chooses an action to execute, and occasionally receives rewards and punishments. Based on these reinforcements from the environment, the task of the learner is to improve its action-choosing module such that it increases the total amount of reinforcement it receives. In our augmentation, an observer watches the learner and periodically provides advice, which RATLE incorporates into the action-choosing module of the RL agent.

In Q-learning (Watkins, 1989) the action-choosing module uses a *utility function* that maps states and actions to a numeric value (the utility). The utility value of a particular state and action is the predicted future (discounted) reward that will be achieved if that action is taken by the agent in that state and the agent acts optimally afterwards. It is easy to see that given a perfect version of this function, the optimal plan is to simply choose, in each state that is reached, the action with the largest utility.

To learn a utility function, a Q-learner typically starts out with a randomly chosen utility function and stochastically explores its environment. As the agent explores, it continually makes predictions about the reward it expects and then updates its utility function by comparing the reward it actually receives to its prediction. In *connectionist* Q-learning, the utility function is implemented as a neural network, whose inputs describe the current state and whose outputs are the utility of each action.

The main difference between our approach and standard connectionist Q-learning is that our agent continually checks for pending advice, and if so, incorporates that advice into its utility function. Table 2 shows the main loop of an agent employing connectionist Q-learning, augmented (in italics) by our process for using advice. The resulting composite system we refer to as RATLE.

Background – Knowledge-Based Neural Networks

In order for us to make use of the advice provided by the observer, we must incorporate this advice into the agent's neural-network utility function. To do so, we extend the KBANN algorithm (Towell & Shavlik, 1994). KBANN is a method for incorporating knowledge, in the form of simple propositional rules, into a neural network. In a KBANN network, the units of the network represent Boolean concepts. A concept is assumed to be true if the unit representing the concept is highly active (near 1) and false if the unit is inactive (near 0). To represent the meaning of a set of rules, KBANN connects units with highly weighted links and sets unit biases (thresholds) in such a manner that the (non-input) units emulate AND or OR gates, as appropriate. Figure 4 shows an example of this process for a set of simple propositional rules.

In RATLE, we use an imperative programming language, instead of propositional rules, to specify advice. In order to map this more complex language, we make use of hidden units that record state information. These units are recurrent and record the activation of a hidden unit from the previous activation of the network (i.e., they "remember" the previous activation value). We discuss how these units are used below.

Table 2. Steps of the RATLE algorithm. Our additions to the standard connectionist Q-learning loop are Step 6 and the subroutine IncorporateAdvice (all shown in italics). We follow Lin's (1992) method exactly for action selection and Q-function updating (Steps 2 and 5). When estimating the performance of a network ("testing"), the action with the highest utility is chosen in Step 2 and no updating is done in Step 5.

Agent's Main Loop	Incorporate Advice
1. Read sensors.	6a. *Parse advice.*
2. Stochastically choose an action, where the probability of selecting an action is proportional to the log of its predicted utility (i.e., its current Q value). Retain the predicted utility of the action selected.	6b. *Operationalize any fuzzy terms.*
	6c. *Translate advice into network components.*
3. Perform selected action.	6d. *Insert translated advice directly into RL agent's neural-network based utility function.*
4. Measure reinforcement, if any.	
5. Update utility function – use the current state, the current Q-function, and the actual reinforcement to obtain a new estimate of the expected utility; use the difference between the new estimate of utility and the previous estimate as the error signal to propagate through the neural network.	6e. *Return.*
6. *Advice pending? If so, call IncorporateAdvice.*	
7. Go to 1.	

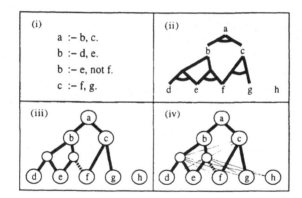

Figure 4. Sample of the KBANN algorithm: (i) a propositional rule set in Prolog notation; (ii) the rules viewed as an AND-OR dependency graph; (iii) each proposition is represented as a unit (extra units are also added to represent disjunctive definitions, e.g., *b*), and their weights and biases are set so that they implement AND or OR gates, e.g, the weights $b \rightarrow a$ and $c \rightarrow a$ are set to 4 and *a*'s bias (threshold) to 6 (the bias of an OR node is 2); (iv) low-weighted links are added between layers as a basis for future learning (e.g., an antecedent can be added to a rule by increasing one of these weights).

Implementing the Five-Step Framework

In the remainder of this section we describe how we implemented the advice-taking strategy presented in the last section. Several worked examples are included.

Step 1. Request/receive the advice. To give advice, the observer simply interrupts the agent's execution and types his or her advice. Advice must be expressed in the language defined by the grammar in Appendix B.

Step 2. Convert the advice into an internal representation. We built RATLE's advice parser using the standard Unix compiler tools *lex* and *yacc* (Levine et al., 1992).

Our advice-taking language has two main programming constructs: IF-THEN rules and loops (both WHILE and REPEAT). The loop constructs also have optional forms that allow the teacher to specify more complex loops (e.g., the REPEAT may have an entry condition). Each of these constructs may specify either a single action or, via the MULTIACTION construct, a "plan" containing a sequence of actions. The observer may also specify that an action should *not* be taken as a consequent (as opposed to specifying an action to take). Examples of advice in our language appear in Table 1 and in Appendix A.

The IF-THEN constructs actually serve two purposes. An IF-THEN can be used to specify that a particular action should be taken in a particular situation. It can also be used to create a new intermediate term; in this case, the conclusion of the IF-THEN rule is not an action, but instead is the keyword INFER followed by the name of the new intermediate term. This allows the observer to create descriptive terms based on the sensed features. For example, the advisor may want to define an intermediate term *NotLarge* that is true if an object is *Small* or *Medium*, and then use the derived term *NotLarge* in subsequent advice.

In order to specify the preconditions of the IF-THEN and looping constructs, the advisor lists logical combinations of conditions (basic "sensors" and any derived features). To make the language easier to use, we also allow the observer to state "fuzzy" conditions (Zadeh, 1965), which we believe provide a natural way to articulate imprecise advice.

Step 3. Convert the advice into a usable form. As will be seen in Step 4, most of the concepts expressible in our grammar can be directly translated into a neural network. The fuzzy conditions, however, require some pre-processing. We must first "operationalize" them by using the traditional methods of fuzzy logic to create an explicit mathematical expression that determines the fuzzy "truth value" of the condition as a function of the sensor values. We accomplish this re-representation by applying the method of Berenji and Khedkhar (1992), adapted slightly (Maclin, 1995) to be consistent with KBANN's mapping algorithm.

Though fuzzy logic is a powerful method that allows humans to express advice using intuitive terms, it has the disadvantage that someone must explicitly define the fuzzy terms in advance. However, the definitions need not be perfectly correct, since we insert our fuzzy conditions into the agent's neural network and, thus, allow their definitions to be adjusted during subsequent training.

At present, RATLE only accepts fuzzy terms of the form:

quantifier object IS/ARE *descriptor*

where the quantifier is a fuzzy term specifying number (e.g., A, No, Few, Many), the object is the type of object being sensed (e.g., Blocks, Trees, Enemies) and the descriptor is a property of the referenced objects (e.g., Near, Big). For example, a fuzzy condition could be "Many Trees ARE Near."

Currently we use only sigmoidal membership functions. To operationalize a fuzzy condition, RATLE determines a set of weights and a threshold that implement the given sigmoidal membership function, as a function of the current sensor readings. The exact details depend on the structure of a given domain's sensors (see Maclin, 1995, for additional details) and have not been a major focus of this research. The result of this process essentially defines a perceptron; hence, operationalized fuzzy conditions can be directly inserted into the agent's neural network during Step 4.

Step 4. Integrate the reformulated advice into the agent's knowledge base. After RATLE operationalizes any fuzzy conditions, it proceeds to insert all of the advice into the agent's current neural-network utility function. To do this, we made five extensions to the standard KBANN algorithm: (i) advice can contain multi-step plans, (ii) it can contain loops, (iii) it can refer to previously defined terms, (iv) it may suggest actions to *not* take, and (v) it can involve fuzzy conditions (discussed above). We achieve each of these extensions by following the general approach illustrated earlier in Figure 2.

Consider, as an example of a multi-step plan, the first entry in Table 1. Figure 5 shows the network additions that represent this advice. RATLE first creates a hidden unit (labeled A) that represents the conjunction of (i) an enemy being near and west and (ii) an obstacle being near and north. It then connects this unit to the action $MoveEast$, which is an existing output unit (recall that the agent's utility function maps states to values of actions); this constitutes the first step of the two-step plan. RATLE also connects unit A to a newly added hidden unit called $State1$ that records when unit A was active in the previous state. It next connects $State1$ to a new input unit called $State1_{-1}$. This *recurrent* unit becomes active ("true") when $State1$ was active for the previous input (we need recurrent units to implement multi-step plans). Finally, it constructs a unit (labeled B) that is active when $State1_{-1}$ is true and the previous action was an eastward move (the network's input vector records the previous action taken in addition to the current sensor values). When active, unit B suggests moving north – the second

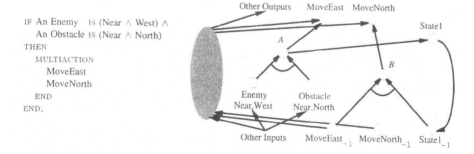

Figure 5. On the left is the first piece of advice from Table 1. On the right is RATLE's translation of this piece of advice. The shaded ellipse represents the original hidden units. Arcs show units and weights that are set to implement a conjunction. RATLE also adds zero-weighted links (not shown here – see Figure 4d) between the new units and other parts of the current network; these links support subsequent refinement.

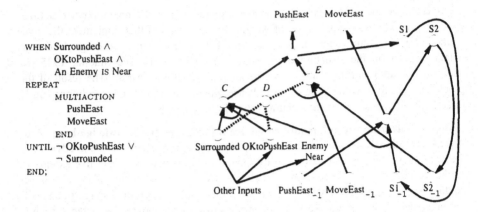

WHEN Surrounded ∧
    OKtoPushEast ∧
    An Enemy IS Near
REPEAT
        MULTIACTION
        PushEast
        MoveEast
    END
UNTIL ¬ OKtoPushEast ∨
        ¬ Surrounded
END;

Figure 6. On the left is the second piece of advice from Table 1. On the right is RATLE's translation of it. Dotted lines indicate negative weights. These new units are added to the existing network (not shown).

step of the plan. (In general, RATLE represents plans of length N using $N - 1$ state units.)

RATLE assigns a high weight[1] to the arcs coming out of units A and B. This means that when either unit is active, the total weighted input to the corresponding output unit will be increased, thereby increasing the utility value for that action. Note, however, that this does not guarantee that the suggested action will be chosen when units A or B are active. Also, notice that during subsequent training the weight (and thus the definition) of a piece of advice may be substantially altered.

The second piece of advice in Table 1 also contains a multi-step plan, but this time it is embedded in a REPEAT. Figure 6 shows RATLE's additions to the network for this advice. The key to translating this construct is that there are two ways to invoke the two-step plan. The plan executes if the WHEN condition is true (unit C) and also if the plan was just run and the UNTIL condition is false. Unit D is active when the UNTIL condition is met, while unit E is active if the UNTIL is unsatisfied and the agent's two previous actions were pushing and then moving east.

A third issue for RATLE is dealing with advice that involves previously defined terms. This frequently occurs, since advice generally indicates new situations in which to perform existing actions. There are two types of new definitions: (i) new preconditions of actions, and (ii) new definitions for derived features. We process the two types differently, since the former involve real-valued outputs while the latter are essentially Boolean-valued.

For new preconditions of *actions*, RATLE adds a highly weighted link from the unit representing the definition to the output unit representing the action. This is done so that in the situations where the advice is applicable, the utility of the action will then be higher that it would otherwise be. When the advisor provides a new definition of a derived feature, RATLE operates as shown in Figure 7. It first creates a new hidden unit that represents the new definition, then makes an OR node that combines the old and

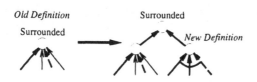

Figure 7. Incorporating the definition of a term that already exists.

new definitions. This process is analogous to how KBANN processes multiple rules with the same consequent.

A fourth issue is how to deal with advice that suggests *not* doing an action. This is straightforward in our approach, since we connect hidden units to "action" units with a highly weighted link. For example, for the third piece of advice shown in Table 1, RATLE would create a unit representing the fuzzy condition "An Enemy IS (Near and East)" and then connect the resulting unit to the action MoveEast with a negatively weighted link. This would have the effect of lowering MoveEast's utility when the condition is satisfied (which is the effect we desire). This technique avoids the question of what to do when one piece of advice suggests an action and another prohibits that action. Currently the conflicting pieces of advice (unless refined) cancel each other, but this simple approach may not always be satisfactory.

Maclin (1995) fully describes how each of the constructs in RATLE's advice language is mapped into a neural-network fragment.

4. Experimental Study

We next empirically judge the value of using RATLE to provide advice to an RL agent.

4.1. Testbed

Figure 8a illustrates the Pengo task. We chose Pengo because it has been previously explored in the AI literature (Agre & Chapman, 1987; Lin, 1992). The agent in Pengo can perform nine actions: *moving* and *pushing* in each of the directions East, North, West and South; and *doing nothing*. Pushing moves the obstacles in the environment. A moving obstacle will destroy the food and enemies it hits, and will continue to slide until it encounters another obstacle or the edge of the board. When the obstacle is unable to move (because there is an obstacle or wall behind it), the obstacle disintegrates. Food is collected when touched by the agent or an enemy.

Each enemy follows a fixed policy. It moves randomly unless the agent is in sight, in which case it moves toward the agent. Enemies may move off the board (they appear again after a random interval), but the agent is constrained to remain on the board. Enemies do not push obstacles.

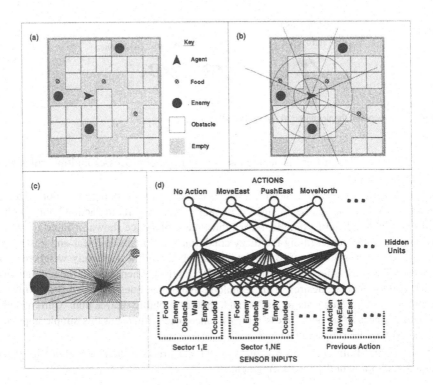

Figure 8. Our sample test environment: (a) sample configuration; (b) sample division of the environment into sectors; (c) distances to the nearest occluding object along a fixed set of arcs (measured from the agent); (d) a neural network that computes the utility of actions.

The initial mazes are generated randomly using a maze-creation program (Maclin, 1995) that randomly lays out lines of obstacles and then creates connections between "rooms." The percentage of the total board covered by obstacles is controlled by a parameter, as are the number of enemies and food items. The agent, enemies, and food are randomly deposited on the board, with the caveat that the enemies are required to be initially at least a fixed distance away from the agent at the start.

The agent receives reinforcement signals when: (i) an enemy eliminates the agent by touching the agent (−1.0), (ii) the agent collects one of the food objects (+0.7), or (iii) the agent destroys an enemy by pushing an obstacle into it (+0.9).

We do not assume a global view of the environment, but instead use an agent-centered sensor model. It is based on partitioning the world into a set of sectors around the agent (see Figure 8b). Each sector is defined by a minimum and maximum distance from the agent and a minimum and maximum angle with respect to the direction the agent is facing. The agent calculates the percentage of each sector that is occupied by each type of object – food, enemy, obstacle, or wall. To calculate the sector occupancy, we assume the agent is able to measure the distance to the nearest occluding object along a fixed set of angles around the agent (see Figure 8c). This means that the agent is only able to represent the objects in direct line-of-sight from the agent (for example, the enemy to the south of the agent is out of sight). The percentage of each object type in a sector is just the number of sensing arcs that end in that sector by hitting an object of the given type, divided by the maximum number of arcs that could end in the sector. So for example, given Figure 8b, the agent's percentage for "obstacle" would be high for the sector to the east. The agent also calculates how much of each sector is empty and how much is occluded. These percentages constitute the input to the neural network (see Figure 8d). Note that the agent also receives as input, using a 1-of-N encoding, the action the agent took in the previous state.[2]

4.2. Methodology

We train the agents for a fixed number of *episodes* for each experiment. An episode consists of placing the agent into a randomly generated, initial environment, and then allowing it to explore until it is captured or a threshold of 500 steps is reached. We report our results by training episodes rather than number of training actions because we believe episodes are a more useful measure of "meaningful" training done – an agent having collected all of the food and eliminated all of the enemies could spend a large amount of time in useless wandering (while receiving no reinforcements), thus counting actions might penalize such an agent since it gets to experience fewer reinforcement situations. In any case, for all of our results the results appear qualitatively similar when graphed by the number of training actions (i.e., the agents all take a similar number of actions per episode during training).

Each of our environments contains a 7x7 grid with approximately 15 obstacles, 3 enemies, and 10 food items. We use three randomly generated sequences of initial environments as a basis for the training episodes. We train 10 randomly initialized networks on each of the three sequences of environments; hence, we report the averaged results of 30 neural networks. We estimate the future average total reinforcement (the average sum of the reinforcements received by the agent)[3] by "freezing" the network and measuring the average reinforcement on a testset of 100 randomly generated environments; the same testset is used for all our experiments.

We chose parameters for our Q-learning algorithm that are similar to those investigated by Lin (1992). The learning rate for the network is 0.15, with a discount factor of 0.9. To establish a baseline system, we experimented with various numbers of hidden units, settling on 15 since that number resulted in the best average reinforcement for the baseline system. We also experimented with giving this system recurrent units (as in the units RATLE added for multi-step and loop plans), but these units did not lead to improved performance for the baseline system, and, hence, the baseline results are for a system without recurrent links. However, recall that the input vector records the last action taken.

After choosing an initial network topology, we then spent time acting as a user of RATLE, observing the behavior of the agent at various times. Based on these observations, we wrote several collections of advice. For use in our experiments, we chose four sets of advice (see Appendix A), two that use multi-step plans (referred to as *ElimEnemies* and *Surrounded*), and two that do not (*SimpleMoves* and *NonLocalMoves*).

4.3. Results

In our first experiment, we evaluate the hypothesis that our approach can in fact take advantage of advice. After 1000 episodes of initial learning, we judge the value of (independently) providing each of the four sets of advice to our agent using RATLE. We train the agent for 2000 more episodes after giving the advice, then measure its average cumulative reinforcement on the testset. (The baseline is also trained for 3000 episodes). Table 3 reports the averaged testset reinforcement; all gains over the baseline system are statistically significant[4].   Note that the gain is higher for the simpler pieces of advice *SimpleMoves* and *NonLocalMoves*, which do not incorporate multi-step plans. This suggests the need for further work on taking complex advice; however, the multi-step advice may simply be less useful.

Each of our pieces of advice to the agent addresses specific subtasks: collecting food (*SimpleMoves* and *NonLocalMoves*); eliminating enemies (*ElimEnemies*); and avoiding enemies, thus surviving longer (*SimpleMoves*, *NonLocalMoves*, and *Surrounded*). Hence, it is natural to ask how well each piece of advice meets its intent. Table 4 reports statistics on the components of the reward.   These statistics show that the pieces of advice do indeed lead to the expected improvements. For example, our advice *ElimEnemies* leads to a much larger number of enemies eliminated than the baseline or any of the other pieces of advice.

Table 3. Testset results for the baseline and the four different types of advice. Each of the four gains over the baseline is statistically significant.

Advice Added	Average Total Reinforcement on the Testset
None (baseline)	1.32
SimpleMoves	1.91
NonLocalMoves	2.01
ElimEnemies	1.87
Surrounded	1.72

Table 4. Mean number of enemies captured, food collected, and number of actions taken (survival time) for the experiments summarized in Table 3.

Advice Added	Enemies Captured	Food Collected	Survival Time
None (baseline)	0.15	3.09	32.7
SimpleMoves	0.31	3.74	40.8
NonLocalMoves	0.26	3.95	39.1
ElimEnemies	0.44	3.50	38.3
Surrounded	0.30	3.48	46.2

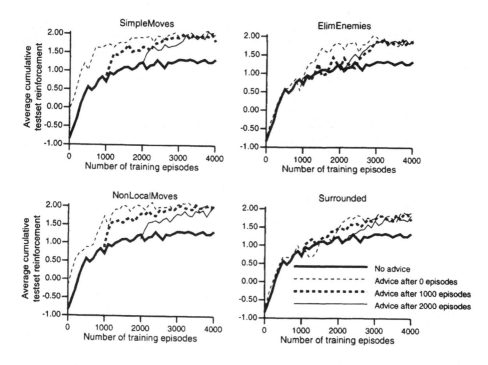

Figure 9. Average total reinforcement for our four sample pieces of advice as a function of amount of training and point of insertion of the advice.

In our second experiment we investigate the hypothesis that the observer can beneficially provide advice at any time during training. To test this, we insert the four sets of advice at different points in training (after 0, 1000, and 2000 episodes). Figure 9 contains the results for the four pieces of advice. They indicate the learner does indeed converge to approximately the same expected reward no matter when the advice is presented.

Our third experiment investigates the hypothesis that subsequent advice will lead to further gains in performance. To test this hypothesis, we supplied each of our four pieces of advice to an agent after 1000 episodes (as in our first experiment), supplied one of the remaining three pieces of advice after another 1000 episodes, and then trained the resulting agent for 2000 more episodes. These results are averaged over 60 neural networks instead of the 30 networks used in the other experiments in order to obtain statistically significant results. Table 5 shows the results for this test.

In all cases, adding a second piece of advice leads to improved performance. However, the resulting gains when adding the second piece of advice are not as large as the original gains over the baseline system. We suspect this occurs due to a combination of factors: (i) there is an upper limit to how well the agents can do – though it is difficult to quantify; (ii) the pieces of advice interact – they may suggest different actions in different situations, and in the process of resolving these conflicts, the agent may use one piece of advice less often; and (iii) the advice pieces are related, so that one piece may cover situations

Table 5. Average testset reinforcement for each of the possible pairs of our four sets of advice. The first piece of advice is added after 1000 episodes, the second piece of advice after an additional 1000 episodes, and then trained for 2000 more episodes (total of 4000 training episodes). Shown in parentheses next to the first pieces of advice are the performance results from our first experiment where only a single piece of advice was added. All of the resulting agents show statistically significant gains in performance over the agent with just the first piece of advice.

First Piece of Advice		Second Piece of Advice			
		SimpleMoves	*NonLocalMoves*	*ElimEnemies*	*Surrounded*
SimpleMoves	(1.91)	-	2.17	2.10	2.05
NonLocalMoves	(2.01)	2.27	-	2.18	2.13
ElimEnemies	(1.87)	2.01	2.26	-	2.06
Surrounded	(1.72)	2.04	2.11	1.95	-

that the other already covers. Also interesting to note is that the order of presentation affects the level of performance achieved in some cases (e.g., presenting *NonLocalMoves* followed by *SimpleMoves* achieves higher performance than *SimpleMoves* followed by *NonLocalMoves*).

In our fourth experiment we evaluate the usefulness of combining our advice-giving approach with Lin's "replay" technique (1992). Lin introduced the replay method to make use of "good" sequences of actions provided by a teacher. In replay, the agent trains on the teacher-provided sequences frequently to bias its utility function towards these good actions. Thrun (1994, personal communication) reports that replay can in fact be useful even when the remembered sequences are not teacher-provided sequences – in effect, by training multiple times on each state-action pair the agent is "leveraging" more value out of each example. Hence, our experiment addresses two related questions: (i) does the advice provide any benefit over simply reusing the agent's experiences multiple times?, and (ii) can our approach benefit from replay, for example, by needing fewer training episodes to achieve a given level of performance? Our hypothesis was that the answer to both questions is "yes."

To test our hypothesis we implemented two approaches to replay in RATLE and evaluated them using the *NonLocalMoves* advice. In one approach, which we will call *Action Replay*, we simply keep the last N state-action pairs that the agent encountered, and on each step train with all of the saved state-action pairs in a randomized order. A second approach (similar to Lin's), which we will call *Sequence Replay*, is more complicated. Here, we keep the last N *sequences* of actions that ended when the agent received a non-zero reinforcement. Once a sequence completes, we train on all of the saved sequences, again, in a randomized order. To train the network with a sequence, we first train the network on the state where the reinforcement was received, then the state one step before that state, then two steps before that state, etc., on the theory that the states nearest reinforcements best estimate the actual utility of the state. Results for keeping 250 state-action pairs and 250 sequences[5] appear in Table 6; due to time constraints we trained these agents for only 1000 episodes.

Table 6. Average total reinforcement results for the advice *NonLocalMoves* using two forms of replay. The advice is inserted and then the network is trained for 1000 episodes. Replay results are the *best* results achieved on 1000 episodes of training (occurring at 600 episodes for Action Replay and 500 episodes for Sequence Replay). Results for the RATLE approach without replay are also shown; these results are for 1000 training episodes.

Training Method	Average Total Testset Reinforcement
Standard RATLE (no replay)	1.74
Action-Replay Method	1.48
Sequence-Replay Method	1.45

Surprisingly, replay did not help our approach. After examining networks during replay training, we hypothesize this occurred because we are using a single network to predict all of the Q values for a state. During training, to determine a target vector for the network, we first calculate the new Q value for the action the agent actually performed. We then activate the network, and set the target output vector for the network to be equal to the actual output vector, except that we use the new prediction for the action taken. For example, assume the agent takes action two (of three actions) and calculates that the Q value for action two should be 0.7. To create a target vector the agent activates the network with the state (assume that the resulting output vector is [0.4,0.5,0.3]), and then creates a target vector that is the same as the output vector except for the new Q value for the action taken (i.e., [0.4,**0.7**,0.3]). This causes the network to have error at only one output unit (the one associated with the action taken). For replay this is a problem because we will be activating the network for a state a number of times, but only trying to correctly predict one output unit (the other outputs are essentially allowed to take on any value), and since the output units share hidden units, changes made to predict one output unit may affect others. If we repeat this training a number of times, the Q values for other actions in a state may become greatly distorted. Also, if there is unpredictability in the outcomes of actions, it is important to average over the different results; replay focuses on a single outcome. One possible solution to this problem is to use separate networks for each action, but this means the actions will not be able to share concepts learned at the hidden units. We plan to further investigate this topic, since replay intuitively seems to be a valuable technique for reducing the amount of experimentation an RL agent has to perform.

Our final experiment investigates a naive approach for using advice. This simple strawman algorithm follows the observer's advice when it applies; otherwise it uses a "traditional" connectionist Q-learner to choose its actions. We use this strawman to evaluate if it is important that the agent refine the advice it receives. When measured on the testset, the strawman employs a loop similar to that shown in Table 2. One difference for the strawman's algorithm is that Step 6 in Table 2's algorithm is left out. The other difference is that Step 2 (selecting an action) is replaced by the following:

Evaluate the advice to see if it suggests any actions:
   If any actions are suggested, choose one randomly,
   Else choose the action that the network predicts has maximum utility.

Table 7. Average testset reinforcement using the strawman approach to using advice compared to the RATLE method.

Advice	STRAWMAN	RATLE
SimpleMoves	1.63	1.91
NonLocalMoves	1.46	2.01
ElimEnemies	1.28	1.87
Surrounded	1.21	1.72

The performance of this strawman is reported in Table 7. In all cases, RATLE performs better than the strawman; all of these reinforcement gains are statistically significant. In fact, in two of the cases, *ElimEnemies* and *Surrounded*, the resulting method for selecting actions is actually worse than simply using the baseline network (whose average performance is 1.32).

4.4. Discussion

Our experiments demonstrate that: (i) advice can improve the performance of an RL agent; (ii) advice produces the same resulting performance no matter when it is added; (iii) a second piece of advice can produce further gains; and (iv) it is important for the agent to be able to refine the advice it receives. In other experiments (not reported here), we demonstrate that an agent can quickly overcome the effects of "bad" advice (Maclin, 1995). We corroborated our Pengo results using a second testbed (Maclin, 1995). A significant feature of our second testbed is that its agent's sensors record the complete state of the environment. Thus, the results in our second testbed support the claim that the value of advice in the Pengo testbed is not due solely to the fact that the teacher sees the complete state, while the learner only has line-of-sight sensors (and, hence, is trying to learn a *partially observable* Markov decision process; Monahan, 1982).

One key question arises from our Pengo results: will the baseline system eventually achieve the same level of performance that the advice-taking system achieves? After all, Q-learning converges to the optimal Q function when a Q table is used to represent the function (Watkins & Dayan, 1992). However, a backpropagation-trained network may only converge to a *local* minimum in the weight space defining the Q function. To further answer the performance-in-the-limit question, we will address a more general one – what effect do we expect advice to have on the agent?

When we introduce "good" advice into an agent, we expect it to have one or more of several possible effects. One possible effect of advice is that the advice will change the network's predictions of some of the Q values to values that are closer to the desired "optimal" values. By reducing the overall error the agent may be able to converge more quickly towards the optimal Q function. A second related effect is that by increasing (and decreasing) certain Q values the advice changes which states are explored by the agent. Here, good advice would cause the agent to explore states that are useful in finding the optimal plan (or ignoring states that are detrimental). Focusing on the states that are important to the optimal solution may lead to the agent converging more quickly to a solution. A third possible effect is that the addition of advice alters the weight space of possible solutions that the learner is exploring. This is because the new weights and hidden units change the set of parameters that the learner was exploring. For example, the advice may construct an intermediate term (represented by a hidden unit) with very large weights, that could not have been found by gradient-descent search. In the resulting altered weight space the learner may be able to explore functions that were unreachable before the advice is added (and these functions may be closer to the optimal).

Given these possible effects of good advice, we can conclude that advice can both cause the agent to converge more quickly to a solution, and that advice may cause the agent to find a better solution than it may have otherwise found. For our experiments, we see the effect of speeded convergence in the graphs of Figure 9, where the advice, generally after a small amount of training, leads the agent to achieve a high level of performance quickly. These graphs also demonstrate the effect of convergence to a better solution, at least given our fixed amount of training. Note that these effects are a result of what we would call "good" advice. It is possible that "bad" advice could have equally deleterious effects. So, a related question is how do we determine whether advice is "good" or not?

Unfortunately, determining the "goodness" of advice appears to be a fairly tricky problem, since even apparently useful advice can lead to poor performance in certain cases. Consider for example, the simple problem shown in Figure 10. This example demonstrates a case where advice, though providing useful information, could actually cause the agent to take longer to converge to an optimal policy. Basically, the good advice "masks" an even better policy. This example suggests that we may want to rethink the stochastic mechanism that we use to select actions. In any case, it appears that defining the properties of "good" advice is a challenging topic for future work. As a first (and admittedly vague) approximation we would expect advice to be "good" when it causes one of the effects mentioned above: (i) it reduces the overall error in the agent's predicted

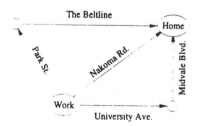

Figure 10. A sample problem where good advice can fail to enhance performance. Assume the goal of the agent is to go from Work to Home, and that the agent will receive a large reward for taking Nakoma Road, and a slightly smaller reward for taking University Avenue followed by Midvale Boulevard. If the agent receives advice that University followed by Midvale is a good plan, the agent, when confronted with the problem of going from Work to Home will likely follow this plan (even during training, since actions are selected proportional to their predicted utility). Thus it may take a long time for the learner to try Nakoma Road often enough to learn that it is a better route. A learner without advice might try both routes equally often and quickly learn the correct utility value for each route.

Q values, (ii) it causes the agent to pick actions that lead to states that are important in finding a solution, or (iii) it transforms the network so that the agent is able to perform gradient descent to a better solution.

5. Future Work

Based on our initial experience, we intend to expand our approach in a number of directions. One important future topic is to evaluate our approach in other domains. In particular, we intend to explore tasks involving multiple agents working in cooperation. Such a domain would be interesting in that an observer could give advice on how a "group" of agents could solve a task. Another domain of interest is software agents (Riecken, 1994). For example, a human could advise a software agent that looks for "interesting" papers on the World-Wide Web.

We see algorithmic extensions as fitting into the three categories explained below.

Broadening the Advice Language

Our experience with RATLE has led us to consider a number of extensions to our current programming language. Examples include:

- **Prefer** *action* – let the teacher indicate that one action is "preferred" in a state. Here the advisor would only be helping the learner sort through its options, rather than specifically saying what should be done.

- **Forget** *advice* – permit the advisor to retract previous advice.

- **Add/Subtract** *condition* **from** *advice* – allow the advisor to add or remove conditions from previous rules, thereby fine-tuning advice.

- **Reward/Punish** *state* – let the teacher specify "internal" rewards that the agent is to receive in certain states. This type of advice could be used to give the agent a set of internal goals.

- **Remember** *condition* – permit the advisor to indicate propositions that the learner should remember (e.g., the location of some important site, like a good place to hide, so that it can get back there again). We would implement this using recurrent units that record state information. This remembered information could then be used in future advice.

We also plan to explore mechanisms for specifying multi-user advice when we explore domains with multiple agents.

Improving the Algorithmic Details

At present, our algorithm adds hidden units to the learner's neural network whenever advice is received. Hence, the network's size grows monotonically. Although recent evidence (Weigend, 1993) suggests overly large networks are not a problem given that one uses proper training techniques, we plan to evaluate techniques for periodically "cleaning up" and shrinking the learner's network. We plan to use standard neural-network techniques for removing network links with low saliency (e.g., Le Cun, Denker, & Solla, 1990).

In our current implementation, a plan (i.e., a sequence of actions) can be interrupted if another action has higher utility (see Figure 5). Recall that the advice only increases the utility of the actions in the plan, and that the learner can choose to execute another action if it has higher utility. Once a plan is interrupted, it cannot be resumed from the point of interruption because the necessary state unit is not active. We intend to evaluate methods that allow plans to be temporarily interrupted; once the higher-utility tasks complete, the interrupted plan will resume. We anticipate that this will involve the use of exponentially decaying state units that record that a plan was being executed recently.

The networks we use in our current implementation have numeric-valued output units (recall that they represent the expected utility of actions). Hence, we need to more thoroughly investigate the setting of the weights between the Boolean-valued advice nodes and the numeric-valued utility nodes, a topic not relevant to the original work with KBANN, since that research only involved Boolean concepts. Currently, advice simply increases the utility of the recommended actions by a fixed amount. Although subsequent training can alter this initial setting, we plan to more intelligently perform this initial setting. For example, we could reset the weights of the network so that the suggested action always has the highest utility in the specified states. This approach would guarantee that the suggested action will have the highest utility, but can be faulty if the action is already considered the best of several bad choices. In this case the alternate approach would simply leave the current network unchanged, since the advised action is already preferred. But the teacher may be saying that the action is not only the best choice, but that the utility of the action is high (i.e., the action is "good" in some sense). Therefore, simply requiring that the action be the "best" choice may not always capture the teacher's intentions. This approach also requires that RATLE find an appropriate set of weights to insure that the suggested action be selected first (possibly by solving a non-linear program).

In another form of reinforcement learning, the agent predicts the utility of a state rather than the utility of an action in a state (Sutton, 1988); here the learner has a model of how its actions change the world, and determines the action to take by checking the utility of the states that are reachable from the current state. Applying RATLE to this type of reinforcement-learning system would be difficult, since RATLE statements suggest actions to take. In order to map a statement indicating an action, RATLE would first have to determine the set of states that meet the condition of the statement, then calculate the set of states that would result by following the suggested action. RATLE would then increase the utility of the states that follow from the suggested action. Other types of advice would be more straightforward under this approach. For example, if the teacher gave advice about a goal the agent should try to achieve (i.e., as in Gordon and Subramanian's, 1994, approach), RATLE could determine the set of states corresponding to the goal and simply increase the utility of all of these states.

Finally, our system maintains no statistics that record how often a piece of advice was applicable and how often it was followed. We intend to add such statistics-gatherers and use them to inform the advisor that a given piece of advice was seldom applicable or followed. We also plan to keep a record of the *original* advice and compare its statistics to the *refined* version. Significant differences between the two should cause the learner to inform its advisor that some advice has substantially changed (we plan to use the rule-extraction techniques described below when we present the refined advice to the advisor).

Converting Refined Advice into a Human-Comprehensible Form

One interesting area of future research is the "extraction" (i.e., conversion to a easily comprehensible form) of learned knowledge from our connectionist utility function. We plan to extend previous work on rule extraction (Craven & Shavlik, 1994; Towell & Shavlik, 1993) to produce rules in RATLE's language. We also plan to investigate the use of rule extraction as a mechanism for transfering learned knowledge between RL agents operating in the same or similar environments.

6. Related Work

Our work relates to a number of recent research efforts. This related work can be roughly divided into five groups: (i) providing advice to a problem solver, (ii) giving advice to a problem solver employing reinforcement learning, (iii) developing programming languages for interacting with agents, (iv) creating knowledge-based neural networks, and (v) refining prior domain theories.

Providing advice to a problem solver

An early example of a system that makes use of advice is Mostow's (1982) FOO system, which operationalizes general advice by reformulating the advice into search heuristics. These search heuristics are then applied during problem solving. In FOO the advice is assumed to be correct, and the learner has to convert the general advice into an executable plan based on its knowledge about the domain. Our system is different in that we try to directly incorporate general advice, but we do not provide a sophisticated means of operationalizing advice. Also, we do not assume the advice is correct; instead we use reinforcement learning to refine and evaluate the advice.

More recently, Laird, Hucka, Yager, and Tuck (1990) created an advice-taking system called ROBO-SOAR. In this system, an observer can provide advice whenever the system is at an impasse by suggesting which operators to explore in an attempt to resolve the impasse. As with FOO, the advice presented is used to guide the learner's reasoning process, while in RATLE we directly incorporate the advice into the learner's knowledge base and then refine that knowledge using subsequent experience. Huffman and Laird

(1993) developed the INSTRUCTO-SOAR system that allows an agent to interpret simple imperative statements such as "Pick up the red block." INSTRUCTO-SOAR examines these instructions in the context of its current problem solving, and uses SOAR's form of explanation-based learning to generalize the instruction into a rule that can be used in similar situations. RATLE differs from INSTRUCTO-SOAR in that we provide a language for entering general advice rather than attempting to generalize specific advice.

Providing advice to a problem solver that uses reinforcement learning

A number of researchers have introduced methods for providing advice to a reinforcement learning agent. Lin (1992) designed a technique that uses advice expressed as sequences of teacher's actions. In his system the agent "replays" the teacher actions periodically to bias the agent toward the actions chosen by the teacher. Our approach differs in that RATLE inputs the advice in a general form; also, RATLE directly installs the advice into the learner rather than using the advice as a basis for training examples.

Utgoff and Clouse (1991) developed a learner that consults a set of teacher actions if the action it chose resulted in significant error. This system has the advantage that it determines the situations in which it requires advice, but is limited in that it may require advice more often than the observer is willing to provide it. In RATLE the advisor provides advice whenever he or she feels they have something to say.

Whitehead (1991) examined an approach similar to both Lin's and Utgoff & Clouse's that can learn both by receiving advice in the form of critiques (a reward indicating whether the chosen action was optimal or not), as well as learning by observing the actions chosen by a teacher. Clouse and Utgoff (1992) created a second system that takes advice in the form of actions suggested by the teacher. Both systems are similar to ours in that they can incorporate advice whenever the teacher chooses to provide it, but unlike RATLE they do not accept broadly applicable advice.

Thrun and Mitchell (1993) investigated a method for allowing RL agents to make use of prior knowledge in the form of neural networks. These neural networks are assumed to have been trained to predict the results of actions. This proves to be effective, but requires previously trained neural networks that are related to the task being addressed.

Gordon and Subramanian (1994) developed a system that is closely related to ours. Their system employs genetic algorithms, an alternate approach for learning from reinforcements. Their agent accepts high-level advice of the form IF *conditions* THEN ACHIEVE *goal*. It operationalizes these rules using its background knowledge about goal achievement. Our work primarily differs from Gordon and Subramanian's in that RATLE uses connectionist Q-learning instead of genetic algorithms, and in that RATLE's advice language focuses on actions to take rather than goals to achieve. Also, we allow advice to be given at any time during the training process. However, our system does not have the operationalization capability of Gordon and Subramanian's system.

Developing robot-programming languages

Many researchers have introduced languages for programming robot-like agents (Chapman, 1991; Kaelbling, 1987; Nilsson, 1994). These systems do not generally focus on programming agents that learn to refine their programs. Crangle and Suppes (1994) investigated how a robot can understand a human's instructions, expressed in ordinary English. However, they do not address correction, by the learner, of approximately correct advice.

Incorporating advice into neural networks

Noelle and Cottrell (1994) suggest an alternative approach to making use of advice in neural networks. One way their approach differs from ours is that their connectionist model itself performs the process of incorporating advice, which contrasts to our approach where we directly add new "knowledge-based" units to the neural network. Our approach leads to faster assimilation of advice, although theirs is arguably a better psychological model.

Siegelman (1994) proposed a technique for converting programs expressed in a general-purpose, high-level language into a type of recurrent neural networks. Her system is especially interesting in that it provides a mechanism for performing arithmetic calculations, but the learning abilities of her system have not yet been empirically demonstrated.

Gruau (1994) developed a compiler that translates Pascal programs into neural networks. While his approach has so far only been tested on simple programs, his technique may prove applicable to the task of programming agents. Gruau's approach includes two methods for refining the networks he produces: a genetic algorithm and a hill-climber. The main difference between Gruau's system and ours is that the networks we produce can be refined using standard connectionist techniques such as backpropagation, while Gruau's networks require the development of a specific learning algorithm, since they require integer weights (-1,0,1) and incorporate functions that do not have derivatives.

Diederich (1989) devised a method that accepts instructions in a symbolic form. He uses the instructions to create examples, then trains a neural network with these examples to incorporate the instructions, as opposed to directly installing the instructions.

Abu-Mostafa (1995) uses an approach similar to Diederich's to encode "hints" in a neural network. A hint is a piece of knowledge provided to the network that indicates some important general aspect for the network to have. For example, a hint might indicate to a network trying to assess people as credit risks that a "monotonicity" principle should hold (i.e., when one person is a good credit risk, then an identical person with a higher salary should also be a good risk). Abu-Mostafa uses these hints to generate examples that will cause the network to have this property, then mixes these examples in with the original training examples. As with Diederich's work, our work differs from Abu-Mostafa's in that RATLE directly installs the advice into the network.

Suddarth and Holden (1991) investigated another form of "hint" for a neural network. In their approach, a hint is an extra output value for the neural network. For example, a neural network using sigmoidal activation units to try to learn the difficult XOR function might receive a hint in the form of the output value for the OR function. The OR function is useful as a hint because it is simple to learn. The network can use the hidden units it constructs to predict the OR value when learning XOR (i.e., the hint serves to decompose the problem for the network). Suddarth and Holden's work however only deals with hints in the form of useful output signals, and still requires network learning, while RATLE incorporates advice immediately.

Our work on RATLE is similar to our earlier work with the FSKBANN system (Maclin & Shavlik, 1993). FSKBANN uses a type of recurrent neural network introduced by Elman (1990) that maintains information from previous activations using the recurrent network links. FSKBANN extends KBANN to deal with *state* units, but it does not create *new* state units. Similarly, other researchers (Frasconi, Gori, Maggini, & Soda, 1995; Omlin & Giles, 1992) insert prior knowledge about a finite-state automaton into a recurrent neural network. Like our FSKBANN work, this work does not make use of knowledge provided after training has begun, nor do they study RL tasks.

Lin (1993) has also investigated the idea of having a learner use prior state knowledge. He uses an RL agent that has as input not only the current input state, but also some number of the previous input states. The difference between Lin's approach and ours is that we use the advice to determine a portion of the previous information to keep, rather than trying to keep all of it, thereby focusing learning.

Refining prior knowledge

There has been a growing literature on automated "theory refinement" (Fu, 1989; Ginsberg, 1988; Ourston & Mooney, 1994; Pazzani & Kibler, 1992; Shavlik & Towell, 1989), and it is from this research perspective that our advice-taking work arose. Our new work differs by its novel emphasis on theory refinement in the context of multi-step problem solving in multi-actor worlds, as opposed to refinement of theories for categorization and diagnosis. Here, we view "domain theories" as statements in a procedural programming language, rather than the common view of a domain theory being a collection of declarative Prolog statements. We also address reinforcement learning, rather than learning-from-examples. Finally, unlike previous approaches, we allow domain theories to be provided piecemeal at any time during the training process, as the need becomes apparent to the advisor. In complex tasks it is not desirable to simply restart learning from the beginning whenever one wants to add something to the domain theory.

7. Conclusions

We present an approach that allows a connectionist, reinforcement-learning agent to take advantage of instructions provided by an external observer. The observer communicates advice using a simple imperative programming language, one that does not require that the observer have any knowledge of the agent's internal workings. The reinforcement learner applies techniques from knowledge-based neural networks to directly insert the observer's advice into the learner's utility function, thereby speeding up its learning.

Importantly, the agent does not accept the advice absolutely nor permanently. Based on subsequent experience, the learner can refine and even discard the advice.

Experiments with our RATLE system demonstrate the validity of this advice-taking approach; each of four types of sample advice lead to statistically significant gains in expected future reward. Interestingly, our experiments show that these gains do not depend on when the observer supplies the advice. Finally, we present results that show our approach is superior to a naive approach for making use of the observer's advice.

In conclusion, we have proposed an appealing approach for learning from both instruction and experience in dynamic, multi-actor tasks. This work widens the "information pipeline" between humans and machine learners, without requiring that the human provide absolutely correct information to the learner.

Acknowledgments

This research was partially supported by Office of Naval Research Grant N00014-93-1-0998 and National Science Foundation Grant IRI-9002413. We also wish to thank Carolyn Allex, Mark Craven, Diana Gordon, Leslie Pack Kaelbling, Sebastian Thrun, and the two anonymous reviewers for their helpful comments on drafts of this paper. A shorter version of this article appeared as Maclin and Shavlik (1994).

Appendix A

Four Sample Pieces of Advice

The four pieces of advice used in the experiments in Section 4 appear below. Recall that in our testbed the agent has two actions (moving and pushing) that can be executed in any of the four directions (East, North, West, and South). To make it easier for an observer to specify advice that applies in any direction, we defined the special term *dir*. During parsing, *dir* is expanded by replacing each rule containing it with four rules, one for each direction. Similarly we have defined a set of four terms {*ahead, back, side1, side2*}. Any rule using these terms leads to *eight* rules – two for each case where *ahead* is East, North, West and South and *back* is appropriately set. There are two for each case of *ahead* and *back* because *side1* and *side2* can have two sets of values for any value of *ahead* (e.g., if *ahead* is North, *side1* could be East and *side2* West, or vice-versa). Appendix A in Maclin (1995) contains the definitions of the fuzzy terms (e.g., Near, Many, An, and East).

SimpleMoves

IF An Obstacle IS (NextTo ∧ *dir*)
THEN INFER OkPush*dir* END;
IF No Obstacle IS (NextTo ∧ *dir*) ∧
No Wall IS (NextTo ∧ *dir*)
THEN INFER OkMove*dir* END;
IF An Enemy IS (Near ∧ *dir*)
THEN DO_NOT Move*dir* END;
IF OkMove*dir* ∧ A Food IS (Near ∧ *dir*) ∧
No Enemy IS (Near ∧ *dir*)
THEN Move*dir* END;
IF OkPush*dir* ∧ An Enemy IS (Near ∧ *dir*)
THEN Push*dir* END

Grab food next to you; run from enemies next to you; push obstacles at enemies behind obstacles. [This leads to 20 rules.]

NonLocalMoves

IF No Obstacle IS (NextTo ∧ *dir*) ∧
No Wall IS (NextTo ∧ *dir*)
THEN INFER OkMove*dir* END;
IF OkMove*dir* ∧ Many Enemy ARE (¬ *dir*) ∧
No Enemy IS (Near ∧ *dir*)
THEN Move*dir* END;
IF OkMove*dir* ∧ No Enemy IS (*dir* ∧ Near) ∧
A Food IS (*dir* ∧ Near) ∧
An Enemy IS (*dir* ∧ {Medium ∨ Far})
THEN Move*dir* END

Run away if many enemies in a direction (even if they are not close), and move towards foods even if there is an enemy in that direction (as long as the enemy is a ways off). [12 rules.]

ElimEnemies

IF No Obstacle IS (NextTo ∧ *dir*) ∧
No Wall IS (NextTo ∧ *dir*)
THEN INFER OkMove*dir* END;
IF OkMove*ahead* ∧ An Enemy IS (Near ∧ *back*) ∧
An Obstacle IS (NextTo ∧ *side1*)
THEN
MULTIACTION
Move*ahead*
Move*side1*
Move*side1*
Move*back*
Push*side2*
END
END

When an enemy is closely behind you and a convenient obstacle is nearby, spin around the obstacle and push it at the enemy. [12 rules.]

Surrounded

IF An Obstacle IS (NextTo ∧ *dir*)
 THEN INFER OkPush*dir* END;
IF An Enemy IS (Near ∧ *dir*) ∨
 A Wall IS (NextTo ∧ *dir*) ∨
 An Obstacle IS (NextTo ∧ *dir*)
 THEN INFER Blocked*dir* END;
IF BlockedEast ∧ BlockedNorth ∧
 BlockedSouth ∧ BlockedWest
 THEN INFER Surrounded END;
WHEN Surrounded ∧ OkPush*dir* ∧ An Enemy IS Near
 REPEAT
  MULTIACTION Push*dir* Move*dir* END
 UNTIL ¬ OkPush*dir*
END

When surrounded by obstacles and enemies, push obstacles out of the way and move through the holes. [13 rules.]

Appendix B

The Grammar for RATLE's Advice Language

The start nonterminal of the grammar is *rules*. Grammar rules are shown with vertical bars (|) indicating alternate rules for nonterminals (e.g., *rules, rules,* and *ante*). Names like IF, THEN, and WHILE are keywords in the advice language. Additional details can be found in Maclin (1995).

A piece of advice may be a single construct or multiple constructs.
  rules ← *rule* | *rules* ; *rule*

The grammar has three main constructs: IF-THENS, WHILEs, and REPEATs.
  rule ← IF *ante* THEN *conc* *else* END
     | WHILE *ante* DO *act* *postact* END
     | *pre* REPEAT *act* UNTIL *ante* *postact* END

  else ← ε | ELSE *act*
  postact ← ε | THEN *act*
  pre ← ε | WHEN *ante*

A MULTIACTION construct specifies a *series* of actions to perform.
  conc ← *act* | INFER Term_Name | REMEMBER Term_Name
  act ← *cons* | MULTIACTION *clist* END
  clist ← *cons* | *cons* *clist*
  cons ← Term_Name | DO_NOT Term_Name | (*corlst*)
  corlst ← Term_Name | Term_Name ∨ *corlst*

Antecedents are logical combinations of terms and fuzzy conditionals.

$$ante \quad \leftarrow \text{Term\_Name} \mid (\ ante\ ) \mid \neg\ ante$$
$$\mid ante \land ante \mid ante \lor ante$$
$$\mid \text{Quantifier\_Name Object\_Name IS } desc$$

The descriptor of a fuzzy conditional is a logical combination of fuzzy terms.

$$desc \quad \leftarrow \text{Descriptor\_Name} \mid \neg\ desc \mid \{\ dlist\ \} \mid (\ dexpr\ )$$
$$dlist \quad \leftarrow \text{Descriptor\_Name} \mid \text{Descriptor\_Name}\ ,\ dlist$$
$$dexpr \quad \leftarrow desc \mid dexpr \land dexpr \mid dexpr \lor dexpr$$

Notes

1. Through empirical investigation we chose a value of 2.0 for these weights. A topic of our future research is to more intelligently select this value. See the discussion in Section 5.
2. The agent needs this information when employing multiple-step plans (see Section 3). We include this information as input for all of the agents used in our experiments so that none will be at a disadvantage.
3. We report the average total reinforcement rather than the average discounted reinforcement because this is the standard for the RL community. Graphs of the average *discounted* reward are qualitatively similar to those shown in the next section.
4. All results reported as statistically significant are significant at the $p < 0.05$ level (i.e., with 95% confidence).
5. We also experimented with keeping only 100 pairs or sequences; the results using 250 pairs and sequences were better.

References

Abu-Mostafa, Y. (1995). Hints. *Neural Computation, 7*, 639–671.

Agre, P., & Chapman, D. (1987). Pengi: An implementation of a theory of activity. In *Proceedings of the Sixth National Conference on Artificial Intelligence*, pp. 268–272 Seattle, WA.

Anderson, C. (1987). Strategy learning with multilayer connectionist representations. In *Proceedings of the Fourth International Workshop on Machine Learning*, pp. 103–114 Irvine, CA.

Barto, A., Sutton, R., & Anderson, C. (1983). Neuronlike adaptive elements that can solve difficult learning control problems. *IEEE Transactions on Systems, Man, and Cybernetics, 13*, 834–846.

Barto, A., Sutton, R., & Watkins, C. (1990). Learning and sequential decision making. In Gabriel, M., & Moore, J. (Eds.), *Learning and Computational Neuroscience*, pp. 539–602. MIT Press, Cambridge, MA.

Berenji, H., & Khedkar, P. (1992). Learning and tuning fuzzy logic controllers through reinforcements. *IEEE Transactions on Neural Networks, 3*, 724–740.

Chapman, D. (1991). *Vision, Instruction, and Action.* MIT Press, Cambridge, MA.

Clouse, J., & Utgoff, P. (1992). A teaching method for reinforcement learning. In *Proceedings of the Ninth International Conference on Machine Learning*, pp. 92–101 Aberdeen, Scotland.

Crangle, C., & Suppes, P. (1994). *Language and Learning for Robots.* CSLI Publications, Stanford, CA.

Craven. M., & Shavlik. J. (1994). Using sampling and queries to extract rules from trained neural networks. In *Proceedings of the Eleventh International Conference on Machine Learning*. pp. 37–45 New Brunswick, NJ.

Diederich, J. (1989). "Learning by instruction" in connectionist systems. In *Proceedings of the Sixth International Workshop on Machine Learning*, pp. 66–68 Ithaca, NY.

Dietterich, T. (1991). Knowledge compilation: Bridging the gap between specification and implementation. *IEEE Expert, 6*, 80–82.

Elman, J. (1990). Finding structure in time. *Cognitive Science, 14*, 179–211.

Frasconi, P., Gori, M., Maggini, M., & Soda, G. (1995). Unified integration of explicit knowledge and learning by example in recurrent networks. *IEEE Transactions on Knowledge and Data Engineering, 7,* 340–346.

Fu, L. M. (1989). Integration of neural heuristics into knowledge-based inference. *Connection Science, 1,* 325–340.

Ginsberg, A. (1988). *Automatic Refinement of Expert System Knowledge Bases.* Pitman, London.

Gordon, D., & Subramanian, D. (1994). A multistrategy learning scheme for agent knowledge acquisition. *Informatica, 17,* 331–346.

Gruau, F. (1994). *Neural Network Synthesis using Cellular Encoding and the Genetic Algorithm.* Ph.D. thesis, Ecole Normale Superieure de Lyon, France.

Hayes-Roth, F., Klahr, P., & Mostow, D. J. (1981). Advice-taking and knowledge refinement: An iterative view of skill acquisition. In Anderson, J. (Ed.), *Cognitive Skills and their Acquisition,* pp. 231–253. Lawrence Erlbaum, Hillsdale, NJ.

Huffman, S., & Laird, J. (1993). Learning procedures from interactive natural language instructions. In *Machine Learning: Proceedings on the Tenth International Conference,* pp. 143–150 Amherst, MA.

Kaelbling, L. (1987). REX: A symbolic language for the design and parallel implementation of embedded systems. In *Proceedings of the AIAA Conference on Computers in Aerospace* Wakefield, MA.

Kaelbling, L., & Rosenschein, S. (1990). Action and planning in embedded agents. *Robotics and Autonomous Systems, 6,* 35–48.

Laird, J., Hucka, M., Yager, E., & Tuck, C. (1990). Correcting and extending domain knowledge using outside guidance. In *Proceedings of the Seventh International Conference on Machine Learning,* pp. 235–243 Austin, TX.

Le Cun, Y., Denker, J., & Solla, S. (1990). Optimal brain damage. In Touretzky, D. (Ed.), *Advances in Neural Information Processing Systems,* Vol. 2, pp. 598–605. Morgan Kaufmann, Palo Alto, CA.

Levine, J., Mason, T., & Brown, D. (1992). *Lex & yacc.* O'Reilly, Sebastopol, CA.

Lin, L. (1992). Self-improving reactive agents based on reinforcement learning, planning, and teaching. *Machine Learning, 8,* 293–321.

Lin, L. (1993). Scaling up reinforcement learning for robot control. In *Proceedings of the Tenth International Conference on Machine Learning,* pp. 182–189 Amherst, MA.

Maclin, R. (1995). *Learning from Instruction and Experience: Methods for Incorporating Procedural Domain Theories into Knowledge-Based Neural Networks.* Ph.D. thesis, Computer Sciences Department, University of Wisconsin, Madison, WI.

Maclin, R., & Shavlik, J. (1993). Using knowledge-based neural networks to improve algorithms: Refining the Chou-Fasman algorithm for protein folding. *Machine Learning, 11,* 195–215.

Maclin, R., & Shavlik, J. (1994). Incorporating advice into agents that learn from reinforcements. In *Proceedings of the Twelfth National Conference on Artificial Intelligence,* pp. 694–699 Seattle, WA.

Mahadevan, S., & Connell, J. (1992). Automatic programming of behavior-based robots using reinforcement learning. *Artificial Intelligence, 55,* 311–365.

McCarthy, J. (1958). Programs with common sense. In *Proceedings of the Symposium on the Mechanization of Thought Processes,* Vol. I, pp. 77–84. (Reprinted in M. Minsky, editor, 1968, *Semantic Information Processing.* Cambridge, MA: MIT Press, 403–409.).

Monahan, G. (1982). A survey of partially observable Markov decision processes: Theory, models, and algorithms. *Management Science, 28,* 1–16.

Mostow, D. J. (1982). Transforming declarative advice into effective procedures: A heuristic search example. In Michalski, R., Carbonell, J., & Mitchell, T. (Eds.), *Machine Learning: An Artificial Intelligence Approach,* Vol. 1. Tioga Press, Palo Alto.

Nilsson, N. (1994). Teleo-reactive programs for agent control. *Journal of Artificial Intelligence Research, 1,* 139–158.

Noelle, D., & Cottrell, G. (1994). Towards instructable connectionist systems. In Sun, R., & Bookman, L. (Eds.), *Computational Architectures Integrating Neural and Symbolic Processes.* Kluwer Academic, Boston.

Omlin, C., & Giles, C. (1992). Training second-order recurrent neural networks using hints. In *Proceedings of the Ninth International Conference on Machine Learning,* pp. 361–366 Aberdeen, Scotland.

Ourston, D., & Mooney, R. (1994). Theory refinement combining analytical and empirical methods. *Artificial Intelligence, 66,* 273–309.

Pazzani, M., & Kibler, D. (1992). The utility of knowledge in inductive learning. *Machine Learning, 9,* 57–94.

Riecken, D. (1994). Special issue on intelligent agents. *Communications of the ACM, 37(7).*

Shavlik, J., & Towell, G. (1989). An approach to combining explanation-based and neural learning algorithms. *Connection Science*, *1*, 233–255.

Siegelmann, H. (1994). Neural programming language. In *Proceedings of the Twelfth National Conference on Artificial Intelligence*, pp. 877–882 Seattle, WA.

Suddarth, S., & Holden, A. (1991). Symbolic-neural systems and the use of hints for developing complex systems. *International Journal of Man-Machine Studies*, *35*, 291–311.

Sutton, R. (1988). Learning to predict by the methods of temporal differences. *Machine Learning*, *3*, 9–44.

Sutton, R. (1991). Reinforcement learning architectures for animats. In Meyer, J., & Wilson, S. (Eds.), *From Animals to Animats: Proceedings of the First International Conference on Simulation of Adaptive Behavior*, pp. 288–296. MIT Press, Cambridge, MA.

Tesauro, G. (1992). Practical issues in temporal difference learning. *Machine Learning*, *8*, 257–277.

Thrun, S., & Mitchell, T. (1993). Integrating inductive neural network learning and explanation-based learning. In *Proceedings of the Thirteenth International Joint Conference on Artificial Intelligence*, pp. 930–936 Chambery, France.

Towell, G., & Shavlik, J. (1993). Extracting refined rules from knowledge-based neural networks. *Machine Learning*, *13*, 71–101.

Towell, G., & Shavlik, J. (1994). Knowledge-based artificial neural networks. *Artificial Intelligence*, *70*, 119–165.

Towell, G., Shavlik, J., & Noordewier, M. (1990). Refinement of approximate domain theories by knowledge-based neural networks. In *Proceedings of the Eighth National Conference on Artificial Intelligence*, pp. 861–866 Boston, MA.

Utgoff, P., & Clouse, J. (1991). Two kinds of training information for evaluation function learning. In *Proceedings of the Ninth National Conference on Artificial Intelligence*, pp. 596–600 Anaheim, CA.

Watkins, C. (1989). *Learning from Delayed Rewards*. Ph.D. thesis, King's College, Cambridge.

Watkins, C., & Dayan, P. (1992). Q-learning. *Machine Learning*, *8*, 279–292.

Weigend, A. (1993). On overfitting and the effective number of hidden units. In *Proceedings of the 1993 Connectionist Models Summer School*, pp. 335–342 San Mateo, CA. Morgan Kaufmann.

Whitehead, S. (1991). A complexity analysis of cooperative mechanisms in reinforcement learning. In *Proceedings of the Ninth National Conference on Artificial Intelligence*, pp. 607–613 Anaheim, CA.

Zadeh, L. (1965). Fuzzy sets. *Information and Control*, *8*, 338–353.

Received November 1, 1994
Accepted February 24, 1995
Final Manuscript October 4, 1995

Machine Learning, 22, 283–290 (1996)

Technical Note
Incremental Multi-Step Q-Learning

JING PENG   jp@vislab.ucr.edu
College of Engineering, University of California, Riverside, CA 92521

RONALD J. WILLIAMS   rjw@ccs.neu.edu
College of Computer Science, Northeastern University, Boston, MA 02115

Editor: Leslie Pack Kaelbling

Abstract. This paper presents a novel incremental algorithm that combines Q-learning, a well-known dynamic-programming based reinforcement learning method, with the TD(λ) return estimation process, which is typically used in actor-critic learning, another well-known dynamic-programming based reinforcement learning method. The parameter λ is used to distribute credit throughout sequences of actions, leading to faster learning and also helping to alleviate the non-Markovian effect of coarse state-space quantization. The resulting algorithm, *Q(λ)-learning*, thus combines some of the best features of the Q-learning and actor-critic learning paradigms. The behavior of this algorithm has been demonstrated through computer simulations.

Keywords: reinforcement learning, temporal difference learning

1. Introduction

The incremental multi-step Q-learning (Q(λ)-learning) method is a new direct (or model-free) algorithm that extends the one-step Q-learning algorithm (Watkins, 1989) by combining it with TD(λ) returns for general λ (Sutton, 1988) in a natural way for delayed reinforcement learning. By allowing corrections to be made incrementally to the predictions of observations occurring in the past, the Q(λ)-learning method propagates information rapidly to where it is important. The Q(λ)-learning algorithm works significantly better than the one-step Q-learning algorithm on a number of tasks and its basis in the integration of one-step Q-learning and TD(λ) returns makes it possible to take advantage of some of the best features of the Q-learning and actor-critic learning paradigms and to be a potential bridge between them. It can also serve as a basis for developing various multiple time-scale learning mechanisms that are essential for applications of reinforcement learning to real world problems.

2. TD(λ) Returns

Direct dynamic-programming based reinforcement learning algorithms are based on updating state values or state-action values according to state transitions as they are experienced. Each such update is in turn based on the use of a particular choice of estimator for the value being updated, which spells out differences among various learning methods.

This section describes an important and computationally useful class of such estimators – the TD(λ) estimators (Sutton, 1988; Watkins, 1989).

Let the world state at time step t be x_t, and assume that the learning system then chooses action a_t. The immediate result is that a reward r_t is received by the learner and the world undergoes a transition to the next state, x_{t+1}. The objective of the learner is to choose actions maximizing discounted cumulative rewards over time. More precisely, let γ be a specified discount factor in $[0, 1)$. The *total discounted return* (or simply *return*) received by the learner starting at time t is given by

$$\mathbf{r}_t = r_t + \gamma r_{t+1} + \gamma^2 r_{t+2} + \cdots + \gamma^n r_{t+n} + \cdots.$$

The objective is to find a policy π, or rule for selecting actions, so that the expected value of the return is maximized. It is sufficient to restrict attention to policies that select actions based only on the current state (called *stationary* policies). For any such policy π and for any state x we define

$$V^\pi(x) = E[\mathbf{r}_0 | x_0 = x, a_i = \pi(x_i) \text{ for all } i \geq 0],$$

the expected total discounted return received when starting in state x and following policy π thereafter. If π is an optimal policy we also use the notation V^* for V^π. Many dynamic-programming-based reinforcement learning methods involve trying to estimate the state values $V^*(x)$ or $V^\pi(x)$ for a fixed policy π.

An important class of methods for estimating V^π for a given policy π is the TD(λ) estimators, which have been investigated by Sutton (1988) and later by Watkins (1989). Following Watkins' notation, let $\mathbf{r}_t^{(n)}$ denote the *corrected n-step truncated return* for time t, given by

$$\mathbf{r}_t^{(n)} = r_t + \gamma r_{t+1} + \gamma^2 r_{t+2} + \cdots + \gamma^{n-1} r_{t+n-1} + \gamma^n \hat{V}_{t+n}^\pi(x_{t+n}) \tag{1}$$

where $\hat{V}_t^\pi$ is the estimate of V^π at time t. If $\hat{V}^\pi$ were equal to V^π, then the corrected truncated returns would be unbiased estimators of V^π. Watkins (Watkins, 1989) shows that corrected truncated returns have the *error-reduction property* in that the expected value of the corrected truncated return is closer to V^π than $\hat{V}^\pi$ is.

Then Sutton's TD(λ) return from time t is

$$\begin{aligned}
\mathbf{r}_t^\lambda &= (1 - \lambda)[\mathbf{r}_t^{(1)} + \lambda \mathbf{r}_t^{(2)} + \lambda^2 \mathbf{r}_t^{(3)} + \cdots] \\
&= r_t + \gamma(1 - \lambda)\hat{V}_t^\pi(x_{t+1}) + \gamma\lambda[r_{t+1} + \gamma(1 - \lambda)\hat{V}_{t+1}^\pi(x_{t+2}) + \cdots] \\
&= r_t + \gamma(1 - \lambda)\hat{V}_t^\pi(x_{t+1}) + \gamma\lambda \mathbf{r}_{t+1}^\lambda.
\end{aligned} \tag{2}$$

The TD(0) return is just $\mathbf{r}_t^0 = r_t + \gamma\hat{V}_t^\pi(x_{t+1})$ and the TD(1) return is

$$\mathbf{r}_t^1 = r_t + \gamma r_{t+1} + \gamma^2 r_{t+2} + \cdots$$

which is the exact actual return. Watkins (1989) argues that, in a Markov decision problem, the choice of λ is a trade-off between bias and variance. Sutton's empirical demonstration (Sutton, 1988) favors intermediate values of λ that are closer to 0. More recent analysis (Sutton & Singh, 1994) suggests that in certain prediction tasks near optimal performance can be achieved by setting λ at each time step to the transition probability of the immediately preceding transitions. For further details, see (Sutton, 1988; Sutton & Singh, 1994; Watkins, 1989).

3. One-Step Q-Learning

One-step Q-learning of Watkins (1989), or simply Q-learning, is a simple incremental algorithm developed from the theory of dynamic programming (Ross, 1983) for delayed reinforcement learning. In Q-learning, policies and the value function are represented by a two-dimensional lookup table indexed by state-action pairs. Formally, using notation consistent with that of the previous section, for each state x and action a let

$$Q^*(x,a) = R(x,a) + \gamma \sum_y P_{xy}(a)V^*(y) \tag{3}$$

where $R(x,a) = E\{r_0|x_0 = x, a_0 = a\}$, and $P_{xy}(a)$ is the probability of reaching state y as a result of taking action a in state x. It follows that $V^*(x) = \max_a Q^*(x,a)$.

Intuitively, Equation (3) says that the state-action value, $Q^*(x,a)$, is the expected total discounted return resulting from taking action a in state x and continuing with the optimal policy thereafter. More generally, the Q function can be defined with respect to an arbitrary policy π as $Q^\pi(x,a) = R(x,a) + \gamma \sum_y P_{xy}(a)V^\pi(y)$ and Q^* is just Q^π for an optimal policy π.

The Q-learning algorithm works by maintaining an estimate of the Q^* function, which we denote by $\hat{Q}^*$, and adjusting $\hat{Q}^*$ values (often just called Q-values) based on actions taken and reward received. This is done using Sutton's prediction difference, or TD error (Sutton, 1988)–the difference between the immediate reward received plus the discounted value of the next state and the Q-value of the current state-action pair:

$$r + \gamma \hat{V}^*(y) - \hat{Q}^*(x,a)$$

where r is the immediate reward, y is the next state resulting from taking action a in state x, and $\hat{V}^*(x) = \max_a \hat{Q}^*(x,a)$. Then the values of $\hat{Q}^*$ are adjusted according to

$$\hat{Q}^*(x,a) = (1-\alpha)\hat{Q}^*(x,a) + \alpha(r + \gamma \hat{V}^*(y)) \tag{4}$$

where $\alpha \in (0,1]$ is a learning rate parameter. In terms of the notation described in the previous section, Equation (4) may be rewritten as

$$\hat{Q}^*(x,a) = (1-\alpha)\hat{Q}^*(x,a) + \alpha \mathbf{r}^0 \tag{5}$$

That is, the Q-learning method uses TD(0) as its estimator of expected returns. Note that the current estimate of the Q^* function implicitly defines a greedy policy by $\pi(x) = \arg\max_a \hat{Q}^*(x,a)$. That is, the greedy policy is to select actions with the largest estimated Q-value.

It is important to note that the one-step Q-learning method does not specify what actions the agent should take at each state as it updates its estimates. In fact, the agent may take whatever actions it pleases. This means that Q-learning allows arbitrary experimentation while at the same time preserving the current best estimate of states' values. Furthermore, since this function is updated according to the ostensibly optimal choice of action at the following state, it does not matter what action is actually followed at that state. For this reason, Q-learning is not *experimentation-sensitive*. On the other hand, because actor-critic learning updates the state value at any state based on the actual action selected, not on what would have been the optimal choice of action, it is experimentation-sensitive.

To find the optimal Q function eventually, however, the agent must try out each action in every state many times. It has been shown (Watkins, 1989; Watkins & Dayan, 1992) that if Equation (4) is repeatedly applied to all state-action pairs in any order in which each state-action pair's Q-value is updated infinitely often, then $\hat{Q}^*$ will converge to Q^* and $\hat{V}^*$ will converge to V^* with probability 1 as long as α is reduced to 0 at a suitable rate.

Finally, Watkins (1989) has also described possible extensions to the one-step Q-learning method by using different value estimators, such as $\mathbf{r}^\lambda$ for $0 < \lambda < 1$, and he has illustrated the use of $\mathbf{r}^\lambda$ returns in Q-learning in his empirical demonstrations by memorizing past experiences and calculating these returns at the end of each learning period, where a learning period specifies the number of experiences occurring in the past the agent needs to store. The following section derives a novel algorithm that enables the value estimation process to be done incrementally.

4.   Q(λ)-Learning

This section derives the Q(λ)-learning algorithm combining TD(λ) returns for general λ with Q-learning in an incremental way. Note that in terms of the notation introduced here, one-step Q-learning is simply Q(0)-learning, making it a special case.

For simplicity, in what follows we drop the superscript π in V^π and assume that the given policy π is the agent's greedy policy. Now let

$$e_t = r_t + \gamma \hat{V}_t(x_{t+1}) - \hat{V}_t(x_t) \tag{6}$$

and

$$e'_t = r_t + \gamma \hat{V}_t(x_{t+1}) - \hat{Q}_t(x_t, a_t) \tag{7}$$

where $\hat{V}(x) = \max_a \hat{Q}(x, a)$. Then, if we use Equation (7) for one step and Equation (6) thereafter, the difference between the TD(λ) return of Equation (2) and the estimated Q-value can be written as

$$\mathbf{r}_t^\lambda - \hat{Q}_t(x_t, a_t) = e'_t + \gamma \lambda e_{t+1} + \gamma^2 \lambda^2 e_{t+2} + \cdots$$

$$+ \sum_{n=1}^{\infty} (\gamma \lambda)^n [\hat{V}_{t+n}(x_{t+n}) - \hat{V}_{t+n-1}(x_{t+n})]. \tag{8}$$

If the learning rate is small, so that Q is adjusted slowly, then the second summation on the right-hand side of the above equation will be small.

The $Q(\lambda)$-learning algorithm is summarized in Figure 1, where $Tr(x, a)$ is the "activity" trace of state-action pair (x, a), corresponding to the "eligibility" trace as described in (Barto, Sutton & Anderson, 1983).

The main difficulty associated with $Q(\lambda)$-learning in a Markov decision process is that rewards received after a non-greedy action cannot be used to evaluate the agent's greedy policy since this will not be the policy that was actually followed. In other words, $Q(\lambda)$-learning is experimentation-sensitive, assuming that $\lambda > 0$ is fixed. One way around this difficulty is to zero λ on each step that a non-greedy action is taken. However, as argued in (Rummery & Niranjan, 1994), zeroing the effect of subsequent rewards on those prior to a non-greedy action is likely to be more of a hindrance than a help in converging to optimal policies since $\max_a Q(x, a)$ may not provide the best estimate of the value of the state x.

Still another difficulty is that changes in $\hat{Q}$ at each time step may affect $\mathbf{r}^\lambda$, which will in turn affect $\hat{Q}$, and so on. However, these effects may not be significant for small α since they are proportional to α^2 (Peng, 1993).

At each time step, the $Q(\lambda)$-learning algorithm loops through a set of state-action pairs which grow linearly with time. In the worst case, this set could be the entire state-action space. However, the number of state-action pairs for which actual updating is required can be kept at a manageable level by maintaining only those state-action pairs whose activity trace $(\gamma\lambda)^n$ is significant, since this quantity declines exponentially when $\gamma\lambda < 1$. For a more elaborate procedure see Cichosz & Mulawka (1995). Another approach is to

---

1. $\hat{Q}(x, a) = 0$ and $Tr(x, a) = 0$ for all x and a

2. Do Forever:

    (A) $x_t \leftarrow$ the current state

    (B) Choose an action a_t according to current exploration policy

    (C) Carry out action a_t in the world. Let the short-term reward be r_t, and the new state be x_{t+1}

    (D) $e'_t = r_t + \gamma \hat{V}_t(x_{t+1}) - \hat{Q}_t(x_t, a_t)$

    (E) $e_t = r_t + \gamma \hat{V}_t(x_{t+1}) - \hat{V}_t(x_t)$

    (F) For each state-action pair (x, a) do

      • $Tr(x, a) = \gamma\lambda Tr(x, a)$

      • $\hat{Q}_{t+1}(x, a) = \hat{Q}_t(x, a) + \alpha Tr(x, a)e_t$

    (G) $\hat{Q}_{t+1}(x_t, a_t) = \hat{Q}_{t+1}(x_t, a_t) + \alpha e'_t$

    (H) $Tr(x_t, a_t) = Tr(x_t, a_t) + 1$

Figure 1. The $Q(\lambda)$-Learning Algorithm.

implement a $Q(\lambda)$-learning system on a parallel machine in which each state-action pair is mapped onto a separate processor. This corresponds directly to the kind of neural network implementation first envisioned for the actor-critic approach (Barto, Sutton & Anderson, 1983).

Finally, it is interesting to note that both $Q(\lambda)$-learning and actor-critic learning use $TD(\lambda)$ returns as their value estimators through a trace mechanism. Therefore, it seems reasonable to expect the $Q(\lambda)$-learning algorithm to exhibit beneficial performance characteristics attributable to the use of $TD(\lambda)$ returns for $\lambda > 0$, as illustrated in (Barto, Sutton & Anderson, 1983; Sutton, 1988). At the same time, both $Q(\lambda)$-learning and one-step Q-learning construct a value function on the state-action space rather than just the state space, making them both capable of discriminating between the effects of choosing different actions in each state. Thus, while $Q(\lambda)$-learning is experimentation-sensitive, unlike one-step Q-learning, it seems reasonable to expect it to be less so than actor-critic learning. Overall, then, $Q(\lambda)$-learning appears to incorporate some of the best features of the Q-learning and actor-critic learning paradigms into a single mechanism. Furthermore, it can be viewed as a potential bridge between them.

5. Discussion

This paper has only examined the $Q(\lambda)$-learning algorithm in which the $TD(\lambda)$ returns are computed by taking the maximum Q values at each state visited. There are other possibilities, however. For example, the algorithm may estimate the $TD(\lambda)$ returns by using the current exploration policy. This is the algorithm, called *sarsa*, described in (Rummery & Niranjan, 1994). In this algorithm, the update rule is

$$\Delta \mathbf{w}_t = \alpha(r_t + \gamma Q_{t+1} - Q_t) \sum_{k=0}^{t} (\gamma\lambda)^{t-k} \nabla_{\mathbf{w}} Q_k \tag{9}$$

where w denotes the weights of connectionist networks, and Q_{t+1} is associated with the action selected. In terms of $Q(\lambda)$ learning, the right hand side of Equations (2D) and (2E) in Figure 1 would be replaced by

$$r_t + \gamma \hat{Q}_t(x_{t+1}, a_{t+1}) - \hat{Q}_t(x_t, a_t).$$

It is demonstrated (Rummery & Niranjan, 1994) that the overall performance of $Q(\lambda)$ learning, including *sarsa*, shows less sensitivity to the choice of training parameters and exhibits more robust behavior than standard Q learning. See also (Pendrith, 1994).

Experiments involving both Markovian and non-Markovian tasks, whose details we omit here, were carried out to validate the efficacy of the $Q(\lambda)$-learning algorithm. The results showed that $Q(\lambda)$-learning outperformed both actor-critic learning and one-step Q-learning on all the experiments. The significant performance improvement of the $Q(\lambda)$-learning system over the simple Q-learning system (including the case where the Q-learner was given the experiences of the $Q(\lambda)$-learner) is clearly due to the use of the $TD(\lambda)$ return estimation process, which has the effect of making alterations to past predictions throughout each trial. If this is the main benefit conferred by TD(λ), one might expect model-based, multiple-update methods like priority-Dyna (Peng, 1993; Peng & Williams, 1993; Moore & Atkeson, 1994), to perform at least as well. However, additional experiments using such techniques showed that they performed significantly worse than Q(λ)-learning. We believe the reason for this is that the coarse state-space quantization used here has the effect of making the environment non-Markovian, and increasing λ makes TD(λ) less sensitive to this non-Markovian effect. Pendrith (1994) made a similar argument on a related algorithm.

It should be noted that both the fixed period learning process of Watkins (1989) for sufficiently long learning periods and the experience-replay process of Lin (1992) produce similar beneficial effects as that of Q(λ)-learning. However, both of these approaches operate in "batch" mode in that they replay, backwards, the memorized sequence of experiences that the learning agent has recently had.

From a computational standpoint, the incrementality of Q(λ)-learning makes it more attractive than Watkins' batch mode learning and Lin's experience replay process since the computation can be distributed over time more evenly, and thus under many circumstances can ease overall demands on the memory and speed. Similar arguments are made in (Sutton, 1988). Furthermore, incrementality speeds up learning. In one experiment where off-line Q(λ) learning was applied to experiences of its own and to those of on-line Q(λ) learning, it was found that the results of off-line Q(λ) learning in both cases were much worse than those obtained using on-line Q(λ) learning. One additional attractive characteristic of the Q(λ)-learning method is that it achieves greater computational efficiency without having to learn and use a model of the world (Peng, 1993; Peng & Williams, 1993; Sutton, 1990) and is well suited to parallel implementation.

Finally, although look-up table representation has been our main focus so far, it can be shown without difficulty that Q(λ) learning can be implemented on-line using connectionist networks, as is done in (Rummery & Niranjan, 1994).

6. Conclusion

The Q(λ)-learning algorithm is of interest because of its incrementality and its relationship to Q-learning (Watkins, 1989) and actor-critic learning (Barto, Sutton & Anderson, 1983). However, this algorithm, unlike the one-step Q-learning algorithm, cannot be expected to converge to the correct Q^* values under an arbitrary policy that tries every action in every state (although the obvious strategies of gradually reducing λ or gradually turning down the Boltzmann temperature as learning proceeds would probably allow such convergence). In spite of this, the Q(λ)-learning algorithm has always outperformed the one-step Q-learning algorithm on all the problems we have experimented with so far.

It is clear that in continuous-time systems, or even systems where time is discrete but very fine-grained, the use of algorithms that propagate information back one step at a time can make no sense or at least be of little value. In these cases the use of TD(λ) methods is not a luxury but a necessity. In general, λ can be viewed as a time scale parameter in such situations, and we argue that better understanding of its use in this regard is an important area for future research.

Acknowledgments

We wish to thank Rich Sutton for his many valuable suggestions and continuing encouragement. We would also like to thank the reviewers of the paper for their insightful comments and suggestions. This work was supported by Grant IRI-8921275 from the National Science Foundation.

References

Barto, A. G., Sutton, R. S. & Anderson, C. W. (1983). Neuronlike elements that can solve difficult learning control problems. *IEEE Transactions on Systems, Man, and Cybernetics* 13:835-846.

Cichosz, P. & Mulawka, J. J. (1995). Fast and efficient reinforcement learning with truncated temporal differences. *Proceedings of the Twelfth International Conference on Machine Learning*, 99-107.

Lin, L. J. (1992). *Reinforcement learning for robots using neural networks.* Ph. D. Dissertation, Carnegie Mellon University, PA.

Moore, A. W. & Atkeson, C. G. (1994). Prioritized sweeping: reinforcement learning with less data and less time. *Machine Learning* 13(1):103-130.

Pendrith, M. (1994). *On reinforcement learning of control actions in noisy and non-Markovian domains.* UNSW-CSE-TR-9410, University of New South Wales, Australia.

Peng, J. (1993). *Efficient Dynamic Programming-Based Learning for Control.* Ph. D. Dissertation, Northeastern University, Boston, MA 02115.

Peng, J. & Williams, R. J. (1993). Efficient learning and planning within the Dyna framework. *Adaptive Behavior* 1(4):437-454.

Ross, S. (1983). *Introduction to Stochastic Dynamic Programming.* New York, Academic Press.

Rummery, G. A. & Niranjan, M. (1994). *On-line Q-learning using connectionist systems.* CUED/F-INFENG/TR 166, Cambridge University, UK.

Sutton, R. S. (1990). Integrated architectures for learning, planning, and reacting based on approximating dynamic programming. In *Proceedings of the Seventh International Conference on Machine Learning*, 216-224.

Sutton, R. S. (1988). Learning to predict by the methods of temporal differences. *Machine Learning* 3:9-44.

Sutton, R S & Singh, S. P. (1994). On step-size and bias in temporal-difference learning. In *Eighth Yale Workshop on adaptive and Learning Systems*, pages 91-96, New Haven, CT.

Watkins, C. J. C. H. & Dayan, P. (1992). Q-learning. *Machine Learning* 8:279-292.

Watkins, C. J. C. H. (1989). *Learning from delayed rewards.* Ph. D. Dissertation, King's College, UK.

Received November 2, 1994
Accepted March 10, 1995
Final Manuscript October 4, 1995

INDEX